11

PHYSICS

PNG UPPER SECONDARY

Roberto Soto
John Boereboom
Athol Binns
Denis Burchill
David Housden
Peter Kinsler

OXFORD

OXFORD
UNIVERSITY PRESS

Oxford University Press is a department of the University of Oxford. It furthers the University's objective of excellence in research, scholarship, and education by publishing worldwide. Oxford is a registered trademark of Oxford University Press in the UK and in certain other countries.

Published in Australia by
Oxford University Press
Level 8, 737 Bourke Street, Docklands, Victoria 3008, Australia.

First published 2012
Reprinted 2013, 2014, 2015, 2017 (twice), 2018, 2019, 2020, 2021, 2022, 2024

This book was originally published by ESA Publications, Auckland, New Zealand. This edition, specially adapted for the Grade 11 syllabus in Papua New Guinea, is published by arrangement with ESA Publications. Authors of the original work were John Boereboom, Athol Binns, Denis Burchill, David Housden, Peter Kinsler, and this edition has been produced by Roberto Soto.

ISBN 978 0 19 557881 2

Illustrations by diacriTech and Birdwing PNG
Typeset by diacriTech
Printed in China by Golden Cup Printing Co. Ltd

Oxford University Press Australia & New Zealand is committed to sourcing paper responsibly.

Contents

Unit 11.4 Work, Power and Energy

Unit 11.5 Electricity Principles

Unit 11.6 Electronics

Introduction

This book has been published to provide the information required by students in order to successfully complete the Upper Secondary course in Physics at Grade 11 in Papua New Guinea.

The book is written in a manner that best develops an overall understanding of the physics concepts and processes set out in the PNG Grade 11 Syllabus. The order of Units in this book follows the order of Units presented in the Physics syllabus for Grade 11:

Unit 11.1 Measurement

Unit 11.2 Motion (Kinematics)

Unit 11.3 Forces and Motion (Dynamics)

Unit 11.4 Work, Power and Energy

Unit 11.5 Electricity Principles

Unit 11.6 Electronics

Within each Unit there are a number of Topics which follow the main subheadings and bullet points set out within each Unit in the syllabus. The aim is to provide structure and content in a concise and compact format for students to use as an effective resource to support the classroom experience. It is acknowledged that Physics is more interesting and meaningful when students are exposed to a variety of resources and materials and we encourage students and teachers not to rely on this book as a sole source of information.

If you have any suggestions about how this book might be improved in future editions, please make contact with Oxford University Press:

Fax: 00 61 3 9934 9100

Customer Service Email: exportsales.au@oup.com

We wish you every success in your studies.

The authors

Acknowledgments

There are many people to be thanked for their help and assistance in enabling the publication of this series to happen. First, we acknowledge the cooperation and generosity of Mark Sayes at ESA Publications in New Zealand, who responded with interest and support when the proposal to adapt his Study Guide series was put to him.

We would also like to acknowledge many other individuals who have been happy to advise and assist in different ways: Joy Sahumlal, Greg Kapanombo, Anne Sangi, Safak Deliismail.

Authorship

Authors of the original editions were John Boereboom, Athol Binns, Denis Burchill, David Housden and Peter Kinsler.

This edition has been produced with the assistance of:

> Roberto Soto MSEE at the University of Houston, Houston, Texas. Major in Control Theory: Robotics Control and Minor in Robotics Vision; BSEE at the Instituto Tecnologico y de Estudios Superiores in Monterrey, Mexico. Electronics and Electronic Communications Engineering; Diploma in Aircraft Engine Turbines at The Allied Signal Aerospace Engineering; more than 10 years experience in Aerospace and Research and Development in Electronic Controls; Teaching Physics and its applications since 1994 at UPNG and at UNITECH since 2006, teaching Applied Physics: Electronic Instrumentation, Classical Mechanics, Sensors and Measurement Systems. Chief Examiner and Exam writer for Grade 12 since 1998.

The aim of this book is to provide Grade 11 Physics students in PNG with a book that they can use as a compact summary of content and skills related to the PNG Grade 11 syllabus.

Preliminary Unit
Investigating and research in physics

Physics is important because of its relevance to the present and future technology-driven lifestyle in Papua New Guinea. In all Units, students engage in practical tasks and investigations that will enable them to build the basic physics principles, concepts and skills needed for higher education or self-employment. This Preliminary Unit offers advice and guidelines on:

- Focusing and planning an investigation.
- Information gathering and processing and interpreting that information.
- Reporting and explaining the relationship between physics and technology.

Investigations in physics typically involve focusing and planning; followed by information gathering, processing and interpreting; ending with reporting.

Focusing and planning

When an investigation is started, a vague idea of the problem to investigate is generally known. During the focusing and planning stage, the problem is clarified and an aim written for the investigation. It may help to do some research to learn more about the physics principles relevant to the investigation. (This may involve visiting a library, an Internet search or consulting an expert.) The physics investigation problem should be stated in the form of *a question about the relationship between the variables*. Decide which variable will be chosen as the independent variable and over what sort of range it will be changed and measured. The dependent variable and how it will be measured must be determined.

After the dependent and independent **variables** have been decided, a **hypothesis** can be formulated. The hypothesis is usually expressed as a probable **relationship** between the dependent and independent variables. Some preliminary trialing of the proposed experiment may need to be carried out to determine if it is feasible. What variables to keep constant or control throughout the investigation will need to be considered.

In a **fair test**, one variable (the independent variable) is changed to determine the effect this has on another variable (the dependent variable). All other variables that may have an effect on the dependent variable should be kept constant throughout the investigation.

A plan for the investigation can now be written. The plan should contain a list of the apparatus needed and an outline of the method that will be used to investigate the problem.

Information gathering

During the information-gathering phase of an investigation, the equipment needed to carry out the experiment will be assembled and /or constructed. After setting up the equipment the data needed is collected. The data must be recorded in a systematic format, such as a table of results. Carefully label each column with the name of the variable and the unit it is measured in. Think about possible sources of uncertainty in the data and try to control or minimise these.

Processing and interpreting

Processing of data typically involves drawing graphs to explore the relationship between the dependent and independent variables. This may lead to a mathematical relationship. **Computer spreadsheets** can be useful at this stage. Answers to a number of important questions should be sought.

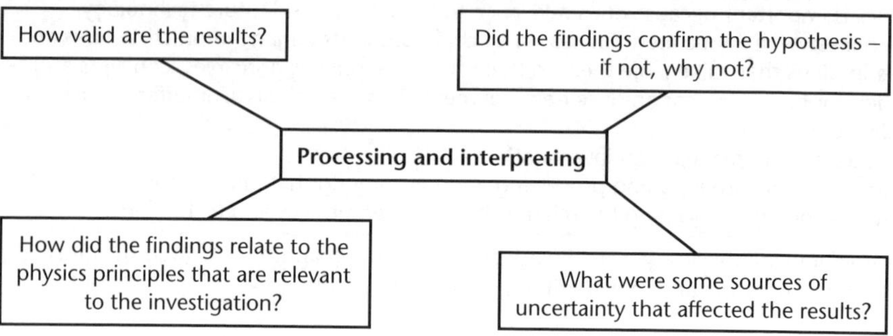

Reporting

The final report may take a variety of forms – a formal report, a poster or video or the findings presented to a class. Whatever format the report may take, it should clearly state:

- The aim of the investigation.
- The hypothesis.
- The plan for the investigation.
- The results and how the data was processed.
- A conclusion and evaluation of the findings.

Example A

A physics investigation

James wanted to investigate how changing the vertical height of a ramp from which a toy car is released would affect the stopping distance of the toy car. He released the car at the top of a ramp and measured the distance the car travelled after leaving the ramp and before coming to a stop.

James changed the angle of the ramp to release the car from different heights. For each height measured, James measured the distance the car travelled before coming to a stop. He carried out repeated trials and averaged his results. All other factors that may have had an impact on the stopping distance – such as the car used and the surface on which the car travelled – were kept constant throughout the investigation. James processed his results by plotting a graph of height versus stopping distance.

The variables in James' investigation are summarised in the following table:

Variable to be changed (independent variable)	Variable to be measured (dependent variable)	Variables to be controlled (kept the same)
Height of the ramp or speed	Stopping distance	Ramp surface
		Floor surface
		Type of toy car

Preliminary Unit Activity A: Force on a current-carrying wire

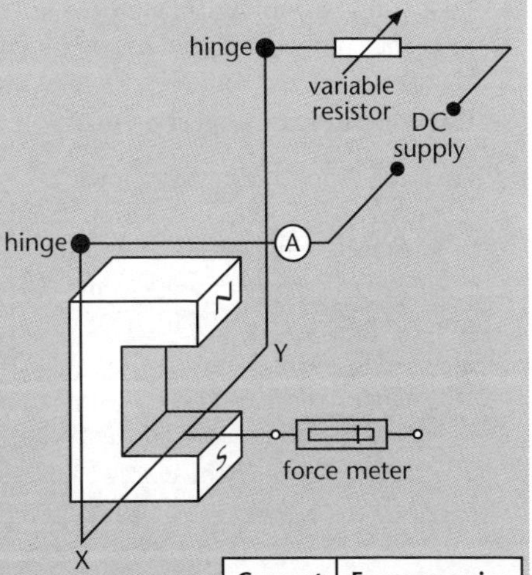

1. Rose and Mary are investigating how the magnetic force on a current-carrying wire in a magnetic field changes as the current moving through the wire is changed. They set up their apparatus as shown, using the variable resistor to change the current in the circuit.

a. Plot the results they found, shown in the table below.

b. Calculate the gradient of the graph.

c. What factors did Rose and Mary keep constant throughout their experiment?

d. Mary and Rose now want to carry out an experiment to investigate how one of the factors you listed in **c** affects force. State which factor they will change.

e. State which factors they will keep constant.

f. List two possible sources of error which may have affected the accuracy of Rose's and Mary's results.

Current (A)	Force on wire XY (N)
0.15	0.65×10^{-2}
0.30	1.25×10^{-2}
0.40	1.70×10^{-2}
0.65	1.75×10^{-2}
0.85	3.60×10^{-2}
0.90	3.80×10^{-2}

2. Rose and James carried out an experiment to find out how the strength of an electromagnet depends on the size of the current through the electromagnet. They were supplied with the following materials:

- 12 V fixed DC power supply.
- Iron rods.
- Variable resistor.
- Long length of insulated wire.
- Ammeter.
- Box of paper clips.

a. List three factors that were kept constant throughout this experiment.

Rose and James obtained the results shown in the table.

Current (A)	Number of paper clips
0.5	1
1.0	2
1.5	3
2.0	4
2.5	5

b. Draw a graph of the number of paper clips versus the current.

c. Their teacher asks them to investigate how changing another factor will affect the strength of the electromagnet.

i. What factor could be changed?

ii. What factors are to be kept constant?

iii. What would be a suitable title for the graph they plot?

iv. What would be the label on the x-axis for the graph they plot?

3. Lucy is investigating the heating coils of a bar heater. She is interested in how the resistance of the wire used to conduct the heating coil varies when its length is changed.

Carrying out the experiment

Lucy constructs the circuit shown alongside. She connects different lengths of resistance wire between A and B.

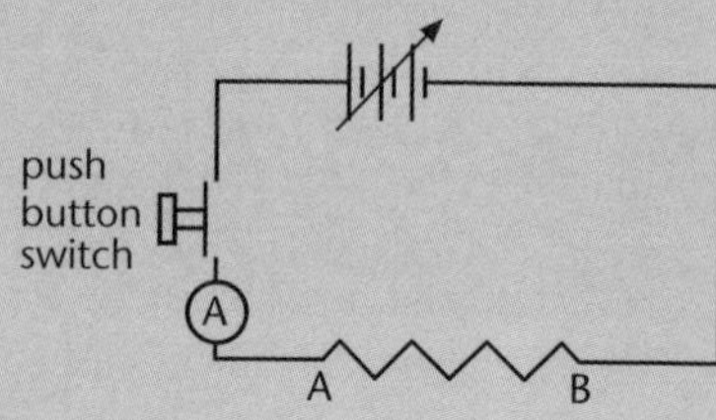

Lucy makes a range of voltage and current measurements for each length of wire.

a. During the investigation, Lucy changes one aspect of the wire but keeps the other facts constant.

Lucy will change the ______ **i**. ______ of the wire connected in the circuit.

Lucy will keep the ______ **ii**. ______ and ______ **iii**. ______ of the wire constant.

Analysing the results

A graph of Lucy's results for a 60 cm length of wire is shown opposite.

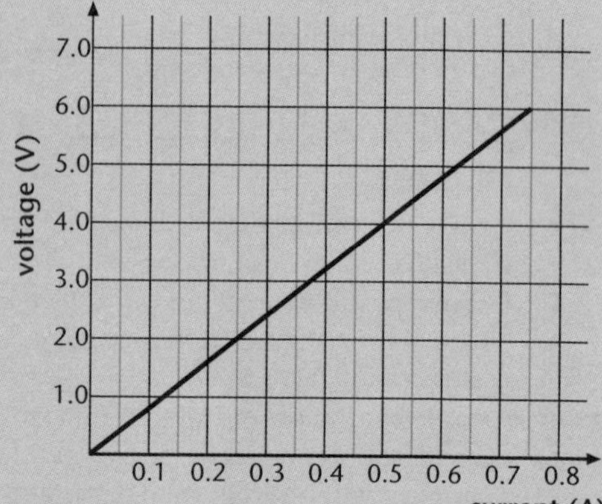

Before she started her investigation, Lucy wrote the following hypothesis:

Hypothesis: I expect the resistance of the wire will increase steadily as the length of the wire is increased.

b. Do Lucy's results support her hypothesis? Give a reason for your answer.

4. Hiris and Paul carried out an experiment to investigate how the volume of a gas varies with temperature. They used the apparatus shown.

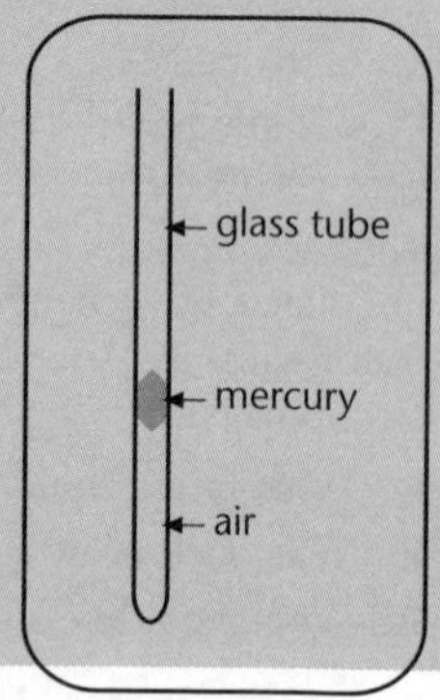

a. Complete the following aim of their experiment.

'To find out how the ______ **i**. ______ of air increases with ______ **ii**. ______ when the ______ **iii**. ______ is kept constant.'

b. Copy and complete the diagram to show how Hiris and Paul set up their experiment. Draw and label any additional equipment needed.

c. Hiris and Paul obtained the set of readings shown in the table.

Temperature (°C)	Length of air column (cm)
0	12.2
25	13.3
50	14.4
75	15.7
100	16.7

Plot the readings on a graph, using the sort on the grid provided.

d. What is the temperature at which the volume of air is zero?

e. How could the apparatus be used to find the outside air temperature?

Research into how physics and technology are related

Technology often involves application of physics principles. For example, the design of an electric steam iron involves application of physics principles ranging from **resistance**, Ohm's law and changes of state to circuit theory. Investigating how technology works often leads to a better understanding of physics and an appreciation of the relevance of physics to everyday life.

To carry out research into the relationship between physics and technology, a range of reference sources needs to be consulted. These may include books, journal articles, the Internet, newspapers, or interviews with people such as technicians or electricians.

The material that is collected needs to be carefully processed to select only those aspects that are directly relevant to the chosen research topic. A research report should be written in the author's own words. Material that is copied from other sources must be clearly acknowledged. All sources of information that have been used must be acknowledged in a list of references (*bibliography*) at the end of the research report – author's name, year of publication, title, details of publisher and page reference. For material sourced from the Internet, a web address should be given.

Example B

Soto, R., (2012) *Save Buk* Grade 11 Physics, Melbourne: Oxford University Press, p. 5.
http://www.oup.com.au

A research report should include:

- A title.
- A research aim or purpose statement.
- An explanation of how the technology works, including appropriate diagrams.
- An explanation of the physics principles involved in the design and operation of the technological application.
- A clear description of the links between physics and the technological application.
- A list of references or bibliography.

Physics, technology and society

The link between society, technology and physics can be seen in many examples. The discipline of thermodynamics, for example, arose from the need to understand and improve the working of heat engines. The steam engine is inseparable from the Industrial Revolution in England during the 18th century, which had a great impact on the course of human civilisation.

Sometimes technology gives rise to new physics; at other times, physics generates new technology. An example of this is the wireless communications technology that followed the discovery of the basic laws of electricity and magnetism in the 19th century.

Another important example of physics giving rise to technology is the silicon 'chip' that triggered the computer revolution in the last three decades of the 20th century.

Physics has contributed enormously to the advancement of astronomical studies and discovery, medical instrumentation, the diagnosis and treatment of cancer and other diseases, information technology and broadband internet communication.

The most significant area to which physics has contributed, and will continue to contribute, is the development of alternative energy sources. The fossil fuels of the Earth are dwindling fast, so there is a need to discover and develop new, affordable sources of energy. Considerable progress has already been made in this area of technology (for example, in conversion of solar energy, geothermal energy, etc. into electricity), but more needs to be accomplished.

Scope of physics

We can get an idea of the scope of physics by looking at its various sub-disciplines. There are basically two major areas of interest, or domains: macroscopic and microscopic.

In the macroscopic domain, we find phenomena at the laboratory, terrestrial and astronomic scales. 'Classical physics' deals mainly with macroscopic phenomena and includes subjects like mechanics, electrodynamics, optics and thermodynamics.

The microscopic domain of physics deals with the constitution and structure of matter at the very small scales of atoms and nuclei and even lower scales of length, and their interaction with different probes such as electrons, photons and other elementary particles. Classical physics is inadequate to handle this domain of knowledge and **quantum theory/** physics is currently accepted as the appropriate framework for explaining microscopic phenomena.

Unit 11.1 Measurement

Topic 1: Quantities and units

In line with the Grade 11 Syllabus (p. 7) Unit 11.1 provides students with an understanding of the parameters of measurement including quantities, units and errors. This Topic introduces 'quantities and units':

- Using the SI system and changing from one unit to another.
- Scientific notation and significant figures.
- Solving physics problems.
- Graphing.

A physical quantity is anything that can be measured, such as time, length, voltage and **current**. To measure a physical quantity a **unit** of measurement is needed. To make communication between scientists around the world easier, most countries use an international system of units called the **SI system** (from the French Système Internationale). The system is based on three **fundamental units**.

Physical quantity	Unit	Abbreviation
Mass (m)	kilogram	kg
Length (l)	metre	m
Time (t)	second	s

The three fundamental SI units.

All other units are derived from the three fundamental units, eg the unit for **speed** is metres per second (m s^{-1}).

Sometimes a single unit replaces a combination of units. These derived units are often named after a famous scientist. The unit for **force** is the kg m s^{-2}. This combination of basic units can be replaced by the **newton** (N) (1 N = 1 kg m s^{-2}).

The SI system is a **metric** system and uses prefixes to express measurements with suitably-sized numbers. The table below shows prefixes commonly used in physics:

Prefix	Symbol	Multiplier	
mega	M	1 000 000	10^6
kilo	k	1 000	10^3
hecto	h	100	10^2
centi	c	0.01	10^{-2}
milli	m	0.001	10^{-3}
micro	μ (greek letter mu)	0.000001	10^{-6}
nano	n	0.000000001	10^{-9}

Prefixes used in the SI system.

Changing from one unit to another

Changing from a large to a smaller unit involves multiplication. Changing from a small to a larger unit involves division. The flow charts overleaf show how to change from kilo to centi and milli and vice versa.

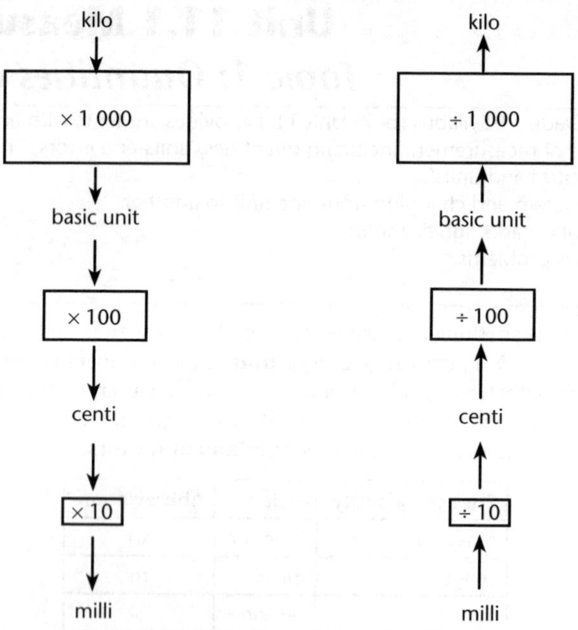

Flow charts of unit conversions.

Example A

154 m to cm:
154 m = 154 × 100 = 15 400 cm
16 500 mg to kg:
16 500 mg = (16 500 ÷ 1 000) = 16.5 g (changing from mg to g)
= (16.5 ÷ 1 000) = 0.0165 kg (changing from g to kg)
1 950 mN to N
1 950 mN = (1 950 ÷ 1 000) = 1.950 N

Scientific notation

Scientific notation or index form is convenient for writing large or small numbers. The distance from the Earth to the Sun is 150 000 000 km. This is more conveniently written as 1.5×10^8 km. The measurement 0.002 m can be written as 2×10^{-3} m or using a prefix, ie 2 mm (the first 'm' in 2 mm refers to milli).

Example B

Expressing in scientific notation:

- 0.0000 0035 m = 3.5×10^{-7} m
- 0.0035 g = 3.5×10^{-3} g or 3.5 mg
- 358 000 000 J = 3.58×10^8 J

Significant figures

A **significant figure** (*sig fig*) is any digit not used as a placeholder. Zeroes are not significant if they are used only to indicate the position of the decimal point.

Example C

534 – all 3 digits are significant.

0.046 – the first two zeroes are not significant, the first significant figure is 4 and the second is 6.

420.05 – the zeroes are significant because they are not placeholders.

21.00 – there are 4 significant figures.

Problem solving in physics typically involves calculations using values that have different numbers of significant figures. As a general rule, always round the final answer to the same number of significant figures as the value in the calculation that has the least number of significant figures. It is good practice to state in brackets how many significant figures the answer has been rounded to.

Example D

205.8 ÷ 5.9 = 34.88 = 35 (2 sig fig) – *5.9 has only 2 sig figs.*

Solving physics problems

There are a number of steps to follow to make solving physics problems easier.

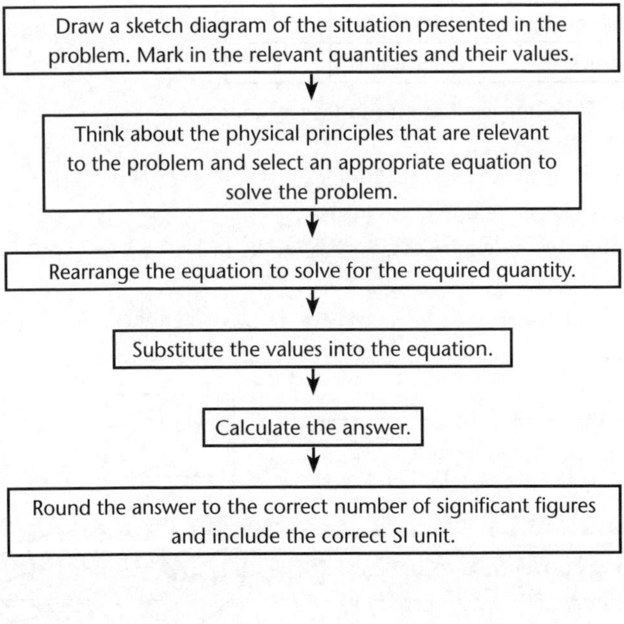

Draw a sketch diagram of the situation presented in the problem. Mark in the relevant quantities and their values.

↓

Think about the physical principles that are relevant to the problem and select an appropriate equation to solve the problem.

↓

Rearrange the equation to solve for the required quantity.

↓

Substitute the values into the equation.

↓

Calculate the answer.

↓

Round the answer to the correct number of significant figures and include the correct SI unit.

Unit 11.1 Activity 1A: Units and measurement

1. For each of the quantities in the table shown, state the name and symbol for the SI unit it is measured in.

Quantity	SI Unit	
	Name	Unit
Force	newton	N
Velocity		
Mass		
Voltage		
Charge		
Heat energy		
Pressure		
Magnetic field		
Energy		
Density		

2. Express the following using SI units. *The first line has been done for you.*

Power of a heater	*2.5 kW*	*2 500 W*
Height of a student	0.0018 km	**a**
Time for a short journey	0.1 hour	**b**
Current in a circuit	250 mA	**c**
Human daily energy needs	10 MJ	**d**
Mass of a truck	1.2 tonne	**e**

3. Convert the following values to the indicated units:
   a. 330 mA = ____ A **b**. 35 276 g ____ = kg **c**. 1.7 MJ ____ = J
   d. 5.3 km = ____ m **e**. 17.4 mg ____ = g **f**. 2 hr 13 min ____ = s
   g. A car travelling at 50 km h^{-1} to m s^{-1}.
   h. Aluminium has a **density** of 2.7 g cm^{-3} to kg m^{-3}.
4. What is the SI unit of:
   A. force **B**. work **C**. power **D**. acceleration?
5. Convert 0.018 km to m.

6. Tom set up the circuit shown to determine the resistance of an unknown resistor which had the resistance colour code rubbed off. The voltmeter measures potential difference in volts and the ammeter measures current in amperes.
   a. For meter A, state what quantity it measures and the reading shown on it.
   b. For meter B, state what quantity it measures and the reading shown on it.

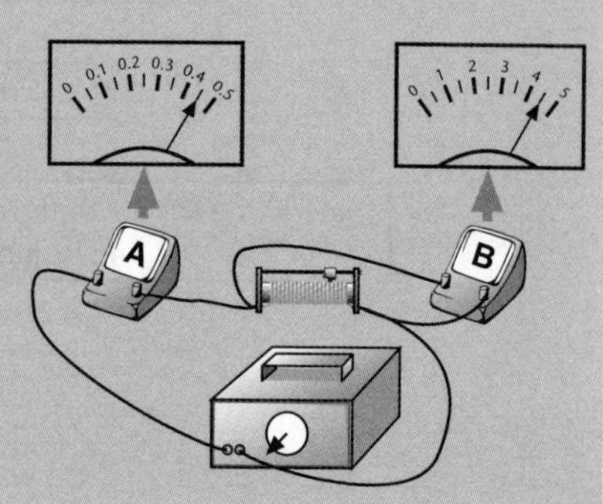

7. Steven and Jenni are selling vegetables at an open-air market. Steven sets up a device for measuring the mass of the vegetables. The vegetables are placed in a bucket. Steven slides a slider weight on a rod until it balances the bucket. The mass is worked out from the position of the slider. To help him work out the mass, Steven records the slider position with some known masses.

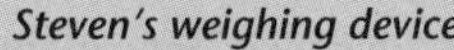
Steven's weighing device

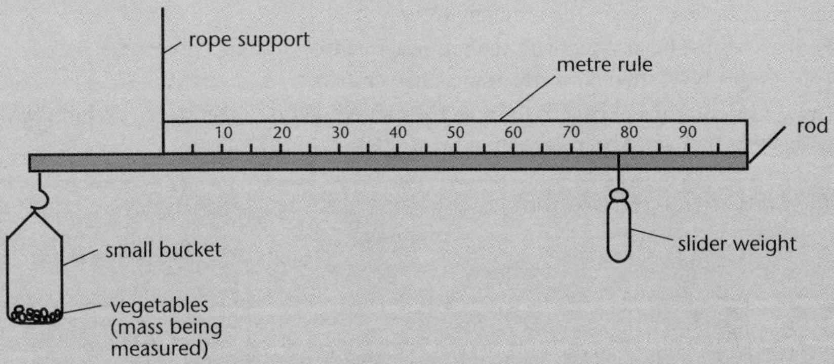

Steven's data is:

Mass (kg)	Slider position (cm)
0	4
0.5	19
1.0	34
2.0	64
3.0	94

a. A customer wants 2.2 kg of tomatoes. Where should Jenni place the slider?

b. Work out the mass of the kumara shown (in kilograms).

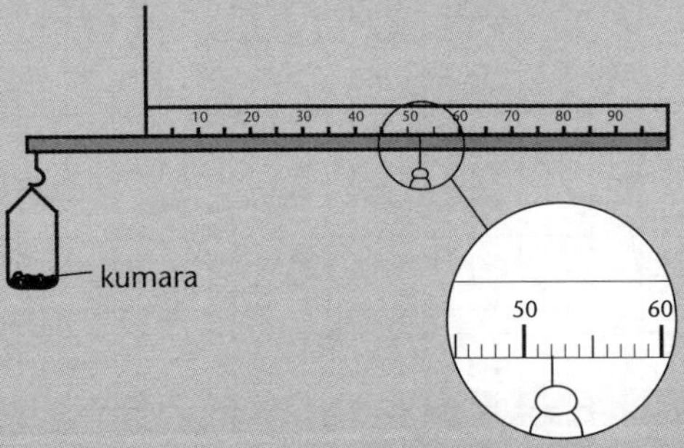

c. The device is not useful for measuring very heavy masses. What is the largest mass Steven could measure with this device?

d. Jenni suggests that if they used a slider twice as heavy as the one they are using at present, they could measure much larger masses as well as lighter ones. Discuss whether or not she is correct.

Graphing

In physics, **graphs** are often drawn to study the relationship between variables. Graphs must be drawn correctly, using the following rules.

- Write a title of the form 'A versus B', where A is the **dependent variable** and B is the **independent variable**.
- Unless instructed otherwise, place the dependent variable on the vertical axis and the independent variable on the horizontal axis.
- Both axes must be labelled with the physical quantity and unit (in brackets).
- Points must be plotted correctly, using small crosses.
- Draw a smooth line of best fit through the plotted points. Take care to exclude wayward points that may have been measured incorrectly.

Unit 11.1 Activity 1B: Graphing

1. Dipa and Maria set up an experiment to investigate how the volume of a gas changes with temperature. Air is trapped inside the glass tubing by a bead.
   The length of the air column can be used to represent the volume of air.
   The data table shows their measurements.

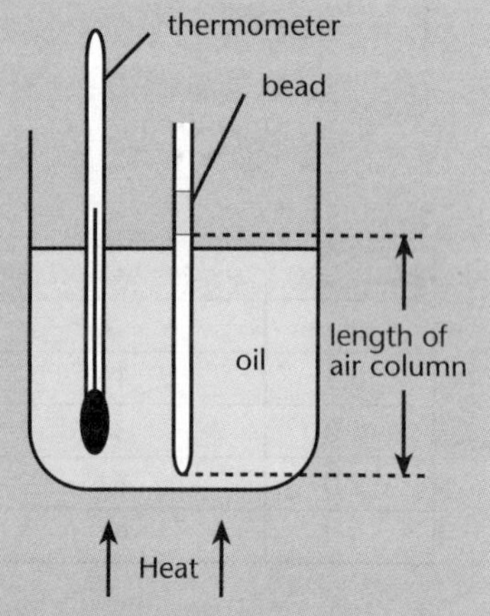

Temperature (°C)	Length (cm)
0	12.9
20	13.9
40	14.8
60	15.7
80	16.7
100	17.6
120	18.6

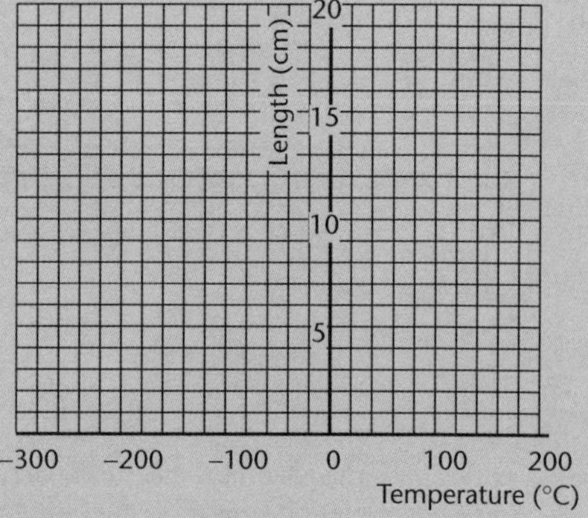

   a. Draw a graph of column length versus temperature.
   b. Extend the graph line to find the temperature at which the gas would have zero volume.

c. When the tubing is placed in a freezer, the length of the air column contracts to 11.5 cm. What is the temperature inside the freezer?

d. Susan and Kila set up a similar experiment. At 0°C, the length of their air column is 10.0 cm. On your graph, draw a line to show how the length of the air column will change.

2. Graph the table of results that shows the speed of a Formula One racing car as it accelerates away from the pit stop at a Grand Prix.

Speed (m s^{-1})	Time (s)
0	0
19	2
36	4
48	6
57	8
64	10
69	12

3. Limu pulls a trolley along the bench and measures the acceleration for various forces. Her results are recorded in the table.
   - **a.** Draw a graph of acceleration versus force.
   - **b.** Find the slope of the graph.
   - **c.** Describe how the mass of the trolley can be found from the slope of the graph. *You do not need to calculate the mass.*

A (m s^{-2})	F (N)
1.5	2
3.0	3
4.5	4
6.0	5
7.5	6

4. The following equation is suggested for the variation of two quantities x and t:

$$x = at^b$$

where a and b are constants. Use the data below to plot a suitable linear graph.

x	1.86	2.45	3.16	4.07	5.37	6.61	8.71
t	1.60	2.50	4.00	6.40	10.0	15.9	25.1

5. The following set of data was obtained from a **radioactive decay** experiment:

N	6630	3650	2000	1000	600	330	165	90	50	30	15
T	0	5	10	15	20	25	30	35	40	45	50

N is the number of counts per second and T is the time in seconds. The equation for such a decay is:

$$N = N_0e^{-kt}$$

where N_0 and N represent the initial activity and the activity after a time t, and k is a constant. Plot a graph of ln N against t.

Unit 11.1 Activity 1C: Multiple choice

1. Suppose the two lines in the following diagram represent successive displacements of 12 km to the North and 16 km to the East. What is the magnitude, in kilometres, of the resultant displacement?

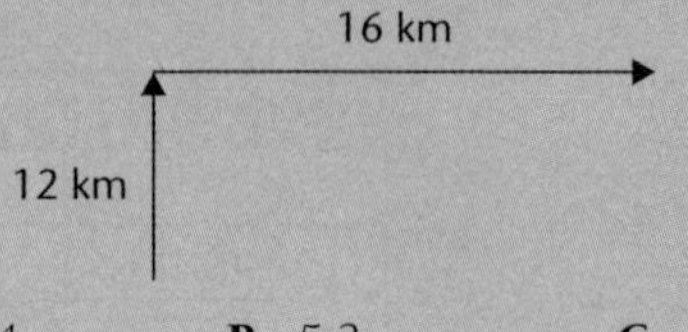

A. 4 **B**. 5.3 **C**. 20 **D**. 28 **E**. 192

2. The following diagram shows the field of vision of a microscope. The scale shown is in micrometres. What is the length of the microbe shown in the field of vision?

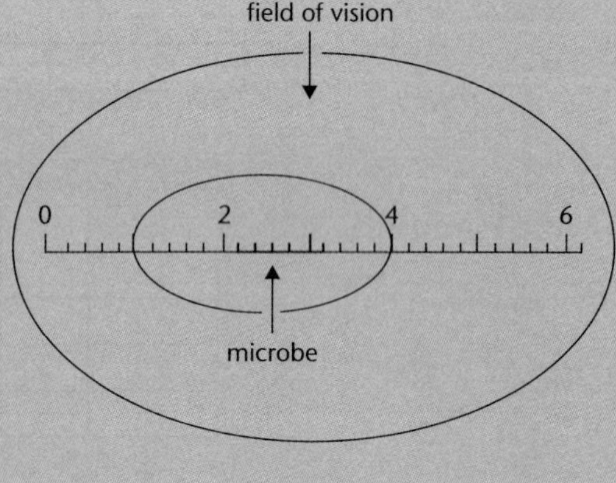

A. 1.6 μm **B**. 2.6 μm **C**. 3.0 μm **D**. 2.0 μm **E**. 4.0 μm

Unit 11.1 Measurement
Topic 2: Vectors and scalars

Topic 2 continues the introduction of 'quantities and units', following on from Topic 1 and in line with the syllabus (pp. 7–8), by looking at:

- Scalar quantities and vectors.
- The International System of Units (SI).
- Scientific notation.
- Use of prefixes.

Introduction

Certain quantities can be described completely when their magnitudes are stated in appropriate units:

For example:

- A speed of 50 km/hr.
- A mass of 10 kg.
- A temperature of 30°C.
- A time of 3 seconds.

These quantities are called **scalar** quantities.

Other quantities need to be described by specifying two of their properties or characteristics; their **magnitude** and their **direction**. Therefore, they are not completely defined unless both properties are specified.

For example:

- A velocity of 5 m/s vertically upward.
- A force of 10 N vertically downward.
- A displacement of 8 km due East.

These quantities are called **vectors**.

Representation of vectors

Because a vector has both magnitude and direction, we can represent it by a segment of a line, where the **length** of the line represents the **magnitude** of the vector quantity and the **direction** of the line show us which way the quantity goes.

Thus the line AB, in the following figure, can be used to represent a **displacement** vector of 3 m North East.

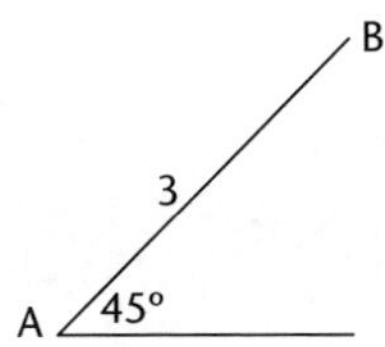

We can indicate that a line segment represents a vector, by using any of the following vector symbols: **AB**, $\overrightarrow{\mathbf{AB}}$, **r**.

In the first two cases, the sense of the vector is given by the order of the letters, but in the case of the symbol r, since it does not include any indication of sense, it must be accompanied by an arrow on the diagram or alternatively, an arrow on top of it ($\vec{\mathbf{r}}$).

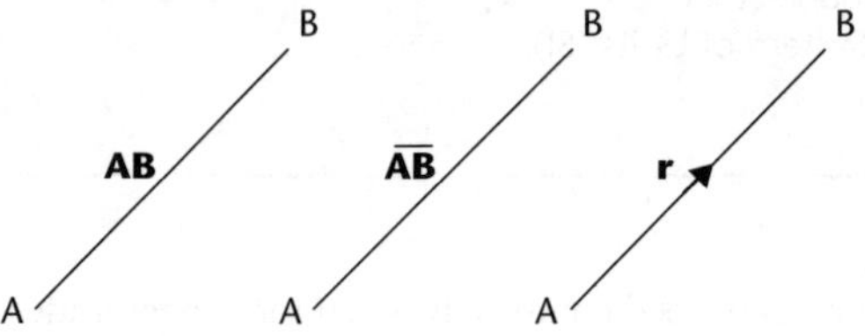

Vectors and scalars

A **scalar** is a physical quantity that has size (magnitude) but no direction.

A **vector** is a physical quantity that has both size and direction.

Scalar quantities	Vector quantities
distance	displacement
speed	velocity
time	acceleration
temperature	force

When drawing vectors, the length represents the magnitude and the arrow represent the direction.

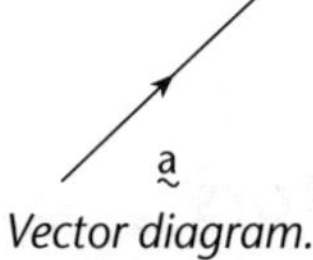

Vector diagram.

Vectors can be added graphically by placing the tail of the second vector on the head of the first. The vector drawn from the tail of the first vector to the head of the second vector is the **resultant**. It is identified by a double arrow.

Example E

Tom walks 6 m in an easterly direction followed by 8 m in a northerly direction. Find:

a. The total distance Tom has travelled.

b. The shortest distance from the starting point.

Solutions:

a. The total distance travelled is 6 + 8 = 14 m.

b. The shortest distance from the starting point is calculated using a vector diagram.

$$\text{Shortest distance} = \sqrt{6^2 + 8^2} = \sqrt{36 + 64} = \sqrt{100} = 10 \text{ m}$$

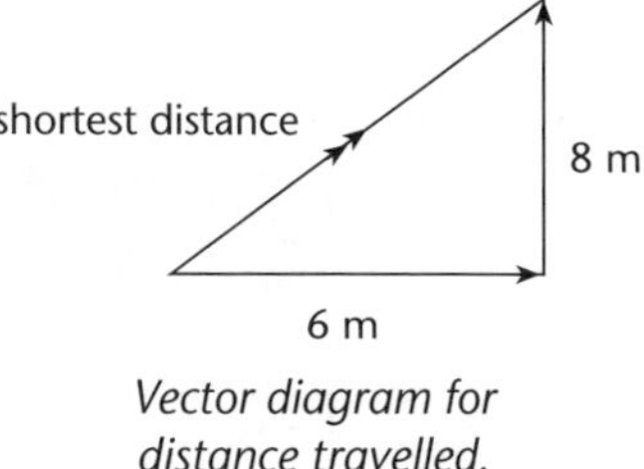

Vector diagram for distance travelled.

Unit 11.1 Activity 2A: Vectors

1. Which of the following gives three vector quantities?

Weight, mass, acceleration. Work, force, distance.

Velocity, acceleration, force. Voltage, resistance, current.

2. State whether each of the following is a vector or a scalar.

A. Speed. **B.** Temperature. **C.** Force. **D.** Work.

3. In a soccer training session, player **A** passes the ball 25 metres due East to player **B**. Player **B** passes the ball 60 metres due North to player **C**.

a. Copy the grid alongside, then draw vectors that represent the displacement for each of the two stages of the soccer ball's path.

It takes the ball 5 seconds to travel from player **A** to player **C**.

b. What is the magnitude of the displacement from **A** to **C**? Draw it on your grid.

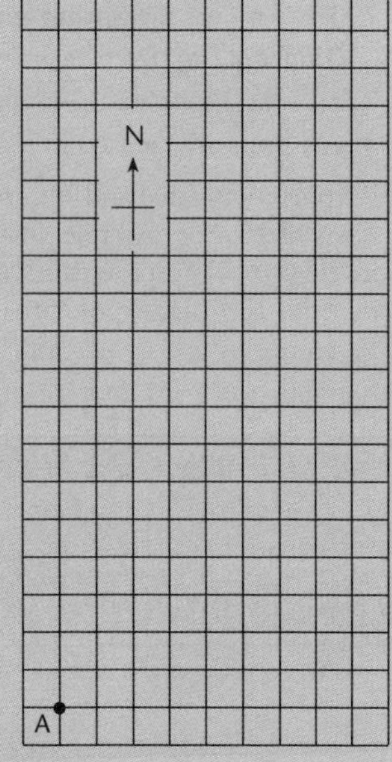

Scale:
For distance 1 square = 5 m
For velocity 1 square = 5 ms^{-1}

4. Allan wants to row his canoe across a river. He rows so that his velocity in a southerly direction relative to the water is 2.5 m s^{-1}. The river is flowing East at 1.5 m s^{-1}.

a. Draw a scale diagram of these velocities, and draw in the resultant velocity of the canoe.

b. Calculate the resultant velocity of the canoe relative to the river bank.

System of units and changing units

In 1971 seven quantities were selected as base quantities to form the basis of the International System of Units, abbreviated as SI, and known as the metric system.

This table gives the basic or fundamental physical quantities measured in everyday life and their International System (SI) of standard units and symbols.

SI base units		
Base quantity	**Name**	**Symbol**
length	metre	m
mass	kilogram	kg
time	second	s
electric current	ampere	A
thermodynamic temperature	kelvin	K
amount of substance	mole	mol
luminous intensity	candela	cd

The units for the three base quantities you will be studying in this Unit are:

- Metre (m) for measuring length.
- **Kilogram** (kg) for measuring mass.
- Second (s) for measuring time.

Many SI units are defined in terms of these three base units.

To express very large and very small quantities we use scientific notation, which employs powers of 10. For example, instead of writing 0.0000000056, we write 5.6×10^{-9}. So, how does this work? You can think of 5.6×10^{-9} as the product of two numbers: 5.6 (the digit term) and 10^{-9} (the exponential term). Here are some other examples:

3,560,000,000 m = 3.56×10^{9} m
0.000 000 492 s = 4.92×10^{-7} s.

Unit 11.1 Activity 2B: Scientific notation

Write the following numbers in scientific notation.

1. 125 000 kg

2. 0.000 932 m

When using very large or very small quantities to express results in measurements, we use the following SI prefixes to form decimal multiples and submultiples of SI units. The 20 commonly used SI prefixes are given in the table below:

SI prefixes					
Factor	Name	Symbol	Factor	Name	Symbol
10^{24}	Yotta	Y	10^{-1}	deci	d
10^{21}	zetta	Z	10^{-2}	centi	c
10^{18}	exa	E	10^{-3}	milli	m
10^{15}	peta	P	10^{-6}	micro	μ
10^{12}	tera	T	10^{-9}	nano	n
10^{9}	giga	G	10^{-12}	pico	p
10^{6}	mega	M	10^{-15}	femto	f
10^{3}	kilo	k	10^{-18}	atto	a
10^{2}	hecto	h	10^{-21}	zepto	z
10^{1}	deka	da	10^{-24}	yocto	y

Attaching a prefix to an SI unit has the same effect as multiplying by the associated factor. For example:

$$2.35 \times 10^{-9}\ \text{s} = 2.35\ \text{nanoseconds} = 2.35\ \text{ns}$$

Sometimes you will need to change the units in which a physical quantity is expressed. The way of doing this is by using a conversion factor that multiplies the original measurement. This conversion factor is usually a **ratio** of units that is equal to unity (since multiplying something by 1 leaves the quantity unmodified). For example, because 1 metre and 100 centimetres are identical length measurements, you can write:

$$\frac{1\ \text{metre}}{100\ \text{cm}} = 1 \text{ and } \frac{100\ \text{cm}}{1\ \text{m}} = 1$$

Note that this is not the same as writing 1/100 = 1 or 100 = 1; you must treat the number and its units together, as a whole entity.

Example F

Q. A vehicle is travelling at a speed of 1.11 metres per second. Express this speed in light-years per year.

A. A light-year (ly) is the distance that light travels in 1 year, 9.46×10^{15} km. Therefore,

$$1.11\,\frac{\text{m}}{\text{s}} = 1.11\,\frac{\text{m}}{\text{s}}\;\frac{1\ \text{light year}}{9.46 \times 10\text{km}} \times \frac{1\ \text{km}}{1000\ \text{m}}\;\frac{3.16 \times 10^{7}\,\text{s}}{1\ \text{year}}$$

$$= 3.71 \times 10^{7}\ \text{ly/y}.$$

Unit 11.1 Activity 2C: Scientific notation, errors and units in measurements

1. How many significant figures are there in each of the following measurements?

a. 3.15 mm **b.** 0.00315 s **c.** 6.025 cm **d.** 36 km

e. 3.34×10^5 km **f.** 36.00 s

2. A very good watch is fast or slow by not more than 1 minute in a month. Another watch is found to be accurate to 0.01%. Which is the more accurate?

3. Express each of the following numbers to the nearest power of ten:

a. 1.21×10^2 **b.** 9.01×10^4 **c.** 0.0008

d. 0.0201

4. A watch ticks 5 times each second. Find to the nearest power of ten the number of times the watch ticks:

a. In a day. **b.** In a year.

5. There are 2.45×10^8 people living in the United States, and 1.81×10^7 of these people live in New York City. How many live in the rest of the country?

6. Write the following numbers using appropriate prefixes.

a. 0.0000063 s

b. 56 000 000 hz

Unit 11.1 Measurement

Topic 3: Measuring instruments

In this Topic we look at various measuring instruments used in the laboratory (Syllabus p. 8), in particular:

- Vernier caliper.
- Micrometer screw gauge.
- Digital multimeter.
- Other instruments such as timers, thermometers and barometer.

This Topic also looks at methods for increasing accuracy in measurement.

Taking measurements

When taking measurements it is important to:

- Choose the most suitable measuring instrument.
- Know how to use the measuring instrument.
- Know methods for increasing the accuracy of measurements.

Measuring instruments

A **vernier calliper** is often used when measuring small lengths and diameters. It has special jaws to measure inside diameters.

A vernier calliper used to measure the diameter of a disk

Vernier callipers have a vernier scale that enables them to indicate an extra one or two significant figures in the reading.

A vernier is an extra scale that slides down the main scale. Nine divisions on the main scale are divided into ten divisions on the vernier. The extra digit is read off the vernier where it coincides nearest to a division mark on the main scale.

Example A

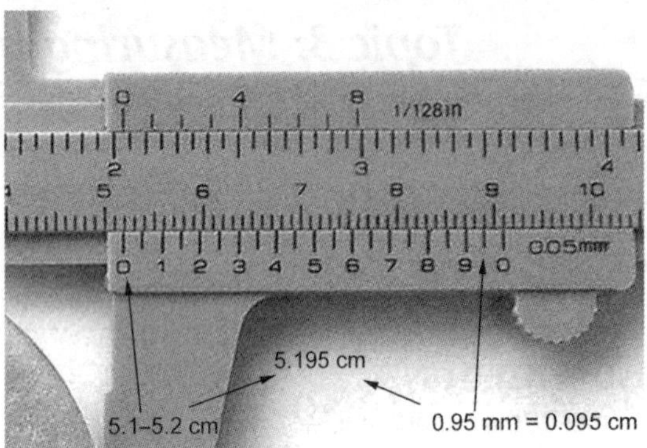

A magnified view of the vernier calliper scale

The first part of the reading is made on the main scale, where the '0' mark of the vernier touches it. This reading is between 5.1 and 5.2 cm and is very close to 5.2 cm.

Next, the vernier scale is inspected to find where it most closely meets a division on the main scale. This happens at the $9^1/_2$ mark on the vernier. Using the vernier scale label, '0.05 mm', $9^1/_2$ divisions represents 0.95 mm.

The final reading is 5.1 + 0.095 = 5.195 cm.

This kind of vernier scale has added 2 significant figures to what would have been read as 5.2 cm without the vernier scale.

Measurement of small lengths can also be made using a **micrometer screw gauge**.

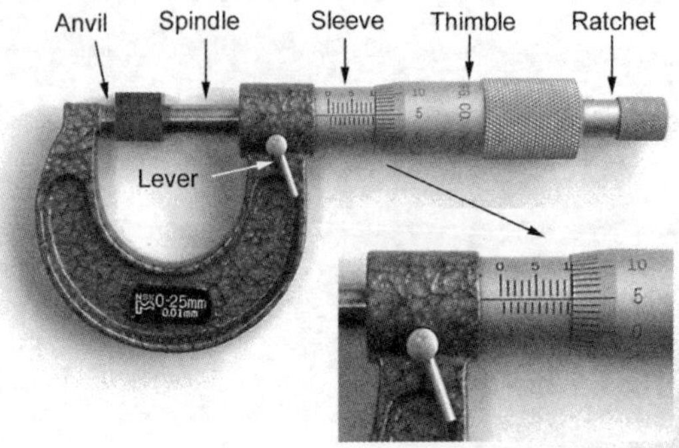

Turning the spring ratchet clockwise screws the thimble and attached spindle along the sleeve towards the anvil. The object to be measured (eg a small metal cube) becomes gripped gently between the anvil and spindle. The ratchet applies a small amount of force and prevents the micrometer from squeezing the object too tightly. The whole micrometer mechanism can be locked by turning the lever.

Whole millimetres are indicated above the line along the sleeve and half millimetres are shown below. The scale on the thimble indicates hundredths of a millimetre.

Micrometer and magnified view of its scale

Example B

The reading on the micrometer shown in the figure on the previous page is:
Sleeve – 10 mm (and no $^1/_2$ mm)
Thimble – 5 divisions = 0.05 mm
Total reading = 10.05 mm

A **digital multimeter** can make different kinds of electrical measurements and has several ranges for each kind.

The meter is of medium quality and can measure to four significant figures. It is connected as an ohmmeter to measure the electrical resistance of a resistor and is switched to the 20K ohm range (ie the meter is able to measure resistances up to 20 kΩ). If the resistance is higher than 20 kΩ, the meter displays an 'over-range' symbol and the central range switch would need to be moved clockwise until a reading is obtained.

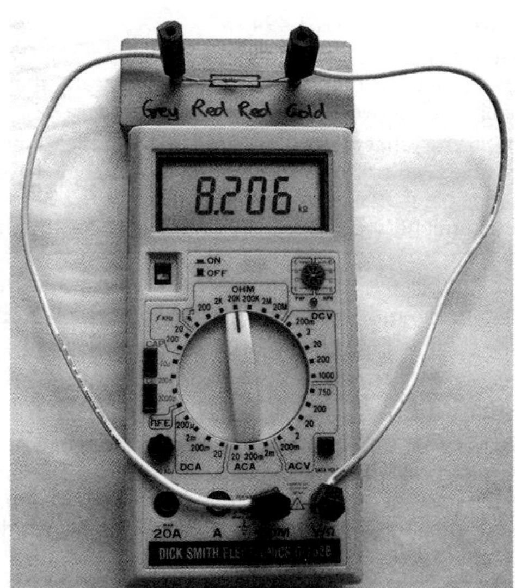

A digital multimeter used as an ohmmeter

Multimeters are usually designed to measure voltage, current and resistance. (The meter shown in the figure above can also measure frequency, capacitance and some transistor values.)

The four sockets below the range switch are used to connect leads to parts of a **circuit**. First select the appropriate range using the range switch and connect two leads to the correct sockets before switching on the circuit being measured. If the range is not known, switch to the highest range and switch to lower, more sensitive ranges as needed. Never switch from one kind of measurement to another (eg voltage to current) while the leads are connected to a circuit, as this could damage the multimeter.

In general, a meter with a digital display is used to measure steady values or ones that change slowly.

Meters with moving pointers (analogue meters) or ones with a bar graph display are better able to indicate readings that fluctuate more rapidly.

Example C

Vehicle speedometers tend to use an analogue pointer or a bar graph.

Methods for increasing accuracy

Usually a measurement can be made more accurate by taking repeated readings and averaging them. The variation in the readings can be used to guide how many significant figures should be written in the measurement.

Example D

A toy car rolls down a slope. Five repeated readings taken with a stopwatch are 3.45 s, 3.55 s, 3.57 s, 3.42 s, 3.48 s.

The average of these is 3.494 s.

There is significant variation in the readings at the second decimal place, so it is probably best to round the number to 3.5 s. This indicates that the time to roll down the slope most probably lies between 3.45 and 3.55 seconds.

Other factors that affect the accuracy of this measurement are the way in which the stopwatch was started and stopped. Most people have a delayed reaction time of about 0.2 s. It is possible that all these readings could be 0.2 s too short if the person using the stopwatch waited until he/she saw the toy car move or heard someone say "Go".

Taking a *multiple measurement* of some quantity and then dividing by the number of items can give a much more accurate measurement, since **errors** in taking the measurement would be smaller in proportion to the larger reading.

Example E

A micrometer is used to measure the thickness of a page in a book. Pages 1 to 200 are selected (giving 100 thicknesses). The micrometer reading for these pages is 10.30 mm. The thickness of one page is 10.30/100 = 0.1030 mm (4 sig figs). However, the reading on the micrometer when it is used to measure only one thickness is 0.10 mm, which is only accurate to 2 sig figs.

Most measuring instruments need to be checked for zero error – do they read zero when nothing is being measured? Many measuring instruments (eg weight scales) can be adjusted to read zero, but some can not. If an instrument has a zero error it must be subtracted from all readings made with that instrument.

Example F

a. A measurement of the length of a microscope slide taken with a vernier calliper reads 7.56 cm. What is the actual measurement, if the vernier calliper has a zero error of +0.03 cm?

b. The 20 volt range on a digital multimeter has a zero error of –0.02 volts. A reading is taken of the voltage of a 9 volt battery. This gives a reading of 9.04 volts. What is the actual measurement of this voltage?

Solutions:

Subtracting the zero error from the reading gives:

a. 7.56 – (+0.03) = 7.53 cm.

b. 9.04 – (–0.02) = 9.06 volts.

Errors due to **parallax** can happen when using some measuring instruments. Care needs to be taken to place your eyes directly in front of the measuring scale. This is so that no error is made due to distortions in reading the scale from an angle.

Example G

Parallax error can be present when using a metre ruler.

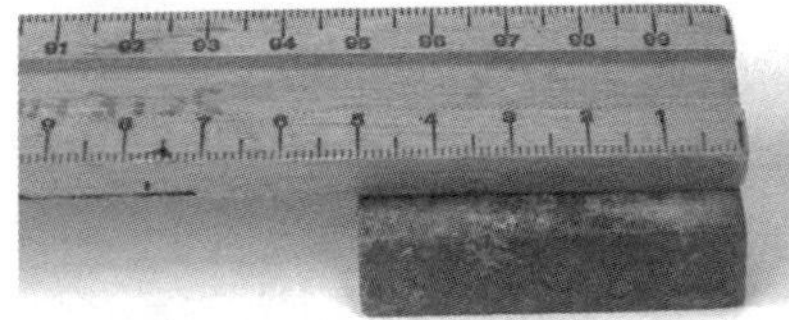

The ruler is positioned so that its scale is not near the length being measured. The left-hand end of the metal block is aligned with the 5.0 cm mark of the ruler. However, it is difficult to see if the right-hand end is aligned with the ruler's zero mark.

Parallax error in reading a ruler

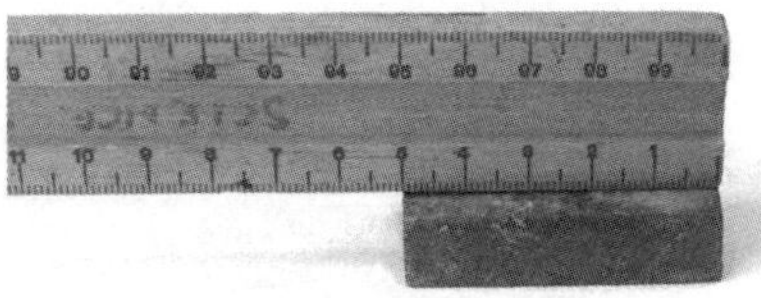

A better method of positioning the scale is next to the length being measured. This reduces parallax error considerably.

Reducing parallax error

Example H

The analogue current meter shown in the figure to the right uses a mirror next to the measuring scale.

A person taking a reading places their eyes so that the **reflection** of the pointer is hidden by the pointer itself. When this happens, the person's eyes are directly over the pointer and a reading can then be taken.

The meter is 'zeroed' correctly (ie it has no zero error).

This particular meter measures alternating current (shown by the ~ symbol beneath the A), so it does not matter which way round the leads are connected to the meter.

The range of possible measurements for this meter is from 0.2 to 1.0 A. The scale is non-linear from zero to 0.2 A and is kept blank.

Ensuring there is no parallax error in reading a meter

Unit 11.1 Activity 3A: Taking measurements

1. Determine the readings indicated on the following scales:

a.

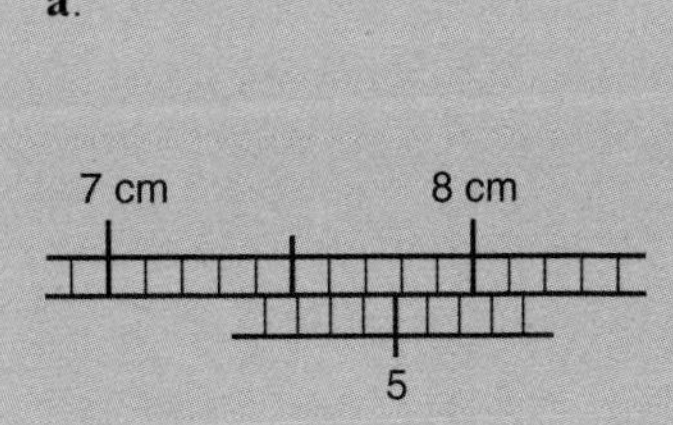

b.

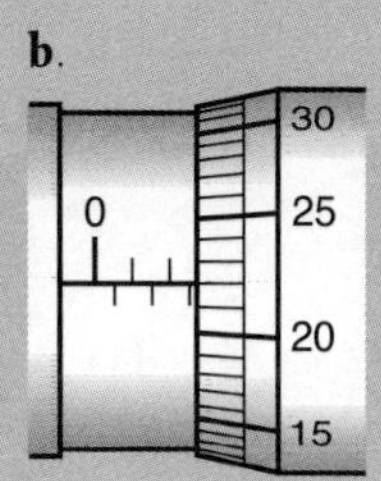

c.

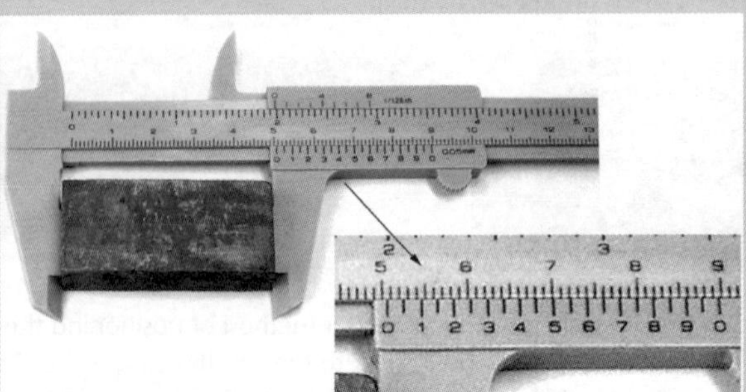

2. The diagram shows a close-up picture of the scale reading of a micrometer.

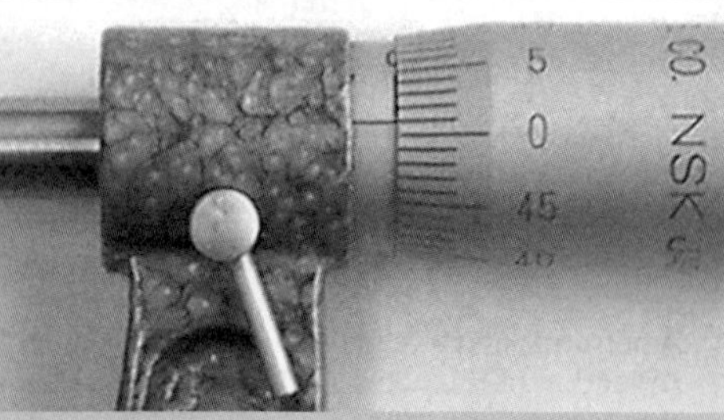

a. Estimate the micrometer's zero error.

b. The micrometer reading for pages 1–200 of a book is 10.30 mm; the thickness of one page is thus 10.30/100 = 0.1030 mm (4 sig figs). When the micrometer is used to measure only one page, it gives a reading of 0.10 mm, which is only to 2 sig figs.

How would the zero error affect the measurement of the paper thickness by both methods?

3. The colour code for the resistor (grey, red, red, gold) shown in the figure on p. 23 indicates the resistor's value should be 8 200 Ω ± 5%. Show whether or not the multimeter measurement lies within the manufacturer's tolerance range.

4. a. Discuss the accuracy in measuring a length of about 30 cm using a wooden metre ruler.

b. i. Estimate how accurately a metre ruler could measure the length of a room which is between 9 and 10 metres long.

ii. List one advantage and one disadvantage in using a long tape measure to measure the length of the room.

5. A heavy mass attached to a spring bounces up and down. Discuss ways that a stopwatch can be used to measure the period of the mass (the time for one complete up-and-down motion).

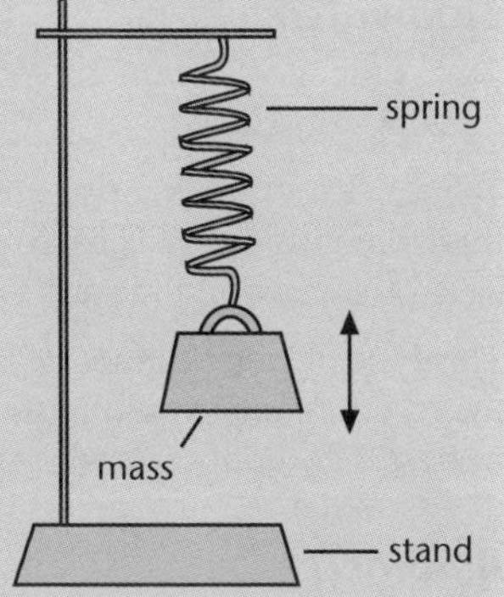

6. A reading on a balance was recorded as 98.76 g, but on checking, the balance required 0.43 g to be added before the balance read zero. What was the actual value of the mass measured?

Other measuring instruments

A description of these instruments is provided on the following page.

Stopwatch

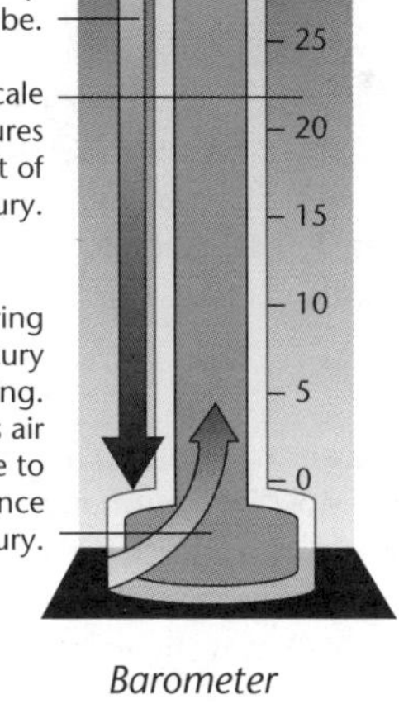

Barometer

Ammeter

Voltmeter

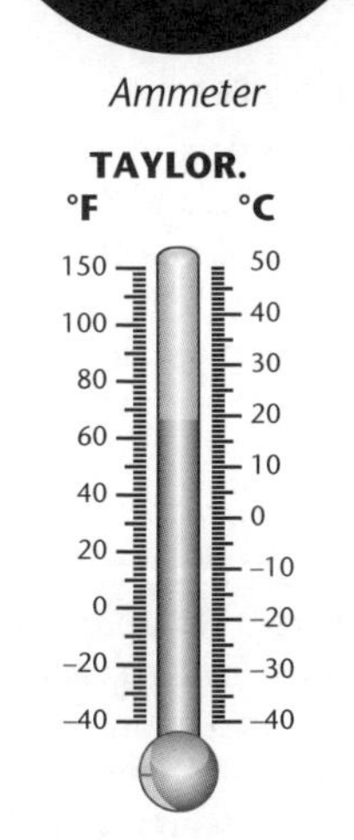

Thermometer

Stopwatch

This is the most common type of clock which is used in science laboratories. It is also called a stop-clock.

To measure a time interval, the clock must be started and stopped by hand. An error is therefore introduced because it takes a certain time to start or stop the clock. This error is called the reaction time and is on average about 0.3 seconds.

This type of clock cannot be used to time an interval of less than about one second since the reaction time would introduce too great an error. A typical stopwatch can be read to the nearest 0.2 s and a typical stop-clock to the nearest 0.5 seconds.

Barometer

This is a modified kind of mercury manometer with the end closed. It is used to measure atmospheric pressure and is called a mercury barometer.

The glass tube is filled with mercury and then inverted into the bowl of mercury. If the tube is long enough, the level of the mercury will drop, leaving a vacuum at the top of the tube, since atmospheric pressure can only support a column of mercury 76 cm high. That is, a column of mercury 76 cm high exerts the same pressure as the whole atmosphere.

Note that $P = \rho gh$, with $\rho = 13.6 \times 10^3$ kg/m^3 for mercury and $h = 76$ cm gives us $P = 1.013 \times 10^5$ N/m^3 = 1 atmosphere.

Household barometers are usually of the aneroid type.

Ammeter

These are laboratory and industrial electromechanical devices that are used to measure electric currents from 1 μA (10^{-6} A) to several hundred amperes. There are ammeters to measure DC currents and ammeters to measure AC currents. The figure above illustrates a DC ammeter. A mechanical type of motion called D'Arsonval movement is used in most DC ammeters as the current detector. Typical laboratory bench meters of this type have accuracies of about 1 percent of their full-scale readings due to inaccuracies of the meter movement.

Voltmeter

This is a device used to measure electric potential or voltages in electric circuits. There are also DC and AC voltmeters. Most DC voltmeters also use D'Arsonval movements. The D'Arsonval movement itself can be considered to be a voltmeter if we consider and note that the current flowing in it, multiplied by its internal resistance, causes a voltage drop. For example, a 1-mA full scale, 50Ω movement has a 50 mV drop across it when 1 mA is flowing in the movement.

Thermometer

There are many types of thermometer, but each makes use of a particular thermometric property (i.e. a property whose value changes with **temperature**) of a particular thermometric substance. For example: a mercury-in-glass thermometer makes use of the change in length of a column of mercury confined in a capillary tube (like the one shown in the above figure), of uniform bore; a platinum resistance thermometer makes use of the increase in the electrical resistance of platinum with increasing temperature. The most common and popular is the mercury thermometer.

Unit 11.1 Measurement

Topic 4: Dimensional analysis

Continuing the exploration of measurement, Topic 4 examines dimensional analysis (Syllabus p. 8) and the concept of fundamental units and the relationship of formula to units. This Topic looks at:

- Dimensions of physical quantities.
- Dimensional formula and dimensional equations.
- Methods of dimensions.

Dimensions of physical quantities

The nature of a physical quantity is described by its dimensions. The dimension is the qualitative nature of a physical quantity (length, mass, time). All the physical quantities represented by derived units can be expressed in terms of some combination of seven fundamental or base quantities.

The dimensions of a physical quantity are the powers (or exponents) to which the base quantities are raised to represent that quantity.

Quantity	Dimension
Area	$[L]^2$
Velocity	$[L]\ [T]^{-1}$
Force	$[M]\ [L]\ [r]^{-2}$
Energy	$[M]\ [L]^{-2}\ [T]^{-2}$
Power	$[M]\ [L]^{-2}\ [T]^{-3}$
Volume	$[L]^3$
Acceleration	$[L]\ [T]^{-2}$
Pressure	$[M]\ [L]^{-1}\ [T]^{-2}$
Momentum	$[M]\ [L]\ [T]^{-1}$

In mechanics, all the physical quantities can be written in terms of the dimensions [L], [M] and [T], as illustrated above.

Note than in this type of representation, the magnitudes are not considered. It is the quality of the type of the physical quantity that enters. Thus, a change in velocity, initial velocity average velocity, final velocity, and speed are all equivalent in this context. Since all these quantities can be expressed as length/time, their dimensions are [L/T] or $[LT^{-1}]$.

Dimensional formulae and dimensional equations

The expressions which show how and which of the base quantities represent the dimension of the physical quantity is called the dimensional formula of the given physical quantity. For example, the dimensional formula of the volume is $[M^0LT^{-1}]$, as shown in the section above.

An equation obtained by equating a physical quantity with its dimensional formula is called the dimensional equation of that physical quantity.

Dimensional analysis and its applications

A thorough understanding of dimensional analysis helps us in deducing certain relations among different physical quantities and checking the derivation, accuracy and dimensional consistency or homogeneity of various mathematical expressions.

Usefulness: Dimensional analysis can be used to derive or check formulas by treating dimensions as algebraic quantities. Quantities can be added or subtracted only if they have the same dimensions, and quantities on two sides of an equation must have the same dimensions.

Example A

Show that the equation for impulse $Ft = mv - mu$ is dimensionally correct.

Writing it in dimensional form: $[M]\,[L]\,[T]\,[T]^{-2} = [M]\,[L]\,[T]^{-1} + [M]\,[L]\,[T]^{-1}$

Therefore $[M]\,[L]\,[T]^{-1} = [M]\,[L]\,[T]^{-1}$ and the equation is correct (both sides of it have the dimensions of momentum).

Method of dimensions

Dimensions can be used as a preliminary test of the consistency of an equation, when there is some doubt about the correctness of the equation. However, the dimensional consistency does not guarantee correct equations.

Dimensions can also be used sometimes to deduce relations among the physical quantities, (ie derive equations).

Example B

Consider the oscillation of a simple pendulum.

Assumption:

Period of the pendulum (t) depends in some way of the following quantities:

i. mass of the pendulum bob (m)
ii. length of the string of the pendulum (l)
iii. the gravitational intensity g.

Then we can write:

$t = km^x\, l^y\, g^z$

x, y, z are unknown powers and k is a dimensionless constant, $T = M^x\, L^y\, L^z\, T^{-2z}$

Equating indices for M, L and T on both sides we have

M: $0 = x$

L: $0 = y + z$ $\qquad$ $x = 0;\ y = \tfrac{1}{2};\ z = -\tfrac{1}{2}$

T: $1 = -2z$

Original equation becomes $t = k\left(\dfrac{l}{g}\right)^{1/2}$

Unit 11.1 Measurement

Topic 5: Error analysis

Topic 4 covers error analysis and the content summarised in the bullet points on p. 8 of the Syllabus. In summary:

- Errors.
- Taking measurements.
- Methods of improving accuracy.
- Graphical techniques.

Errors in measurement

When people measure different quantities, there is always some error in the measured value – a measurement is only an *estimation* of the true value.

The two main sources of error in measurements are *random* and *systematic errors*. Together, these contribute to the *experimental error* in a measurement or in a whole experiment.

A **random error** is one for which the measurement is just as likely to be larger or smaller than the true value. This type of error shows up when the same quantity is measured several times. The measured values are spread randomly around an average value.

Measurements are described as **precise** if the *spread* of their values is small. If several measurements of the same thing are taken, the result can be taken as the average of all the measurements.

Random errors can be caused by:

- The measuring instrument not being particularly sensitive, eg random errors occur when measurements are estimated to the nearest division on a measuring instrument.

Example A

The ammeter shown measures an electrical current. Should the current shown by the ammeter be read as 3 A, 3.4 A or 3.5 A?

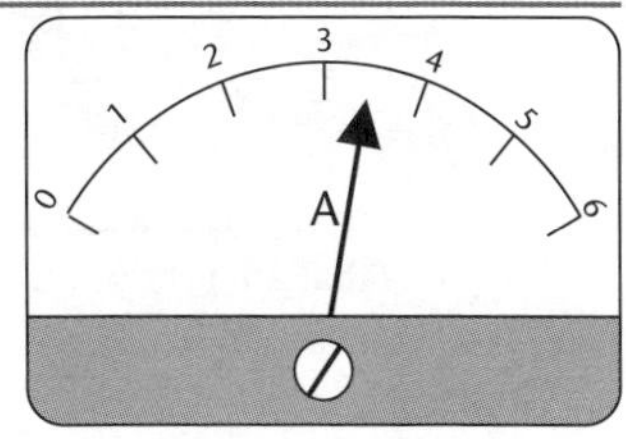

An ammeter

- Lack of perfection in the observer.

Example B

Several people timing the same running event with stopwatches will produce a range of times randomly spread around an average value.

- Random fluctuations in the object being measured.

Example C

The light level may vary when the voltage across a solar cell is being measured.

A **systematic error** is one that consistently makes the measurement either larger or smaller than the true value. This effect can result from:

- An incorrectly adjusted measuring instrument (such as a stopwatch that runs too fast and so always gives a greater time). This is called a **calibration error**.
- An instrument that has a **zero error,** ie it does not read zero for zero measurement (such as a voltmeter needle may point to 0.1 V even when it is not connected in a circuit).
- A measuring procedure carried out by the observer that consistently gives a biased reading (such as reading a metre rule from the right-hand side rather than from directly above). An uncertainty due to viewing the measuring scale from an angle is called a **parallax error**.

Systematic errors are difficult to eliminate. It is not always possible to use other measuring instruments and observers.

An **accurate** measurement (or experiment) is one that has both a small systematic error (ie it closely measures the desired quantity) *and* a small random error. Precision indicates a small random error. It is possible to be precise but inaccurate.

Example D

The figure below illustrates the difference between precision and accuracy.

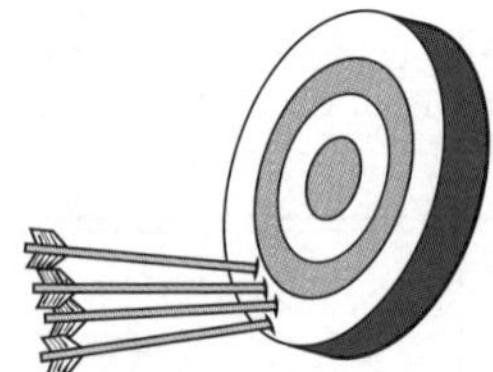

A cluster of arrows that are precise but inaccurate

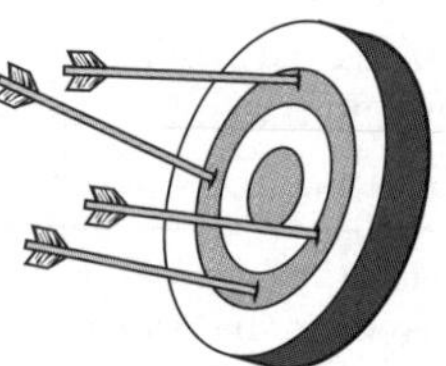

A cluster of arrows that are less precise and so less accurate

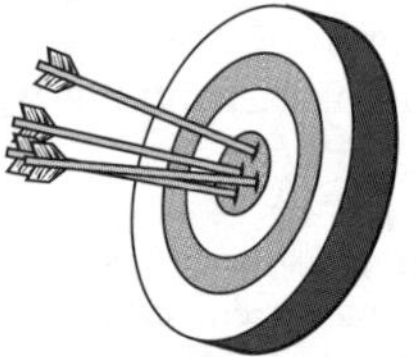

A cluster of arrows that are accurate (and precise)

Precision and accuracy

A race measured using a stopwatch precise to hundredths of a second is likely to be less accurate than using electronic timing. This is due to the delayed reaction of the person starting the stopwatch at the beginning of the race and would make the measured time consistently too small.

The number of **significant figures** (sig figs or sf) in a measurement expresses the precision of the measurement. The larger the number of significant figures, the more precise the measurement.

A significant figure is any digit (0, 1, 2, ..., 9) in a number that is not being used as a place-holder. In any number, the first significant figure is the first non-zero digit. The next digit is the second significant figure and so on.

Example E

0.0104 has 3 significant figures [the first two zeros are place-holders].

23.56×10^{-3} has 4 significant figures.

For some measurements, *trailing* zeros may be significant.

Example F

The length and width of an A4 sheet of paper is measured using a 30 cm ruler able to measure down to tenths of a centimetre (0.1 cm) and are recorded as 29.8 cm and 21.0 cm. Each of these measurements is to 3 sig figs, including 21.0 cm. The trailing zero in 21.0 cm is significant because the ruler is able to measure tenths of a centimetre; 21.0 cm shows that there were zero tenths of a centimetre.

The measurement 29.8 cm means that the actual length lies somewhere between 29.75 cm and 29.85 cm. The ruler is not sensitive enough to distinguish between distances within this range. Similarly, 21.0 cm means an actual length somewhere between 20.95 and 21.05 cm.

It is not always clear how many significant figures there are in some measurements. This difficulty can be overcome by using **scientific notation** (also called **standard form**), to express each measurement.

Example G

It would be difficult to say how many significant figures are in a measurement written as 300 s. The measurement may be to 1 sig fig if it was estimated by a person without a watch, or it may be to 3 sig figs if it was measured using a watch able to be read to the nearest second.

$300 \text{ s} = 3 \times 10^2 \text{ s}$ (1 sig fig)

$300 \text{ s} = 3.0 \times 10^2 \text{ s}$ (2 sig figs)

$300 \text{ s} = 3.00 \times 10^2 \text{ s}$ (3 sig figs)

The only type of measurement that should not involve error is the direct *counting* of whole objects. Even then, a **mistake** could be made in the counting. Mistakes can contribute to the overall error in the result of an experiment.

Mistakes can also arise through:

- Misreading scales.
- Faulty arithmetic or incorrect use of calculators.
- Using a wrong equation.
- Copying (correct) numbers incorrectly (called a *transcription error*).

Mistakes can usually be avoided by careful *double-checking* of measurements and calculations.

Error analysis: Rules for combining errors

When doing error analysis, after performing an experiment involving several measurements, you must know how the errors in all the measurements combine. Therefore, you must know the following rules to combine errors.

- Error of a sum of a difference: when two quantities are added or subtracted, the absolute error in the final result is the sum of the absolute errors in the individual quantities.

Example H

Suppose the temperatures of two objects measured by a thermometer are $t_1 = 20°C \pm 0.5°C$ and $t_2 = 50°C \pm 0.5°C$. You need to calculate the temperature difference and the error.

Solution:

$t' = t_2 - t_1 = (50°C \pm 0.5°C) - (20°C \pm 0.5°C) = 30°C \pm 1°C$

- Error of a product or a quotient: when two quantities are multiplied or divided, the relative error in the result is the sum of the relative errors in the multipliers.

Example I

Suppose the resistance $R = V/I$ where $V = (100 \pm 5)V$ and $I = (10 \pm 0.2)$A. You need to find the percentage error in R.

Solution:

The percentage error in V is 5% and in the current I it is 2%. Therefore, the total error in R would be 5% + 2% = 7%.

- Error in the case of a measured quantity raised to a power: the relative error in the physical quantity raised to the power k is k times the relative error in the individual quantity.

Example J

Suppose you need to find the relative error in Z, if $Z = (A^4B^{1/3})/CD^{3/2}$

Solution:

In this case the relative error in Z is $\Delta Z/Z = 4(\Delta A/A) + (1/3)(\Delta B/B) + (\Delta C/C) + (3/2)(\Delta D/D)$.

Minimising errors

Errors unfortunately cannot be eliminated but can only be minimised. This can be done by averaging the results of several independent readings, choosing an appropriate measuring instrument for that particular task, carrying out the procedure carefully step by step and checking each result regularly as recorded.

Taking a number of readings

When you perform an experiment, taking a number of readings will reduce the error in the final answer. Consider the measurement of the thickness of a hacksaw blade made at a number of places with a micro-meter, in such a way that each reading is accurate to ±0.01 mm.

Suppose that six measurements are taken and they are: 0.65, 0.66, 0.63, 0.66, 0.64 and 0.65 mm. The mean of these, being the sum divided by 6, is 0.65 mm.

To calculate the final error you work out the difference between each reading and the mean value (without regard to sign) and then divide by the number of readings. Therefore,

$$\text{Final error} = (0.00 + 0.01 + 0.02 + 0.01 + 0.01 + 0.00)/6 = 0.008$$

This is the likely error in your answer, and the thickness should therefore be quoted as 0.65 ± 0.008 mm.

A graphical method can also be used to identify which result is not within the accepted range so that it can be isolated from the rest of the readings. The average of a constant can be determined from the graph.

Errors in graphs

If you plot a graph, then you can find the error in the result as follows. Consider the graph shown in the following figure. The best fit line to the points is drawn and its slope found. The average value of all the *x*- and *y*-coordinates is found; this gives you the **centroid** of the line. The lines of greatest and least slope through the centroid are then drawn and their respective slopes found (m_1 and m_2).

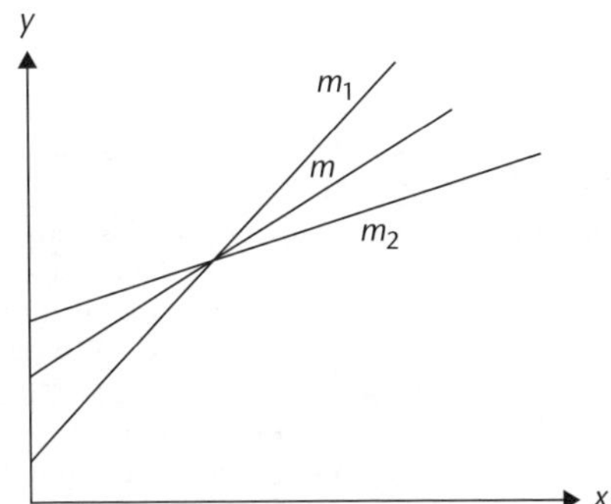

The final error in the slope is then given by $\Delta m/m$ where Δm is the difference between either m_1 or m_2 (whichever is the greater) and m.

Combining measurements

When measurements are combined, significant figures play an important role. When two or more measurements are multiplied or divided, the result should be **rounded off** to the smallest number of significant figures in the original measurements.

Example K

The area of an A4 sheet of paper is:

Area = length × width
= 29.8 × 21.0
= 625.8 cm^2
= 626 cm^2 [rounding to 3 sig figs]

It would be misleading to leave the calculated area as 625.8 cm^2 because:

The maximum possible area = 29.85 × 21.05
= 628.3425 cm^2
= 628 cm^2 (3 sig figs)

The minimum possible area = 29.75 × 20.95
= 623.2625 cm^2
= 623 cm^2 (3 sig figs)

626 cm^2 is midway between these maximum and minimum values and shows that the error is several square centimetres rather than several tenths of a square centimetre.

Example L

A dress designer measures a rectangular piece of a skirt using a tape measure, recording the measurements length = 107 cm and width = 53 cm.

a. The area of the segment in square centimetres is:
area = length × width
= 107 × 53
= 5 671 cm^2
= 5 700 cm^2 [rounded to 2 significant figures because 53 cm has the least number of sig figs]

b. The area of material needed for 5 segments is: [the number 5 is a *counting number* and not a measurement; for this reason it does not contribute any error to the final area]

total area = 5 671 × 5 [the *unrounded* number 5 671, not 5 700, is used – this eliminates **rounding error**, which occurs when partial results are rounded and then used in further calculations]
= 28 355 cm^2
= 28 000 cm^2 (rounded to 2 sig figs) [the total area is rounded to 2 sig figs due to the 53 cm measurement in the original measurements]

5 700 × 5 = 28 500
= 29 000 cm^2 (2 sig figs) which is incorrect.

The final results should be written in standard form, as 5.7×10^3 cm^2 and 2.8×10^4 cm^2.

When measurements are added or subtracted, the procedure above does not apply – the final result is rounded off to the last decimal place of the measurement with the least number of decimal places.

Example M

A box is measured and found to be 41.4 cm high. It is stacked on top of a wardrobe which is measured by a different ruler, as 178 cm in height. Their combined height is:

$$\begin{array}{rl} & 41.4 \\ + & 178 \\ \hline & 219.4 \text{ cm} \\ = & 219 \text{ cm} \end{array}$$

[since 178 cm has the least number of decimal places and ends in the cm column]

Subtraction often reduces the number of significant figures in a calculated result.

Example N

The masses of two kahawai caught in a fishing contest are 1.253 kg and 1.238 kg respectively. The difference in mass is:

$$\begin{array}{rl} & 1.253 \\ - & 1.238 \\ \hline & 0.015 \text{ kg} \end{array}$$

This result does not need to be rounded because both measurements end in the third decimal place.

The difference in mass of the two fish, 0.015 kg, has only 2 sig figs, while the original mass measurements were more precise, each having 4 sig figs.

It is quite acceptable to ignore the rules above and to work to 3 or 4 significant figures when solving 'theoretical' or example problems. As a guide, work to at least one extra significant figure than the number of significant figures in the final answer. This will reduce the effect of rounding errors.

Key words or phrases to look for when deciding whether or not to use the rules for combining measurements are 'measured' and 'in an experiment'. If these words or some like them appear in a problem, then the rules should be applied.

Unit 11.1 Activity 5A: Combining measurements

1. Solve the following, paying attention to how the measurements are combined:
   - **a**. Use the formula, $A = \pi r^2$, to find the area, A, of a circle with radius $r = 2.12$ m.
   - **b**. Two sticks have lengths 12.132 cm and 12.4 cm. What is their total length when placed end to end?
   - **c**. If the two sticks in **b**. are placed side by side, what is the difference in their lengths?

2. The measurements of a rectangular block are:
length = 6.3 cm, width = 12.1 cm and height = 0.84 cm.
   - **a**. What is the volume of this block?
   - **b**. What is the surface area of this block?

3. A bag contains 25 identical marbles. A student measures the diameter of one marble to be 1.94 cm. What is the total volume of all the marbles?
Use the formulas: volume, $V = \frac{4}{3}\pi r^3$ and radius, $r = \frac{\text{diameter}}{2}$

Unit 11.1 Measurement

Topic 6: Graphs

Topic 6 covers graphs and the content summarised in the bullet points on p. 8 of the Syllabus:

- Dependent and independent variables.
- Line of best fit and error bar.
- Extrapolation.

A plot of a set of coordinates which represent experimental results or theories is called a graph. In physics you often draw graphs to study the relationship between variables. You can, in a general sense, say that a graph is a plot of a dependent variable versus the independent variable.

A graph should have the following features.

- Both vertical and horizontal axes are labelled.
- It is given a title, eg 'Position against time graph'.
- It is set to fit into the size of the graph paper.
- Suitable scales are chosen so that the intervals are constant for each axis.
- Points are plotted clearly and accurately.
- The points are joined by a line of best fit which must be drawn clearly but finely.

The presentation of theories or any experimental measurements in the form of a graph has usually two main advantages:

i. The variation of one quantity with another may be seen easily.

ii. The average value of a constant may be determined from the graph.

Probably the most useful form of graph is one in the form of a straight line and we can start by considering this type of graph, which has the equation $y = mx + c$ and is shown in the following figure.

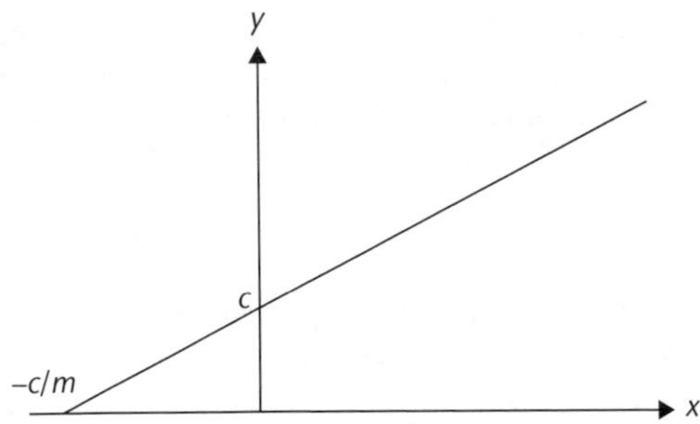

There are many types of graph, but let us consider only two or three more with their relevant equations, as follows.

The following graph shows an exponential increase in y with respect to x; and k is a constant. (An example of this would be the increase in pressure of air with depth.)

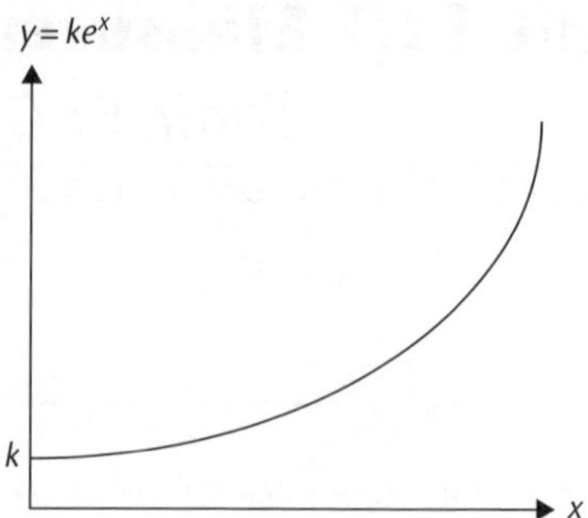

A more common form of graph is the exponential decrease of y in respect to x. Again k is a constant. This behaviour applies, for example, to **radioactive decay**, the discharge of a capacitor and many other physical phenomena. The equation is $y = ke^{-x}$.

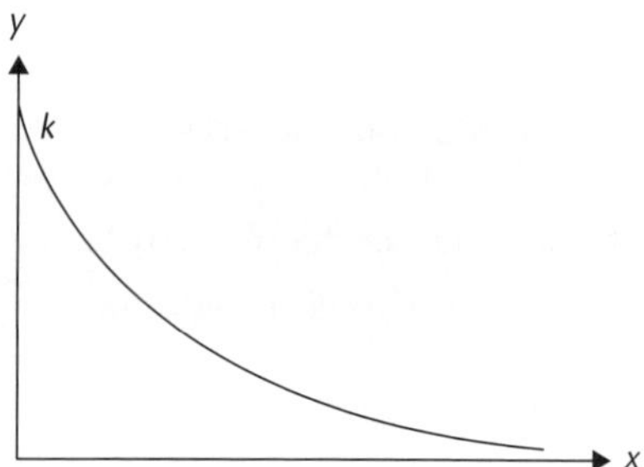

It is much more useful to plot the results of an experiment in the form of a straight line and so a means has to be found by which the above equations can be altered to give a linear relation between a function of y and a function of x.

A graph can be used to estimate the value of a variable. But if you need to estimate the value of a variable, outside the range of your measurements, this is called **extrapolation**. This type of estimation is less certain than **interpolation** because it is not known whether or not the relationship is still valid for larger or smaller values.

More about graphs

A useful way to display the relationship between two sets of measurements is to use a **graph**. Usually, trends in the relationship between measurements can be easily seen on a graph.

Two sets of numbers are called **variables** because their values vary. One variable is described as being the **dependent variable** and is plotted vertically. The **independent variable** is plotted horizontally. The dependent set of measurements are those measurements that are taken as a result of varying the independent variable.

Time is a quantity that we have no control over, ie it is impossible to make time pass quickly or slowly. For this reason, time is considered an independent variable and is plotted on the horizontal axis.

Example A

The temperature of the air near a flame depends on the distance away from the flame. To investigate this, an experiment could be done to measure the temperature of the air near a flame by placing a thermometer at different distances away from the flame. The measured results could then be displayed on a graph with temperature (the dependent variable) plotted vertically and distance (the independent variable) plotted horizontally.

The most common type of graph used in physics is the **line graph** which displays measurements from two **continuous variables**. A continuous variable is one in which measurements vary smoothly. Between any two measurements of a continuous variable there can be another measurement; eg the heights of penguins in the Antarctic.

Variables that are not continuous are called **discrete** variables; eg the number of students in a class; the number of days with a temperature over 18° C. Values in between are not possible. **Bar graphs** are often used to display discrete variables.

The following points should be followed in drawing a graph:

1. Determine the range of values to be plotted on both axes.
2. Choose the scale of the axes so that the graph is as large as possible.
3. The graph should have a title. It is usual to write the dependent variable first and the independent variable second.
4. Each axis should be labelled with the variable quantities. Their units should be placed in parentheses.
5. When plotting by hand, plot the measurements in pencil (a mistake is easy then to correct). Use crosses, or circled dots for the points, since these are easy to see.
6. Where possible, include the origin (0, 0) on the graph. Do not concertina (-⋀⌐, ⋧) either axis.

Example B

The following set of measurements were recorded by measuring the masses and volumes of different sized pieces of a metal:

Mass (g)	8.4	17.1	25.5	34.2	42.3	51.0
Volume (cm^3)	1.0	2.0	3.0	4.0	5.0	6.0

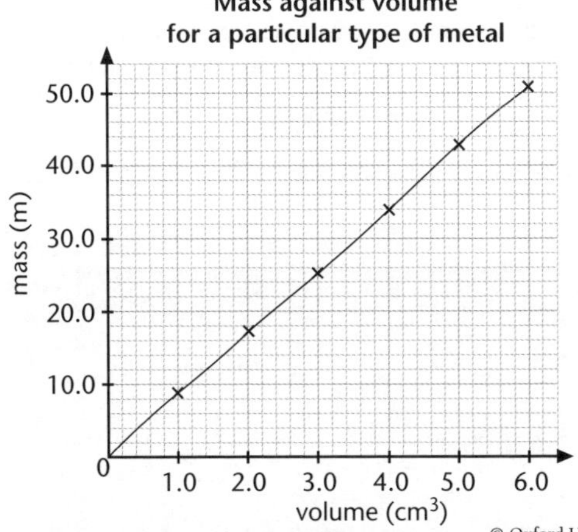

Using the words *mass against volume* in the title of the graph for a particular type of metal shows that mass is the dependent variable and volume is the independent variable.

The graph shows a **linear** relationship between the metal's mass and its volume; ie the graph's points lie in a straight line. When this occurs, a 'straight line of best fit' can be drawn using a clear perspex ruler. The line of best fit may or may not cross over any of the plotted points.

A line's **slope** (or **gradient**) gives the **proportionality constant** between the two variables.

The slope (symbol *m*) is found by choosing two points on the line and finding the change in the *y* axis, Δy, and the change in the *x* axis, Δx, between the two points.

The slope is $m = \frac{\Delta y}{\Delta x}$.

Example C

The slope of the graph in Example B can be found from the co-ordinates of points A and B. Points A and B are chosen as far apart as possible to reduce uncertainty in the slope of a graph caused by uncertainties in reading the values of A and B from the graph.

Point A has co-ordinates (1.4 cm^3, 12.0 g) and point B has co-ordinates (5.4 cm^3, 46.0 g).

$$\Delta y = 46.0 - 12.0 = 34.0 \text{ g}$$

and $\Delta x = 5.4 - 1.4 = 4.0 \text{ cm}^3$

The slope is:

$$m = \frac{\Delta y}{\Delta x} = \frac{34.0}{4.0} = 8.5 \text{ g cm}^{-3} \text{ (to 2 sig figs)}$$

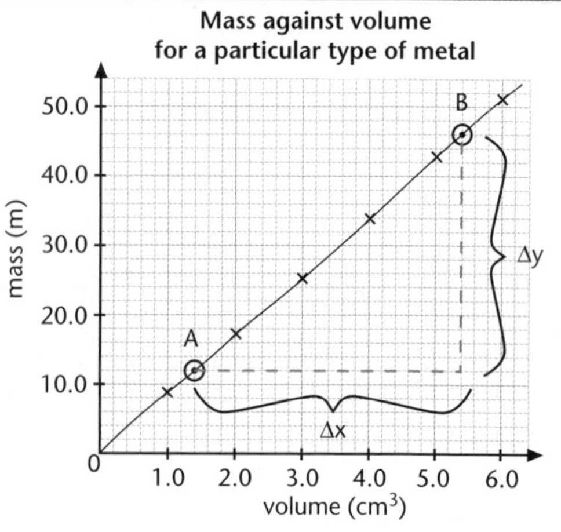

The measurements between mass and volume can be expressed mathematically: mass = slope of graph × volume.

The slope of a graph often has units; in this case, the units g cm^{-3} are those of density. The slope of the graph in Example B is the density of the metal:

mass = density × volume.

When the relationship between two variables is not linear a *curved* graph results.

Example D

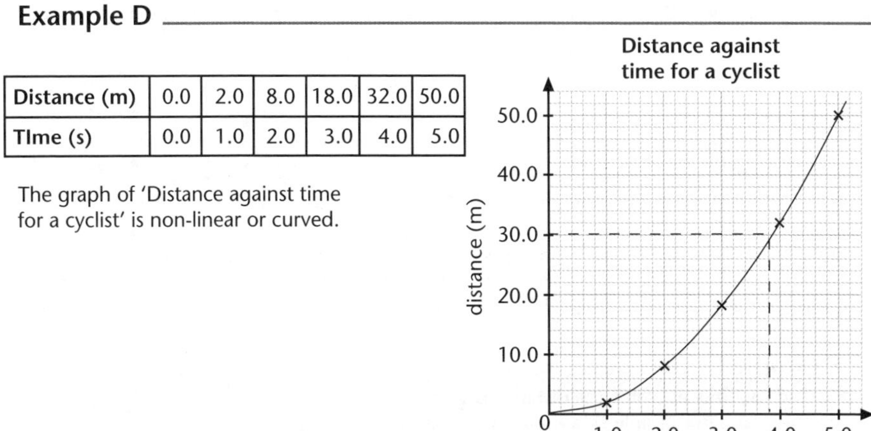

Distance (m)	0.0	2.0	8.0	18.0	32.0	50.0
TIme (s)	0.0	1.0	2.0	3.0	4.0	5.0

The graph of 'Distance against time for a cyclist' is non-linear or curved.

It is difficult to see the relationship between variables on a curved graph, especially if there is an amount of random error present.

Usually the variables are changed (by squaring or taking square roots, etc) until a graph becomes linear. When a linear graph is obtained, the relationship is more clearly seen, since the changed variables are then **proportional** to one another.

Example E

Distance (m)	0.0	2.0	8.0	18.0	32.0	50.0
Time (s)	0.0	1.0	2.0	3.0	4.0	5.0
Time^2 (s^2)	0.0	1.0	4.0	9.0	16.0	25.0

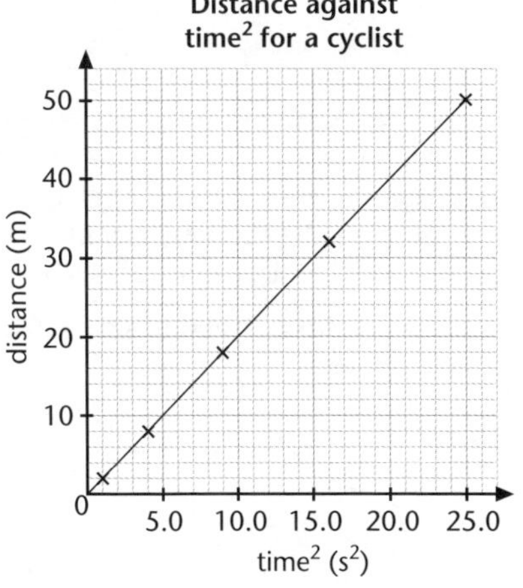

The table and graph use the same data from Example D and square the time measurements. The graph between the the changed variables is a straight line. This means that distance is proportional to time^2.

Taking the origin (0, 0) and the co-ordinates (25.0 s^2, 50.0 m) as the two points to calculate the slope:

$$m = \frac{\Delta y}{\Delta x}$$
$$= \frac{50.0 - 0}{25.0 - 0}$$
$$= 2.00 \text{ m s}^{-2}$$

The mathematical relationship between distance and time is:
distance = 2.0 × time^2 or $d = 2.0t^2$

This slope, 2.0 m s^{-2}, is related to the acceleration of the motion measured.

A graph can be used to estimate the value of a variable.

Example F

To find the time at which the distance is 30 m on the graph in Example D, a horizontal dotted line is drawn across from the 30 m mark on the distance axis to where it intersects the curve. Another dotted line is drawn vertically down to where it intersects the time axis. The time is estimated to be 3.8 s.

This process of estimating values *within* the range of plotted measurements is called **interpolation**.

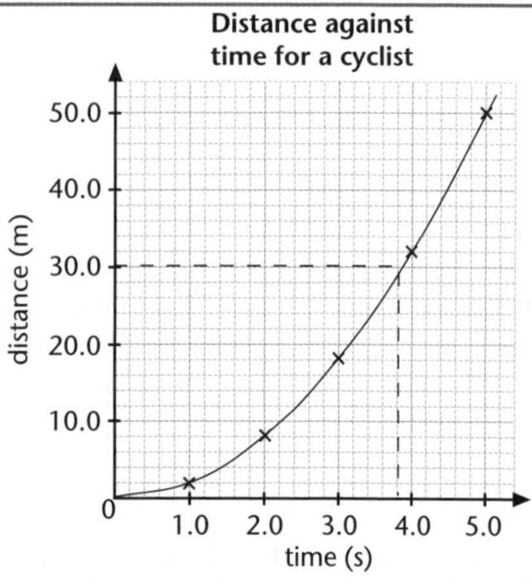

Estimation of values *outside* the range of measurements is called **extrapolation**. This type of estimation is less certain than interpolation because it is not known whether or not the relationship is still valid for larger or smaller values. Weather forecasts are extrapolations of measured weather data and are thus sometimes incorrect.

Linear, square, **inverse** and inverse square relationships have characteristic shapes when plotted as graphs.

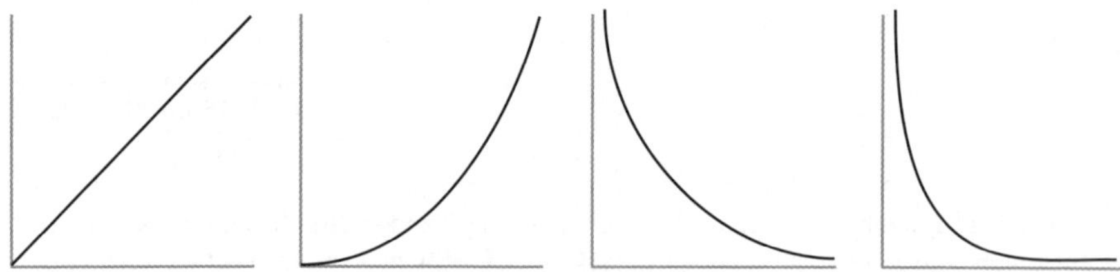

These shapes can be used as a guide when trying to identify the relationship between plotted variables in a graph.

Graphs of linear, square, inverse and inverse square relationships

Unit 11.1 Activity 6A: Graphs

1. Find the relationship between the measurements in each of the following tables and express the relationship as a formula:

   a.

t (s)	0.0	1.0	2.0	3.0	4.0
d (m)	0.0	3.5	7.0	10.5	14.0

   b.

t (s)	5.0	10.0	15.0	20.0	25.0
v (m s^{-1})	2.5	5.0	7.5	10.0	12.5

2. Measurements between *I* (amperes) and *V* (volts) are plotted on the graph shown. A line of best fit is drawn.

   a. Estimate the value of *V* when *I* is 10.0 amperes.

   b. Estimate the value of *I* when *V* is 8.0 volts.

   c. Determine the graph's slope and write down the mathematical relationship between *I* and *V*.

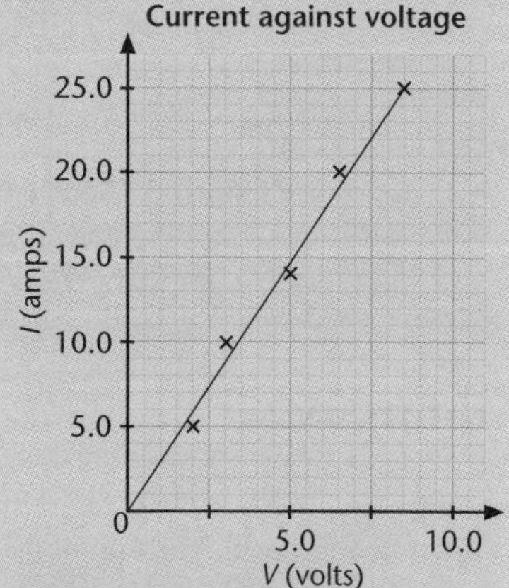

3. Different masses are hung from a spring in an experiment. The resulting extension of the spring (ie the extra distance the spring extends) is measured each time and is shown in the table.

Mass (g)	0.0	200.0	250.0	300.0	400.0	500.0
Extension (cm)	0.0	10.0	12.5	15.0	20.0	25.0

   a. Plot a fully labelled graph of extension against mass.

   b. Estimate how much mass causes an extension of 5.0 cm.

   c. Calculate the graph's slope (remember to include the slope's units).

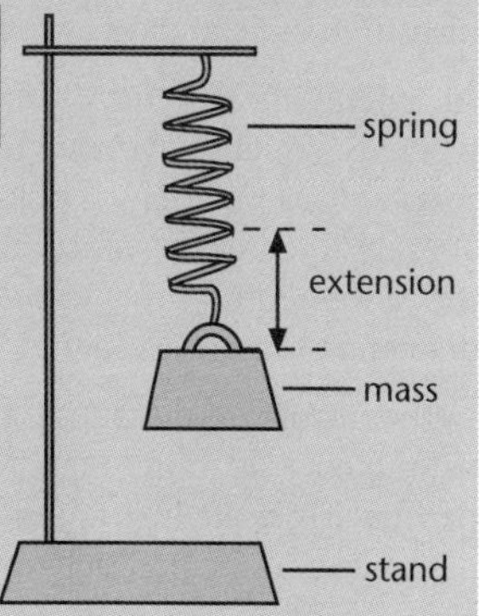

4. The air resistance acting on a car is measured as it travels through still air at different speeds. The table shows the car's speed in km h^{-1} and the corresponding air resistance measurements.

Speed (km h^{-1})	20	30	40	50	60	70	80	90	100
Air resistance (N)	370	833	1 481	2 315	3 333	4 537	5 926	7 500	9 259

Convert each speed from km h^{-1} to m s^{-1} and graphically determine a mathematical relationship between air resistance in newtons and speed in m s^{-1}.

5. The electrical resistances of metre lengths of wire of various diameters are measured. The table shows the various diameters and the corresponding resistance measurements.

Diameter (mm)	0.2	0.5	1.0	1.5	2.0
Resistance (ohms)	37.50	6.00	1.50	0.67	0.38

Graphically determine a mathematical relationship between electrical resistance in ohms and diameter in mm.

Summary

A basic base quantity is one which cannot be derived from other quantities. Mass, length and time are three basic base quantities. They help to build up other quantities to be used to describe the world. The unit of mass is the kilogram, the unit of length is the metre, and the second is the unit of time. These are defined very accurately in such a way that their understanding is beyond most people.

The process of measurement involves two main processes. The first process deals with the assignment of a number – the unit – and the second deals with the statement of the measurement standard against which the number can be compared.

In general, measurements tend to fall into one of two categories. The first category is that where the quantity you are trying to measure has a well defined value, so the aim is to find this value as precisely as possible. On the other hand, quite a number of measurements attempt to quantify something that is not fixed. For instance, measuring the diameter of the Moon requires that the Moon be a perfect sphere (because only then can we define the diameter).

In the process of measuring physical quantities errors are introduced. Errors should usually be quoted only to one significant figure; two significant figures are sometimes justified, particularly if the first figure is a 1. Do not worry about errors in the errors! Errors by their nature cannot be precisely quantified.

Unit 11.2 Motion (Kinematics)

Topic 1: Characteristics of motion – introduction

In line with the Grade 11 Syllabus (p. 10) Unit 11.2 introduces motion and the fundamental principles governing motion that can be used to explain the motion of an object. Topic 1 looks at:

- Distance, displacement, and acceleration of moving objects.
- Solving problems related to speed and velocity.

The study of **motion** is usually divided in two different areas: kinematics and dynamics. **Kinematics** is the description of how the motion takes place, and **dynamics** is the study of the forces involved in the motion of the particles or objects (or the cause of the motion). Therefore, the motion of an object can be described using the terms **distance, speed** and **acceleration**. The vector quantities **displacement, velocity** and **acceleration** describe the motion in more detail.

In this Unit, we will study the details of translational motion only. We will concentrate on linear motion, that is, motion along a straight line. The basic topic of motion in more than one dimension will also be covered briefly.

Distance and displacement

The distance travelled by an object is the length of the path of its journey. Distance is a scalar quantity. When told the distance a car has travelled is 50 km, no information is known about the direction the car travelled in.

The displacement of an object is the shortest distance from the start of the journey to the end of the journey. Since displacement is a vector quantity, its direction must be stated. When told the displacement of a car is 50 km due North from the starting point, enough information to locate the car is given.

The SI unit for distance and displacement is the metre (m), but mm, cm and km are also frequently used.

Example A

A car travels a distance of 30 km East followed by 40 km North.

What is the total distance travelled by the car?

The total distance travelled by the car is 70 km (30 km + 40 km).

What is the displacement of the car? The displacement of the car may be calculated by vector addition:

Displacement = $\sqrt{30^2 + 40^2}$ = 50 km 53° North of East (*using Pythagoras' theorem to get the distance,* $\theta = \sin^{-1}\frac{40}{30} = 53°$).

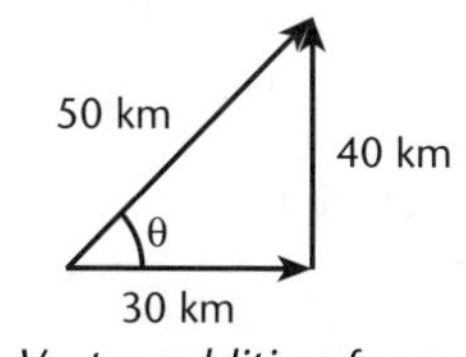

Vector addition for car movement.

Speed

The **speed** of an object is the distance it travels divided by the time it took.

$$\text{speed} = \frac{\text{distance travelled}}{\text{time taken}}$$

$$v_{av} = \frac{s}{t}$$ v = speed, s = distance, t = time taken

Speed is a scalar quantity. The statement 'the speed of a car is 80 km h^{-1}' gives no information about the direction the car is travelling in.

The **instantaneous speed** of an object is its speed at an instant in time. The speedometer in a car shows the instantaneous speed of a car.

Since the speed of an object may change throughout a journey, it can be useful to calculate the **average speed** for the journey.

$$\text{average speed} = \frac{\text{total distance travelled}}{\text{total time taken}}$$

Example B

A car takes 5 hours to travel from Lae to Goroka, a distance of 285 km.

The average speed of the car = $\frac{\text{total distance travelled}}{\text{total time taken}} = \frac{285\text{ km}}{5\text{ h}} = 57\text{ km h}^{-1}$.

If the speed of an object does not change throughout a journey, it is travelling at **constant speed**. This means that every second of its journey, the object travels the same distance.

Velocity is speed in a given direction. Velocity is a vector quantity. The velocity of a car travelling from Lae to Madang may be 60 km h^{-1} due Northwest.

The SI unit for speed and velocity is m s^{-1}. Other commonly used units are cm s^{-1} and km h^{-1}.

Acceleration

An object **accelerates** when it changes its speed. If the object speeds up, it is said to be accelerating. If it slows down, the acceleration is negative. Negative acceleration is called **deceleration**.

$$\text{acceleration} = \frac{\text{change in speed}}{\text{time taken}}$$

$$a = \frac{\Delta v}{t}$$ a = acceleration, v = speed, t = time taken

The symbol Δ is the Greek letter delta and means 'change in'.

Δv equals $v_{final} - v_{initial}$ (final speed – initial speed).

The SI unit for acceleration is m s^{-2}. Other units used are cm s^{-2}, km/h^2 and km h^{-1} s^{-1}.

Example C:

Rowena is riding her bicycle and is being chased by a dog. She speeds up from 1.5 m s^{-1} to 4.8 m s^{-1} in 6 s. Calculate her acceleration.

$$a = \frac{\Delta v}{t} = \frac{v_f - v_i}{t} = \frac{4.8 - 1.5}{6} = 0.6 \text{ m s}^{-1}$$

Unit 11.2 Activity 1A: Motion

1. What is the length of the object shown in cm?

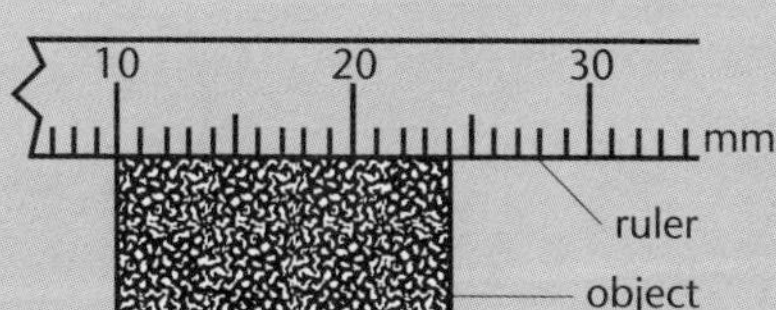

2. Complete the following table by calculating the values **a** to **f**.

Speed	Distance	Time
a	180 m	90 s
b	2.4 km	0.8 h
440 m s^{-1}	c	2 s
2.5 m s^{-1}	d	20 s
50 m s^{-1}	200 m	e
0.5 m s^{-1}	1 m	f

3. Change:

a. 8 km to metres.
b. 90 cm to metres.
c. 180 minutes to hours.
d. 0.5 h to seconds.

4. Complete the table by calculating the values **a** to **h**.

Acceleration, a	Change in speed, Δv	Change in time, Δt
a	180 m s^{-1}	90 s
b	2.4 k m h^{-1}	0.8 h
440 m s^{-2}	c	2 s
2.5 m s^{-2}	d	20 s
50 m s^{-2}	200 m s^{-1}	e
0.5 m s^{-2}	1 m s^{-1}	f

5. A girl rides a bicycle at steady speed of 2.5 m s^{-1}. How long does she take to ride 50 m?

6. A car accelerates at 2 m s^{-2} starting from rest. What is the speed it reaches 10 s after the start of the journey?

Use the following information and the diagram to answer Questions 7 and 8.

A ball is launched horizontally from **T**, the top of the cliff, and it hits the ground some time later at point **Z**. **W**, **X**, **Y** and **Z** are the positions at equal time intervals.

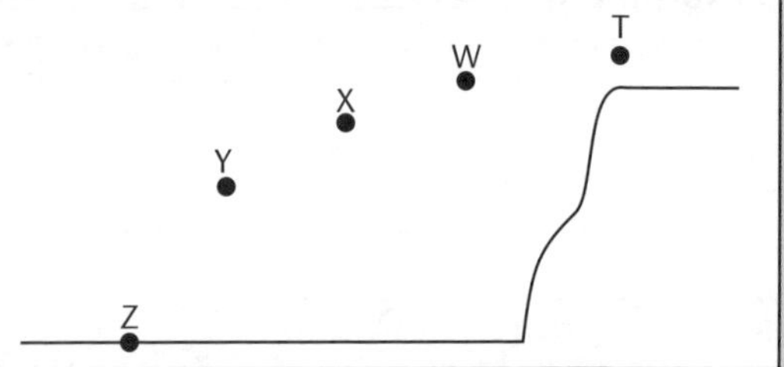

7. Draw a set of vectors to represents the *horizontal* velocity of the ball at **W**, **X**, and **Y**.
8. Draw a set of vectors to represent the acceleration of the ball at **W**, **X**, and **Y**.
9. As part of a science fair project, a student sets up the experiment shown to investigate how long it takes for a mouse to solve the maze shown and reach the cheese.
   a. What is the total distance travelled by the mouse, if it travels directly from **A** to **B** to **C**?
   b. When it follows this path, it takes the mouse 15 s to reach the cheese. Calculate the average speed of the mouse.
   c. Draw a scale vector diagram of the path travelled by the mouse and draw in the resultant displacement. Use a scale of 1 cm = 0.1 m. Begin at **A**.
   d. When the mouse has reached the cheese, calculate the displacement of the mouse from point **A**.
   e. Calculate the size of the average velocity of the mouse.
10. A small plane takes off to fly from **A** due West to **B**. The plane's air speed is 200 km hr^{-1} and the trip normally takes 2 hours. After 2 hours, the pilot finds that a steady wind from the North has blown him off course and he is now 80 kilometres from **B** at point **C**.

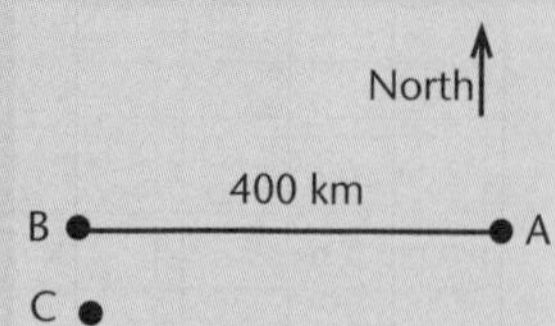

   a. What is the distance from A to C?
   b. What was the speed of the plane relative to an observer on the ground?
   c. At **C** the pilot turns the plane around and flies North into the wind to **B**. How many minutes will it take him to fly there?

Unit 11.2 Motion (Kinematics)

Topic 2: Motion – extension

The material covered in Topic 2 extends the concepts introduced in the previous Topic, by looking at:

- Distance, displacement, velocity and acceleration.
- Constant acceleration in a straight line.
- Kinematic equations.

Introduction

The motion of an object (eg car, satellite, tennis ball) can be described using various quantities such as distance, speed and acceleration.

Motion also involves the *direction* in which an object moves. When direction is important, the quantities of distance, speed and acceleration are replaced by the quantities of displacement, velocity and acceleration.

Distance and displacement

The symbol used for both **distance** and **displacement** is *d*. Distance and displacement both involve a change in position.

Distance is a **scalar** quantity because it only involves the *size* and not the direction of movement.

Displacement is a **vector** quantity, because it involves both *size* of movement *and direction* from a reference or starting point.

The distance an object travels and its displacement are often different.

Example A

A model train travels around the track shown below. The table shows the distance moved by the train, and its displacement at various stages of the journey around the track.

	Distance Travelled	Displacement from Start
At B	2 m	2 m north
At C	4 m	$\sqrt{8}$ m northeast
At D	6 m	2 m east
Back at Start, A	8 m	0 m

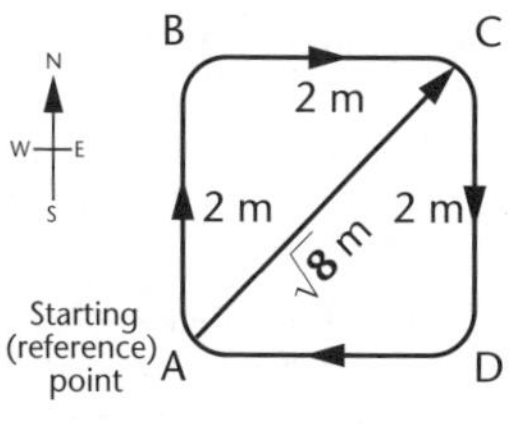

Scalars and vectors

Velocity

The symbol for velocity is the same as for speed, v. The velocity of an object refers to the object's *speed* and the *direction* in which the object moves. Velocity is a **vector** quantity. The units for velocity are the same as for speed.

Velocity is calculated by dividing the *change in an object's displacement* by the *time taken for a change in displacement* to occur.

$$\text{velocity, } v = \frac{\textit{change in displacement}}{\textit{change in time}} = \frac{\Delta d}{\Delta t}$$

Example B

Three cars A, B and C are travelling in the directions shown. The speed of each car is also shown.

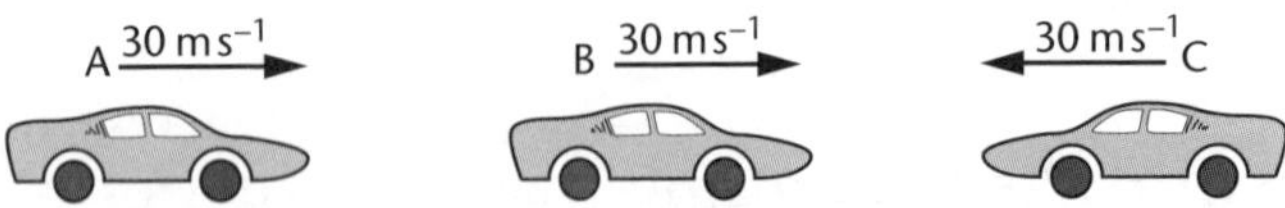

Velocity and speed

All three cars have the same speed. A and B also have the *same* velocity, because both their speed and direction are the same. A and C (and B and C) have *different* velocities because, although their speeds are the same, their direction of travel is different.

As with speed, velocity may refer to **average velocity** or **instantaneous velocity**.

A constant (or uniform) velocity means *both the speed and direction* of an object are constant, ie the speed and direction are not changing.

Example C

A person walks north at constant velocity. At $t = 2$ s the person is 3 m north and at $t = 26$ s the person is 39 m north. The distance travelled is:

(39 – 3) m in a time of (26 – 2) s.

The person's speed is $v = \frac{d}{t}$

$= \frac{36}{24}$ [39 – 3 = 36 and 26 – 2 = 24]

$= 1.5 \text{ m s}^{-1}$

The person's velocity is 1.5 m s^{-1} north.

Average velocity and average speed are not always the same.

Example D

On a trip, a family starts out at 9.00 am and travels 100 km south. They stop for lunch then travel a further 250 km south. Following afternoon tea, they realise they have misread their map and travel 50 km north to reach their motel at 5.00 pm. Determine the average speed and average velocity for the whole journey.

Solution:

Total time taken for the journey = 5.00 pm – 9.00 am

= 8.0 hours

Total displacement for the journey = 100 km S + 250 km S + 50 km N
= 100 + 250 – 50
= 300 km south

Total distance covered during the journey = 100 + 250 + 50
= 400 km

$\therefore$ average velocity $\bar{v} = \frac{300}{8.0}$

$= 37.5 \text{ km h}^{-1}$ south

and average speed, $\bar{v} = \frac{400}{8.0}$

$= 50 \text{ km h}^{-1}$

Acceleration

Acceleration is a *vector quantity* which is defined as 'the rate at which an object changes its velocity'. An object is accelerating if it is changing its velocity.

$$a = \frac{\Delta\text{velocity}}{\text{time}}$$

Constant acceleration

Sometimes an accelerating object will change its velocity by the same amount each second. This is referred to as a **constant acceleration** since the velocity is changing by a constant amount each second. The data tables below depict motions of objects with a constant acceleration and a changing acceleration. Note that each object has a changing velocity.

Accelerating Objects are Changing Their Velocity ...

... by a constant amount each second ...

Time (s)	Velocity (m/s)
0	0
1	4
2	8
3	12
4	16

...in which case, it is referred to as a constant acceleration.

... or by a changing amount each second ...

Time (s)	Velocity (m/s)
0	0
1	1
2	4
3	5
4	7

...in which case, it is referred to as a non-constant acceleration.

Average acceleration

The average acceleration of any object over a given interval of time can be calculated using the equation.

$$\text{Ave. acceleration} = \frac{\Delta\text{velocity}}{\text{time}} = \frac{v_f - v_i}{t}$$

This equation can be used to calculate the acceleration of the object which accelerated from rest to reach 50 m/s in a time of 5 s is as shown below.

$$a = \frac{v_f - v_i}{t} = \frac{50\ \text{m/s} - 0\ \text{m/s}}{5\ \text{s}} = \frac{10\ \text{m/s}}{1\ \text{s}}$$

Acceleration values are expressed in units of velocity/time. Typical acceleration units include the following: m/s/s, mi/hr/s, km/hr/s.

The direction of the acceleration vector

Since acceleration is a vector quantity, it will always have a direction associated with it. The direction of the acceleration vector depends on two things:

- Whether the object is speeding up or slowing down.
- Whether the object is moving in the positive or negative direction.

Now that you have learnt the concept of acceleration and average acceleration, you can define **instantaneous acceleration** in analogy to instantaneous velocity:

$a = \Delta v/\Delta t$ when Δt is very small (approaching to zero).

Therefore,

$$a = dv/dt$$

In other words, acceleration of an object at any instant is the *rate at which its velocity is changing at that instant*. Once you have defined acceleration and you realise that its basic concept involves a **change in velocity**, you know an acceleration can result not only when the magnitude of the velocity changes, but also if the direction changes.

Example E

If you are riding in a car travelling at **constant speed** around a curve in the road or if a child is riding on a merry-go-round, both of you will experience an acceleration because of a change in the direction of the velocity. You will learn more about this type of acceleration when you study motion in more than one dimension.

The concept of acceleration is often confused with velocity. Therefore, to help you make the distinction between the two concepts, you will now analyse and understand the following examples.

Example F

Let us suppose you are observing an object moving to the right along the x-axis (horizontal axis) and you realise it is decreasing its speed, eg a car that is braking. Suppose the initial speed (when you started observing it) was 20.0 m/s and it takes 5.0 s to reach a final speed of 5.0 m/s, then

$$\bar{a} = \frac{5.0\ \text{m/s} - 20.0\ \text{m/s}}{5.0\ \text{s}} = -3.0\ \text{m/s}^2$$

The negative sign in this case is because the final velocity is less than the initial velocity; therefore, it indicates that the speed was decreasing. In other words, the car was **decelerating**. (This case is sometimes considered as a negative acceleration.)

Example G

Suppose you accelerate your car from rest to 80 km/h in 10 s; the magnitude of the average acceleration is then:

$$a = \frac{80\ \text{km/h} - 0\ \text{km/h}}{10\ \text{s}} = \frac{80\ \text{km/h}}{10\ \text{s}} = 8.0\ \text{km/h/s}$$

This result tells you that the average acceleration was *eight kilometres per hour per second* which means that, on average, the speed changed by 8 km/h during each second. In other words, assuming the acceleration was uniform, during the first second your car's speed increased from zero to 8 km/h. During the next second its speed increased by another 8 km/h to 16 km/h, and so on.

From analysing and understanding the examples in this section and the examples in the previous section on speed and velocity, you are now ready to make the following conclusion:

Acceleration is the rate at which the velocity changes, whereas velocity is the rate at which the position of an object changes.

Unit 11.2 Activity 2A: Motion and velocity

An object moves along the x axis. Find out the sign of its acceleration if it is moving:

a. In the positive direction with increasing speed.

b. In the positive direction with decreasing speed.

c. In the negative direction with increasing speed.

d. In the negative direction with decreasing speed.

Unit 11.2 Motion (Kinematics)

Topic 3: Vectors and relative motion

Vector operations is the last bullet point under 'characteristics of motion' on p. 10 of the Syllabus. This Topic covers:

- Scalars, vectors and reference frames.
- Vector addition, subtraction and multiplication by a scalar.
- Velocity vector components and relative motion.

Introduction – Scalars and vectors

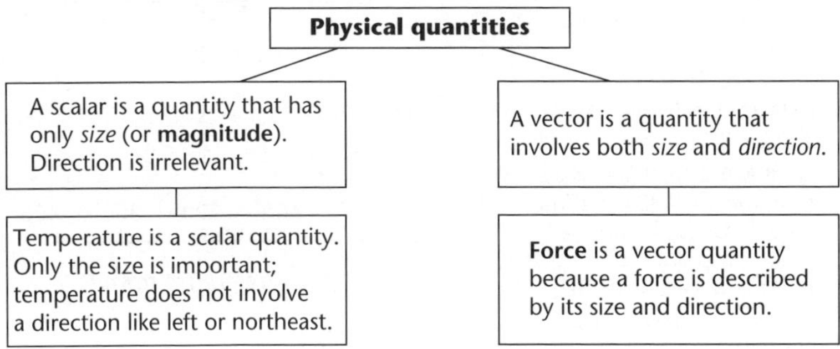

Scalars and vectors

Examples of scalar quantities are mass, time, distance, speed, voltage and **energy**.

Examples of vector quantities are displacement, velocity and momentum.

Vectors

A vector is drawn as a straight, arrowed line. A vector has a **tail** and a **head** (indicated by the arrow). The arrow points in the *direction* of the vector. The *length* of the line represents the vector's size or *magnitude*.

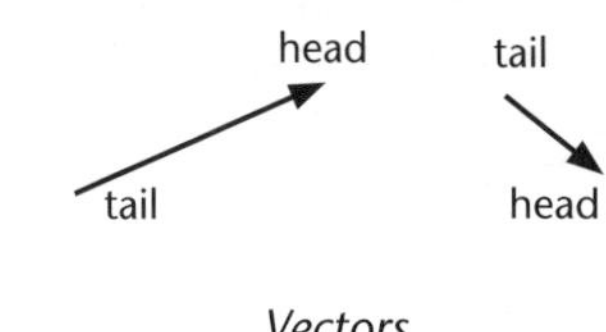

Vectors

Several symbols can be used to show a vector quantity. The notation, $\underset{\sim}{A}$, will be used in this book to show that A is a vector quantity.
Other notations for the vector $\underset{\sim}{A}$ are: $\tilde{A}$, $\underset{\rightarrow}{A}$, $\vec{A}$, or $\mathbf{A}$.

Many quantities in physics are vectors and need to be added together, or multiplied and divided by scalars.

Reference frames

A **reference frame** is a pair of directions at *right angles* to one another. These directions are called **axes**. Reference frames are used to describe the direction of a vector. The three common reference frames are points of the **compass**, compass bearings and the vertical and horizontal.

Points of the compass

The north/south axis is perpendicular to the west/east axis. On this reference frame, a vector's direction is given as an angle to either north, west, east or south.

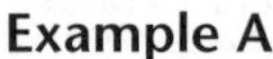

Example A

The figure to the right shows a reference frame with north/south and east/west axes. A vector representing a velocity of 350 km h^{-1}, 32° south of east, is shown on the frame.

The length of the line segment (3.5 cm) represents the size of the vector (350 km h^{-1}). The angle, 32° south of east, is shown.
The scale factor for the diagram is 1 cm : 100 km h^{-1}.
32° south of east is the same as 58° east of south (ie 90° – 32°). It is usual to use the expression having the *smaller* angle, so that 32° south of east is preferred to 58° east of south.
32° south of east can be abbreviated to E 32° S.

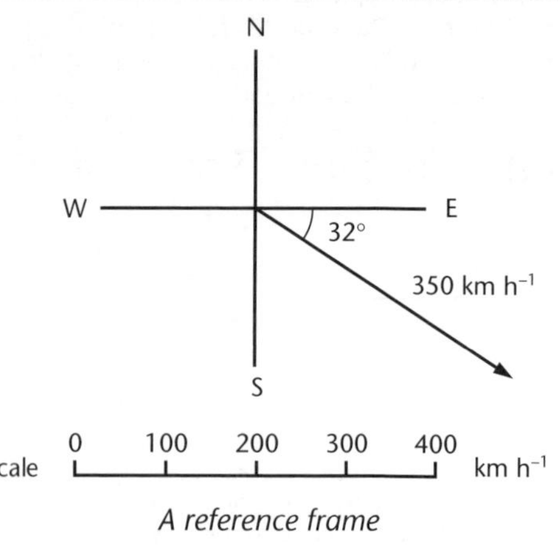

A reference frame

Compass bearings

A **bearing** is a direction given as an angle *clockwise* from *north*. Bearings are used in navigation.

Example B

The direction of the vector, 32° south of east, shown in Example A is the same as the *bearing* 122° (ie 90° + 32°).

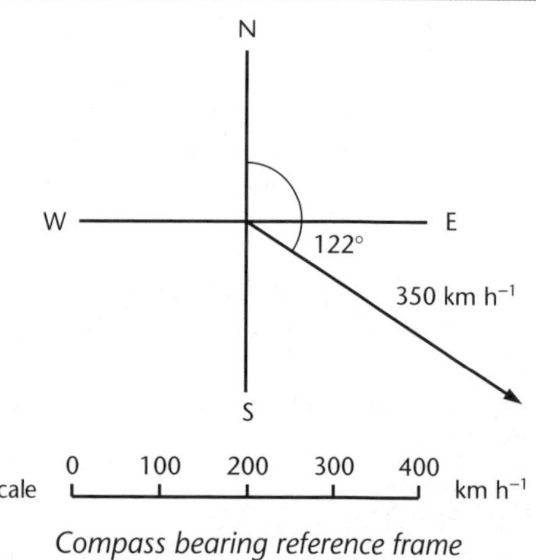

Compass bearing reference frame

All bearing angles are written with three digits, beginning with zero(s) if required, eg 024° means 24° E of N (or N 24° E).

Vertical/Horizontal

When the direction of a vector is in a vertical plane, its direction is given by an angle to either the vertical or the horizontal direction.

Example C

The vector shown might represent a tennis ball moving at 5.0 m s^{-1} *and angled 25° up from the horizontal.*

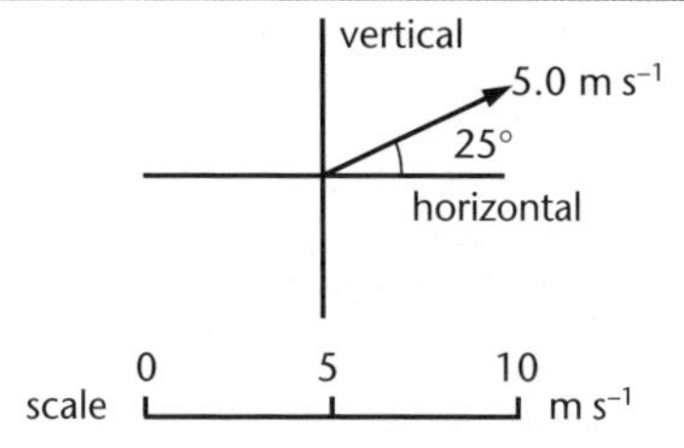

Vertical/horizontal reference frame

If no reference frame is given in a problem involving vectors, it is usual to choose a suitable direction as a reference. The other reference direction will be at right angles to the first.

Example D

If a boat is crossing a river, then a suitable reference frame might be the directions 'along the river bank' and 'directly across the river'.

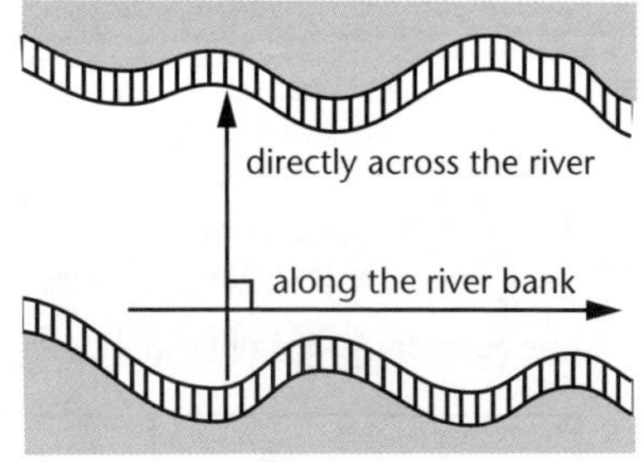

A river reference frame

Multiplication and division of a vector by a scalar

Vectors can be multiplied (or divided) by numbers (called **scalars**). When the scalar is positive, the result is a new vector having the *same direction* but usually a *different* magnitude from the original vector. The magnitude of the new vector is the magnitude of the original vector multiplied by the scalar.

Example E

If the vector 10 m s^{-1} bearing 056° is multiplied by 5, the new vector is 50 m s^{-1} bearing 056°. The new vector is in the same direction and its size is $10 \times 5 = 50$ m s^{-1}.

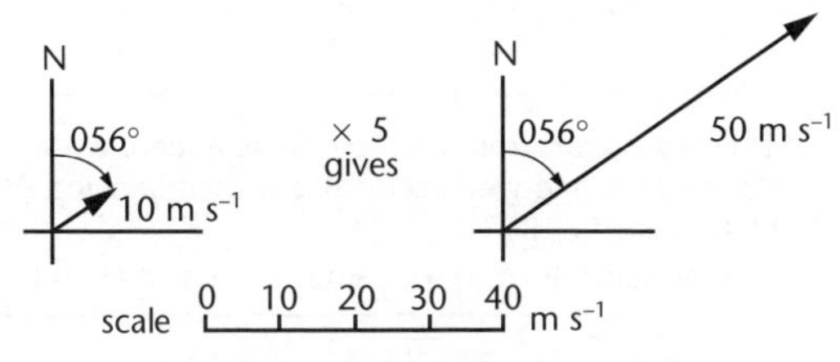

Vector multiplication by a scalar

If the multiplying number is *negative*, the new vector will have the *opposite* direction to the original vector (ie the original vector is rotated by 180°).

Example F

If the vector 10 m s^{-1} bearing 056° is multiplied by –0.5, the new vector is the vector 5 m s^{-1} bearing 236° (ie 056° + 180°).
The new vector is in the opposite direction to the original vector and its magnitude is $10 \times 0.5 = 5$ m s^{-1}.

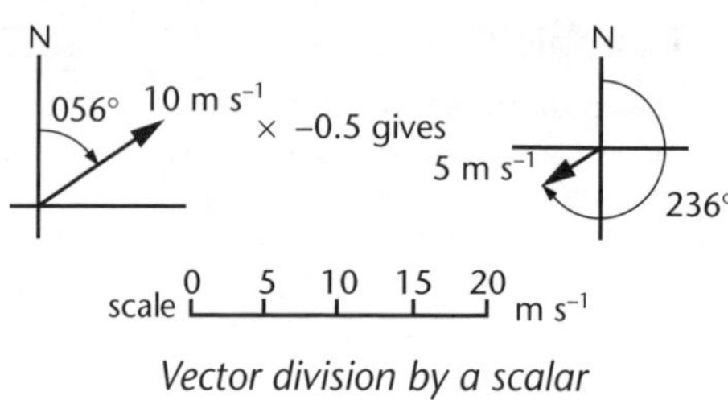

Vector division by a scalar

Often the scalar is a physical quantity and has its own unit. The vector resulting from scalar multiplication or division will have a combination of units different from both the original scalar and vector.

Addition of vectors

Two (or more) vectors are added using a scale drawing called a **vector diagram**. The result of addition is a vector called the **resultant**. The resultant is a vector that has the *same effect* as the combined effect of the vectors being added. Vectors are added 'head to tail'.

Example G

Two vectors $\underset{\sim}{A}$ and $\underset{\sim}{B}$ are added.
$\underset{\sim}{A}+\underset{\sim}{B}$ means that vector $\underset{\sim}{B}$ is *added* to vector $\underset{\sim}{A}$.

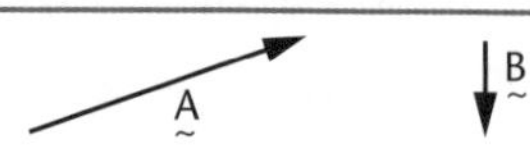

The addition $\underset{\sim}{A}+\underset{\sim}{B}$ is done in the following way:

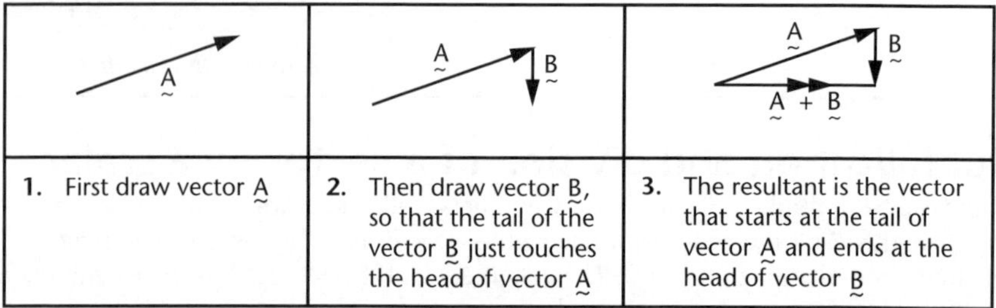

1. First draw vector $\underset{\sim}{A}$	**2.** Then draw vector $\underset{\sim}{B}$, so that the tail of the vector $\underset{\sim}{B}$ just touches the head of vector $\underset{\sim}{A}$	**3.** The resultant is the vector that starts at the tail of vector $\underset{\sim}{A}$ and ends at the head of vector $\underset{\sim}{B}$

The double arrow (—▶▶—) shows that $\underset{\sim}{A}+\underset{\sim}{B}$ is a resultant vector.

It does not matter in what order vectors are added. The resultant is always the same.

Example H

The two vectors from Example G are added, $\underset{\sim}{B}+\underset{\sim}{A}$. This is done by starting with vector $\underset{\sim}{B}$ and then drawing $\underset{\sim}{A}$ (tail of $\underset{\sim}{A}$ at head of $\underset{\sim}{B}$).
The resultant, $\underset{\sim}{B}+\underset{\sim}{A}$, is the same vector as $\underset{\sim}{A}+\underset{\sim}{B}$ in Example G.

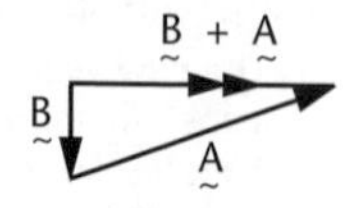

Sometimes more than two vectors are combined.

Example I

A tourist couple visited several places of interest around Lae. They began by travelling 5.0 km east from their hotel, then 3.0 km south, then 4.0 km west and finally 6.0 km north. What was their final displacement, $\underset{\sim}{d}$, from the hotel?

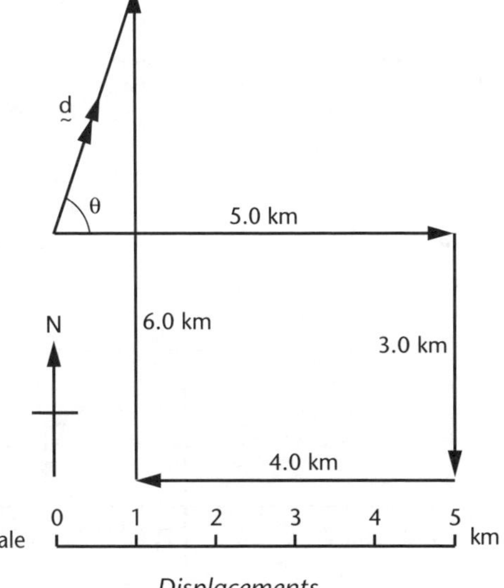

Displacements

Solution:

Displacement is a vector quantity, so the four displacements covered by the tourists can be represented by vectors and added using a vector diagram. The resultant is their final displacement from the hotel.

The size of the resultant displacement, and its direction, θ, can be measured from the diagram as 3.2 km from the hotel at an angle of 72° N of E (bearing 018°).

Rather than using a scale drawing to find the resultant vector, it is common (and often more accurate) to use trigonometry and a *sketched* vector diagram.

Example J

In the north/south direction the couple in Example I have moved 3.0 km S + 6.0 km N = 3.0 km N.

In the east/west direction they have moved 5.0 km E + 4.0 km W = 1.0 km E.

The resultant displacement, $\underset{\sim}{d}$, is found by adding the vectors 1.0 km E and 3.0 km N.

$d = \sqrt{3.0^2 + 1.0^2}$ [Pythagoras' theorem]

$= \sqrt{10}$

$= 3.2\text{ km}$

$\tan\theta = \dfrac{3.0}{1.0}$ $\left[\tan\theta = \dfrac{\text{opposite side}}{\text{adjacent side}}\right]$

$= 3.0$

$\theta = 72^\circ$ [taking inverse tan]

N
d
3.0 km N
θ
E
1.0 km E

Using trigonometry

The displacement is thus 3.2 km, 72° N of E (or bearing 018°).

The answer is more precise than that found in Example I since it does not involve drawing vectors accurately and measuring the resultant.

Only vectors of the *same* physical quantity can be added together; two velocity vectors can be added, but a velocity vector cannot be added to a displacement vector, etc.

Vector subtraction

The *negative* of a vector is a vector with the *same magnitude* but the *opposite direction* of the original vector.

Example K

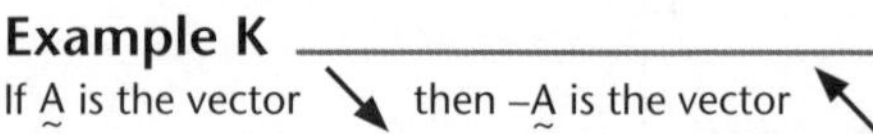
If $\underset{\sim}{A}$ is the vector ↘ then $-\underset{\sim}{A}$ is the vector ↖.

To subtract a vector, *add the negative* vector. To subtract vector $\underset{\sim}{B}$ *from* vector $\underset{\sim}{A}$, begin with vector $\underset{\sim}{A}$ and *add* the vector $-\underset{\sim}{B}$ to $\underset{\sim}{A}$. The subtraction can be written $\underset{\sim}{A} - \underset{\sim}{B} = \underset{\sim}{A} + (-\underset{\sim}{B})$.

Example L

Vector $\underset{\sim}{B}$ is subtracted from vector $\underset{\sim}{A}$.
The subtraction, $\underset{\sim}{A} - \underset{\sim}{B}$, is shown:

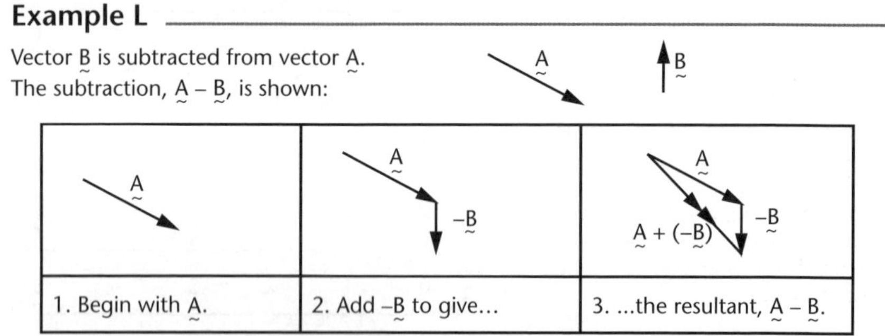

1. Begin with $\underset{\sim}{A}$.	2. Add $-\underset{\sim}{B}$ to give…	3. …the resultant, $\underset{\sim}{A} - \underset{\sim}{B}$.

Unlike vector addition, the order of operations in vector subtraction is important:
$\underset{\sim}{A} - \underset{\sim}{B} \neq \underset{\sim}{B} - \underset{\sim}{A}$.

Vector subtraction is commonly used when calculating a change in a vector quantity.

Example M

To find the change in velocity, $\Delta \underset{\sim}{v}$, from an initial velocity of $\underset{\sim}{v_i}$ to a final velocity of $\underset{\sim}{v_f}$, the *vector* subtraction $\underset{\sim}{v_f} - \underset{\sim}{v_i}$ is done.

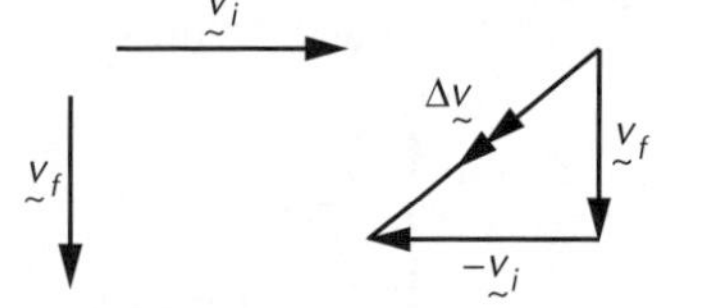

$$\Delta \underset{\sim}{v} = \text{final velocity} - \text{initial velocity}$$
$$\Delta \underset{\sim}{v} = \underset{\sim}{v_f} - \underset{\sim}{v_i}$$
$$= \underset{\sim}{v_f} + (-\underset{\sim}{v_i})$$

Components of a vector

Any vector can be drawn as the sum of two other vectors drawn at right angles to each other. These two vectors are called **components** of the first vector.

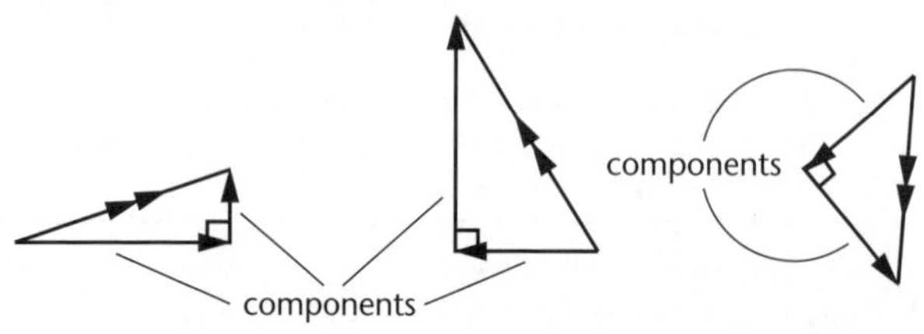

Components

If the directions of the component vectors are chosen carefully, they can be very useful. Finding two component vectors is called **resolving** a vector into components.

Example N

The vertical/horizontal reference frame (Fig. **a** below) shows a vector representing a 10.0 newton force.

The direction of the vector is angled up from the horizontal at 30°. The two force vectors which represent the components of this 10 N force are found by drawing two construction lines, one parallel to the vertical axis and the other parallel to the horizontal axis. The components are labelled (in Fig. **b**).

The 10 N vector and its components form a right-angled triangle (Fig. **c**).

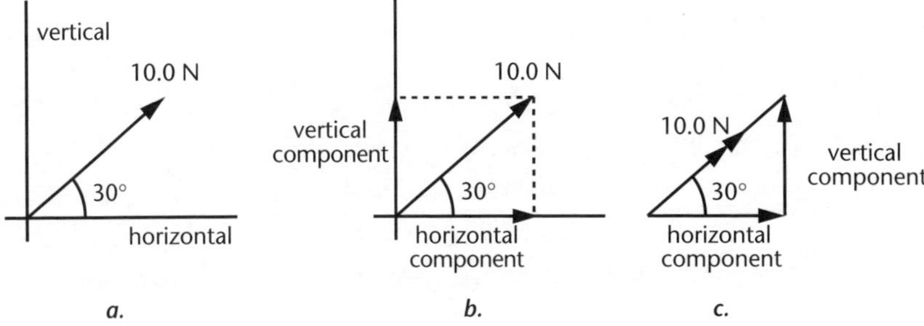

Finding components

Using trigonometry, the vertical and horizontal components can be calculated.

$$\sin 30^\circ = \frac{\text{opposite side}}{\text{hypotenuse}}$$

$$= \frac{\text{vertical component}}{10.0}$$

$$\text{vertical component} = 10.0 \times \sin 30^\circ \qquad \text{[rearranging]}$$

$$= 10.0 \times 0.500$$

$$= 5.00 \text{ N}$$

$$\cos 30^\circ = \frac{\text{adjacent side}}{\text{hypotenuse}}$$

$$= \frac{\text{horizontal component}}{10.0}$$

$$\text{horizontal component} = 10.0 \times \cos 30^\circ \qquad \text{[rearranging]}$$

$$= 10.0 \times 0.866$$

$$= 8.66 \text{ N}$$

The force vector of 10.0 newtons angled 30° up from the horizontal is resolved into components 5.00 N vertically and 8.66 N horizontally.

The 10.0 N force angled 30° up from the horizontal has the same horizontal effect as a force of 8.66 N and the same vertical effect as a force of 5.00 N.

Unit 11.2 Activity 3A: Vector calculations

1. How many vector quantities are in the following list?
   Force, mass, temperature,velocity, density, time, acceleration.
2. Write the equivalent bearing angle for:
   a. 40° south of east b. N 56° E
   c. W 30° S d. 80° north of west
3. Force vectors $\underset{\sim}{a}$ and $\underset{\sim}{b}$ are as shown.

   a. Draw a vector that represents the resultant of $\underset{\sim}{a}$ and $\underset{\sim}{b}$.
   b. Draw a force vector that could act together with $\underset{\sim}{a}$ to produce force $\underset{\sim}{b}$ as a resultant.
4. A cyclist travels 2.5 km east then 10.0 km south and finally 5.0 km west. Determine the cyclist's overall displacement from their starting point.
5. Four forces are applied to an object, as shown below.

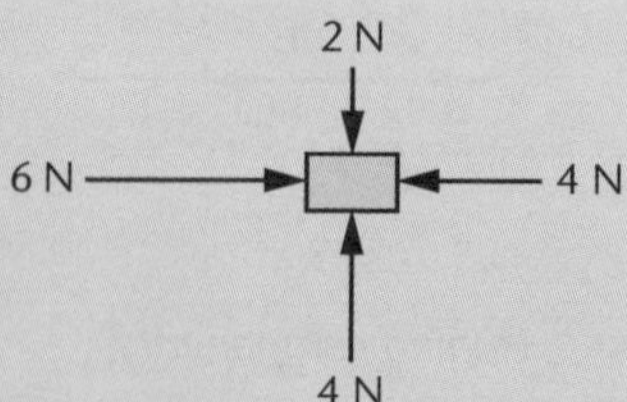

   Determine the direction of the resultant force.
6. A bird flies 3 km to the west and then 4 km to the south. Find the resultant displacement of the bird.
7. The resultant of a 12 m displacement and a 7 m displacement is 5 m. Draw a vector diagram to find the angle between the two displacements.
8. A velocity of 30 km h^{-1} west is added to another velocity of 20 km h^{-1} north. Use a vector diagram to determine the magnitude and direction of the resultant.
9. A vector is as shown in the diagram.
   What is the magnitude of:
   a. The vertical component?
   b. The horizontal component?
   c. The vector?

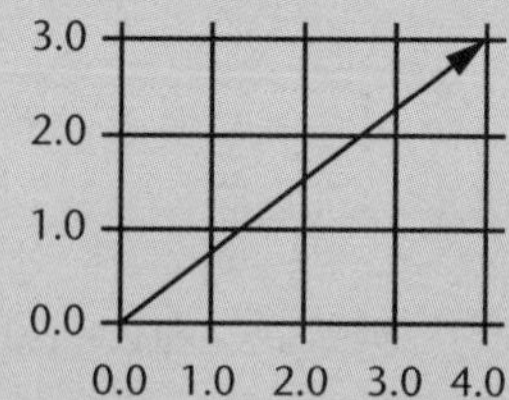

10. Two component forces, 4.0 N vertically upwards and 6.0 N horizontally, act together on the same object. Find their resultant force by drawing a scale vector diagram. Use a protractor to measure the angle which the resultant makes with the horizontal direction.
11. In the diagram alongside, $\underset{\sim}{A}$ represents the initial velocity and $\underset{\sim}{B}$ represents the final velocity. Draw a vector representing the change in velocity.

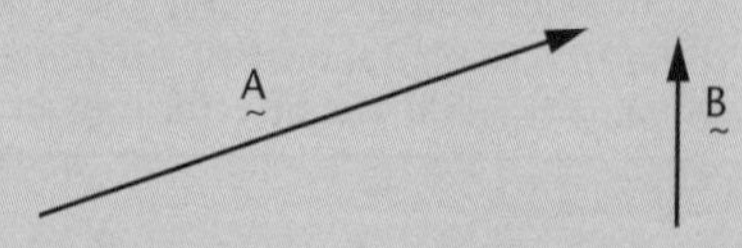

12. A soccer ball collides with the crossbar of a goalpost at 5 m s^{-1}. It rebounds at 4 m s^{-1} in the opposite direction away from the crossbar. Determine the ball's change in velocity.

13. A car travelling east at 40.0 km h^{-1} turns a 90° corner and ends up travelling north at 20.0 km h^{-1}. Find the car's change in velocity (both magnitude and direction).

14. A tennis ball thrown against a wall rebounds without loss in speed, as shown. Draw a scale vector diagram to find the direction of the ball's change in velocity.

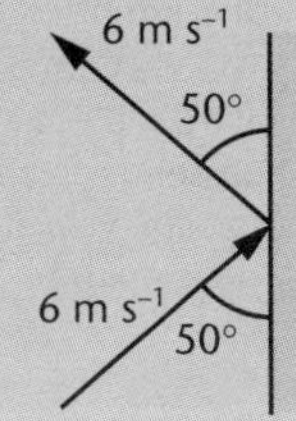

Relative velocity

Relative velocity is the velocity of one object *in relation* to another object. The velocity of an object may appear to be different depending on where it is measured from.

Example O

A woman is standing still on an escalator which is moving at an angle upward at 4 km h^{-1}. This woman's velocity appears to differ relative to where it is measured from.

A physics student *on the ground* would measure the velocity of the woman as being the same as the escalator.

Relative to the ground, the woman's velocity is 4 km h^{-1} angled upward.

A second physics student *standing still on the escalator* would measure the woman's velocity to be zero.

Relative to the escalator, the woman has a velocity of 0 km h^{-1}.

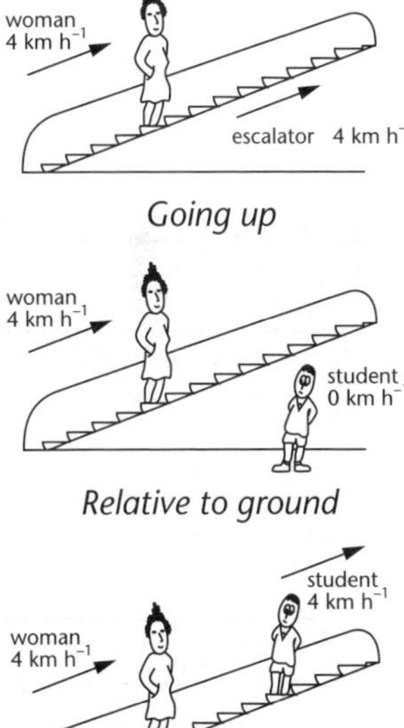

Relative velocity problems involve adding and subtracting vectors.

Example P

A PMV moves forward at 10 km h^{-1} and beside it a car moves in the same direction at 6 km h^{-1}.

The PMV moves forward away from the car at 10 – 6 = 4 km h^{-1}. This means that to a passenger (called an observer) sitting in the car, the velocity of the PMV relative to *the car* is 4 km h^{-1} forward.

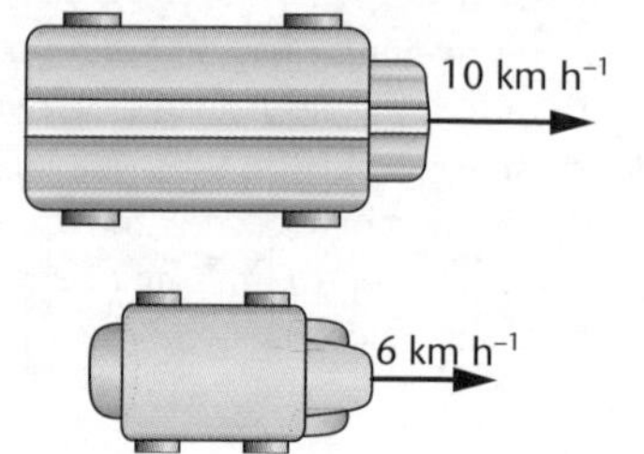

How velocities 'see' each other

To the observer in the car, the situation can appear as if the car is stationary and the PMV is moving forward at only 4 km h^{-1}.

A useful method for calculating relative velocity is illustrated by Example P. If the velocity of the PMV is written $\underset{\sim}{v}_{PMV}$, and the velocity of the car $\underset{\sim}{v}_{car}$, then the velocity of the PMV relative to the car, written $\underset{\sim}{v}_{PMV\ rel\ car}$, is:

$\underset{\sim}{v}_{PMV\ rel\ car} = \underset{\sim}{v}_{PMV} - \underset{\sim}{v}_{car}$.

More generally, if the velocity of an object A is written $\underset{\sim}{v}_A$, and the velocity of an object B is written $\underset{\sim}{v}_B$, then the velocity of A relative to B, written $\underset{\sim}{v}_{A\ rel\ B}$, is:

$$\underset{\sim}{v}_{A\ rel\ B} = \underset{\sim}{v}_A - \underset{\sim}{v}_B$$

Example Q

In Example P,

10 km h^{-1} → 6 km h^{-1} →	10 km h^{-1} → –6 km h^{-1} ←	10 km h^{-1} → 4 km h^{-1} –6 km h^{-1}
$\underset{\sim}{v}_{car}$ and $\underset{\sim}{v}_{bus}$	$\underset{\sim}{v}_{PMV\ rel\ car}$ is found by subtracting $\underset{\sim}{v}_{car}$ from $\underset{\sim}{v}_{PMV}$	giving 4 km h^{-1} forwards

In example P, the velocity of *the car relative to the PMV* is $\underset{\sim}{v}_{car}$ rel $\underset{\sim}{v}_{PMV} = (\underset{\sim}{v}_{car} - \underset{\sim}{v}_{PMV})$ and is 4 km h^{-1} *backwards*. To a person in the PMV, it can *appear* that the nearby car is going backwards at 4 km h^{-1} .

Example R

A motorcyclist rides horizontally at 50 km h^{-1} through the rain. The steady rain falls vertically at 15 km h^{-1}. What is the velocity of the rain relative to the cyclist?

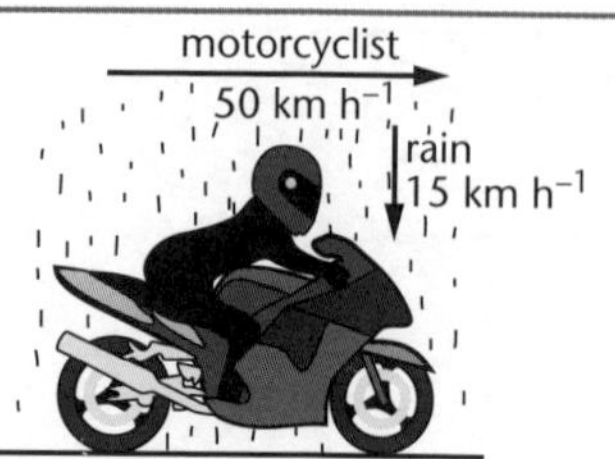

How the rain falls

Solution:

The two velocities are both given *relative to the ground*. Yet experience tells us that *relative to the motorcyclist*, the rain would appear to be coming more towards their face than straight down.

The velocity of the rain relative to the motorcyclist is found by subtracting the motorcyclist's velocity from the velocity of the rain.

Using Pythagoras (where v is the magnitude):

$$v = \sqrt{15^2 + 50^2}$$

$$= 52 \text{ km h}^{-1} \text{ (2 sf)}$$

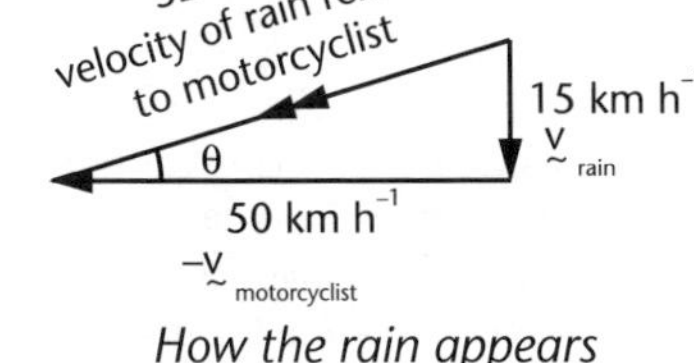

How the rain appears

Using trigonometry (where θ is the angle):

$$\tan\theta = \frac{15}{50}$$

$$= 0.30$$

$$\theta = 17^\circ \text{(2 sf)}$$

The resultant is *the velocity of the rain relative to the motorcyclist* and is 52 km h^{-1} at an angle of 17° to the horizontal.

The rain hits the motorcyclist at 52 km h^{-1}.

Boats crossing rivers in which there is a current give rise to relative velocity problems.

Example S

A boat starts to cross a river which is 2.0 km wide. The boat travels at 20 km h^{-1} *relative* to the surface of the water and points directly across the river. The river is moving at 6.0 km h^{-1} downstream *relative* to the river bank.

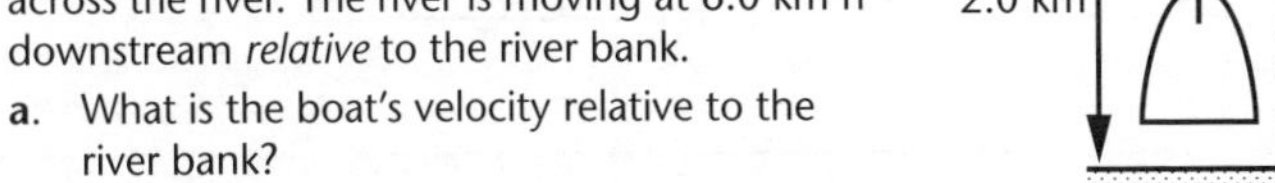

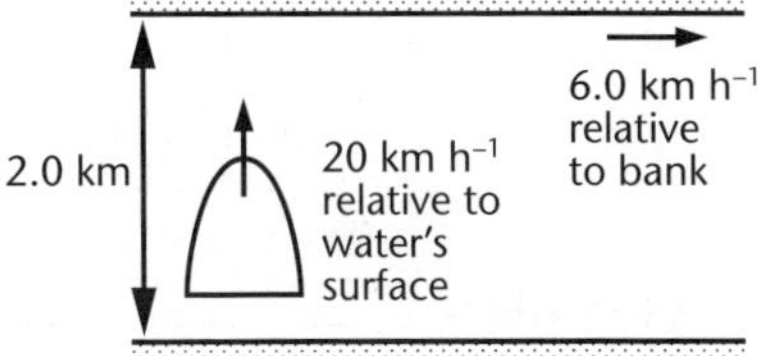

a. What is the boat's velocity relative to the river bank?

b. How long does the boat take to cross the river?

c. At what point does the boat land on the other side?

Solution:

The current is relative to the river bank and the river bank (and any observer standing on the bank) is at rest.

a. The boat's velocity relative to the river bank is the *combined result* of its velocity relative to the river and the velocity of the river relative to the river bank:

$$\underset{\sim}{V}_{\text{boat rel river}} + \underset{\sim}{V}_{\text{river rel bank}} = \underset{\sim}{V}_{\text{boat rel bank}}$$

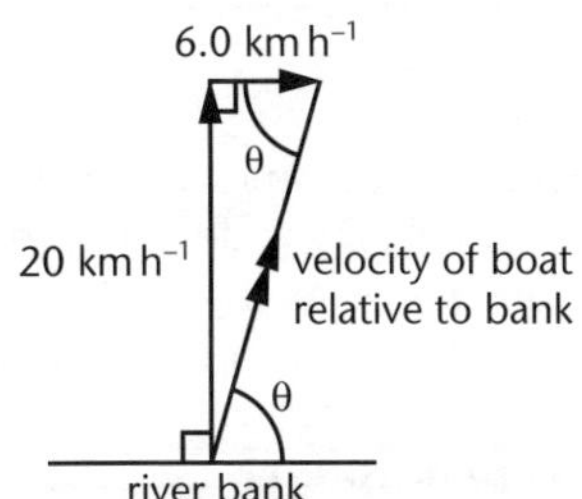

Using Pythagoras and trigonometry, the boat moves with a speed of:
21 km h^{-1} ($\sqrt{20^2 + 6.0^2}$) at an angle downstream of 73° ($\tan\theta = \frac{20}{6.0}$) to the river bank.

b. The boat continues to travel *across* the river at 20 km h^{-1}, despite the fact that it is also being carried downstream. The river's motion *does not affect* the boat's movement *across* the river because the motion of the river is at right angles to the direction in which the boat is pointed.

Time taken to cross the river is $t = \frac{d}{v}$ [rearranging $d = vt$]

$$= \frac{2.0}{20}$$

$$= 0.10\text{h (6 minutes)}$$

c. As well as travelling across the river, the boat is carried at 6.0 km h^{-1} downstream for 0.10 hour (the time taken to cross the river). The distance travelled downstream is:

$$d = vt$$
$$= 6.0 \times 0.10$$
$$= 0.60 \text{ km}$$

The boat touches the opposite river bank 0.60 km downstream from a point directly across the river.

The displacement vector diagram forms a triangle which is *similar* to the velocity vector diagram:

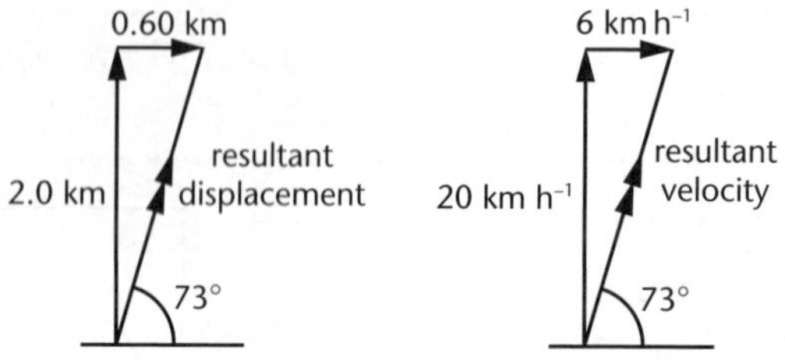

Example T

Consider two cars approaching one another, each with a speed of 90 km/h with respect to ground. The speed of one car relative to the other is 180 km/h. What this means is that, to an observer in one of the cars, the other car seems to be approaching at 180 km/h.

Similarly, when one car travelling at 110 km/h and in the same direction passes a second car travelling at 80 km/h, the first car has a speed relative to the second car of 110 km/h – 80 km/h = 30 km/h.

Note:

From this example, you now understand that when the velocities are along the same line of action, all you do is a simple addition or subtraction to calculate the relative velocity. However, when they are not along the same line, you have to use vector addition (since the velocity is a vector quantity), as in the next example.

Example U

(Adapted from Giancoli D.C. 1980 *Physics Principles and Applications*; Englewood Cliffs, N. J)

Suppose you are by the Sepik River, where your boat can travel 1.20 m/s in still water.

a. If you steer the boat to head directly across the stream where the velocity of the current is 0.75 m/s, what is the velocity (magnitude and direction) of your boat relative to the shore?

b. What will be the position of your boat relative to the point of origin after 3 seconds?

Solution:

a. First you draw a vector diagram of the situation, as follows:

V_{BW} = velocity of the boat relative to the water
V_{WS} = velocity of the water relative to the ground
V_{BS} = velocity of the boat relative to the ground (what you are trying to find)

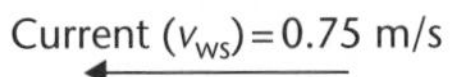

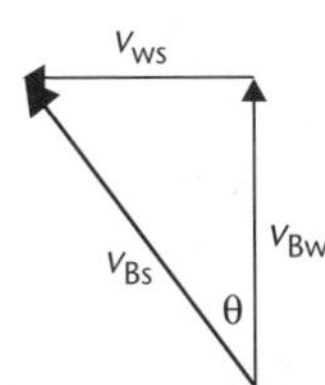

i. As shown in your diagram, your boat will be pulled down the stream by the current. Your boat's velocity with respect to the shore, v_{Bs}, will be the sum of its velocity with respect to the water, v_{Bw}, plus the velocity of the water with respect to the shore, v_{ws}: $v_{BS} = v_{Bw} + v_{Ws}$

ii. You can measure θ and v_{Bs} (magnitude of v_{Bs}) from your diagram to obtain v_{Bs} = 1.4 m/s, and θ = 32°. Or, since v_{Bw} is perpendicular to v_{ws}, you can get v_{Bs} using Pythagoras Theorem; $v_{Bs} = \sqrt{v_{Bw}^2 + v_{ws}^2} = \sqrt{(1.20\ \text{m/s})^2 + (0.75\ \text{m/s})^2}$ = 1.4 m/s.

b. To find your boat's position after 3.0 seconds, you need to determine the displacement vector, *d*. The vector was previously defined as $d = \bar{v}t$, where t is the elapsed time. Since *t* is a scalar, it does not affect directions; so the direction of d in this case will be the same as the direction of V_{BS}.

$d = v_{Bs}t$ = (1.4 m/s)(3.0 s) = 4.2 m. Therefore, your boat after 3.0 s will be 4.2 m from its point of origin in a direction 32° downstream from directly across the river.

Unit 11.2 Activity 3B: Relative velocity

1. In a river 500 m wide, water flows at 8.0 km h^{-1}. A boat, pointing directly across the river, takes 6.0 minutes to cross the river.
   a. How far is the boat carried downstream as it crosses?
   b. What is the speed of the boat relative to the water's surface?
   c. What is the boat's velocity relative to the river bank?
2. A Jumbo jet's air speed is 600 km h^{-1} as it points south. A cross-wind blows at 60 km h^{-1} from the west.
   a. Draw a vector diagram to show the situation.
   b. What is the jet's velocity relative to the ground?
3. Two wharves A and B are directly opposite each other on a 40 m wide river which flows in the direction shown. A boat leaves A and heads at constant speed at right angles to the flow of the river. It lands at point C, with the trip taking 20 seconds.

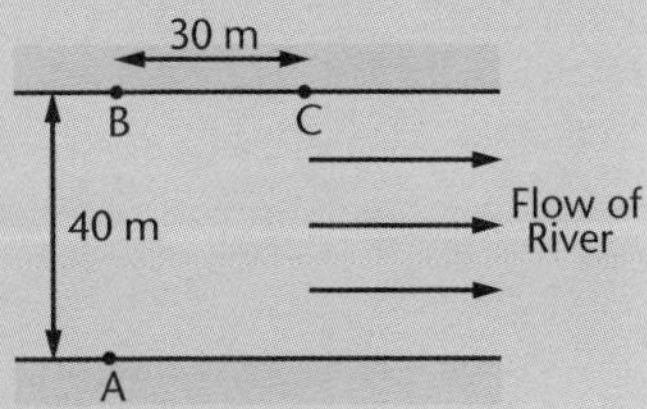

Find:

a. The displacement of C from A.

b. The speed of the boat as seen by people standing at A.

c. The speed of the water in the river.

d. The speed of the boat as seen by a fish drifting with the river.

e. Draw a vector diagram and use it to find the direction the boat should head (at the same constant speed as before) if it is to travel *directly from A to B.*

4. A man in a speedboat wishes to cross a river which is flowing at 0.50 m s^{-1}. He wishes to cross from a point A to a point B which is directly opposite as shown:

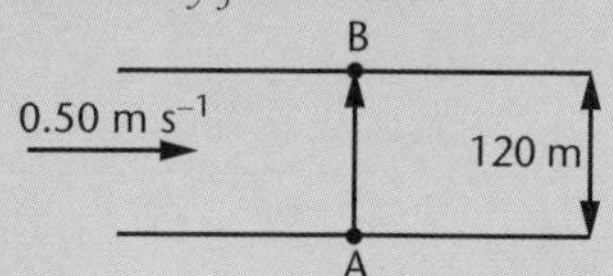

The river is 120 m wide and he wishes to make the crossing in exactly 100 seconds. To achieve this he must angle his boat slightly upstream and adjust his engine speed. With what speed, relative to the water, must the speedboat move in order to travel from A to B in 100 seconds?

5. A woman sitting in a train watches a car approach a level crossing, as shown in the diagram.

What is the velocity of the car relative to the woman in the train?

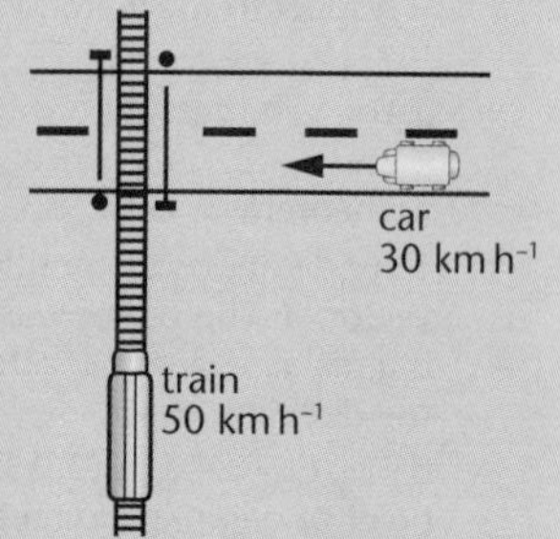

6. A Cessna aeroplane flies with a speed of 150 km h^{-1} relative to the air, and points due east. A wind is blowing at a speed of 20 km h^{-1} and bearing 135°, ie the wind is blowing from the northwest. Use the component method or a scale diagram to find the resultant velocity of the aeroplane relative to the ground.

Unit 11.2 Motion (Kinematics)
Topic 4: Graphs of motion

'Graphs of motion' is the last heading on p. 10 of the Syllabus. This Topic covers the bullet points listed under that heading:

- Distance–time, speed–time, velocity–time, displacement–time, and acceleration–time graphs.
- Interpreting kind of motion from graphs.
- The ticker timer.
- Free fall and acceleration of gravity.

Distance versus time graphs

A **distance versus time graph** provides a picture of the history of the motion of an object. The shape of the distance versus time graph indicates whether or not the object is travelling at constant speed or if it is accelerating.

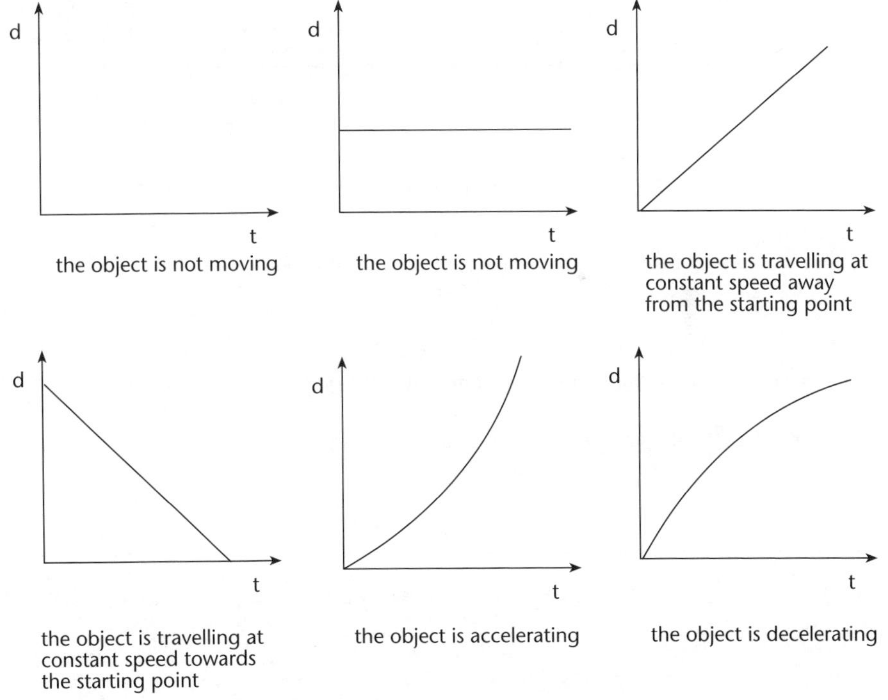

Shapes of distance versus time graphs.

The speed of an object at any point in time is equal to the gradient (slope) of the distance versus time graph at that point in time.

If the graph is curved, a tangent should be drawn – the speed is calculated by finding the gradient of the tangent.

Example A

John is training for a running race. The following distance versus time graph represents his motion during a training run. Use the graph to calculate his speed at:

a. 2 s b. 5 s c. 8 s

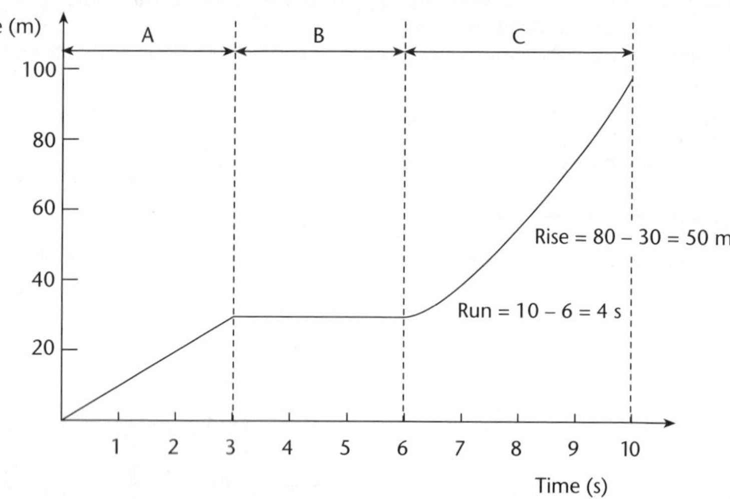

Distance versus time graph for John's race.

Solutions:

a. Speed 2 s after the start of the race = gradient of the graph at $t = 2$ s

$$= \frac{30 \text{ m}}{3 \text{ s}}$$

$$= 10 \text{ m s}^{-1}$$

b. Speed 5 s after the start of the race = gradient of the graph at $t = 5$ s

$$= 0 \text{ m s}^{-1}$$

c. Speed 8 s after the start of the race = gradient of the graph at $t = 5$ s

$$= \frac{50 \text{ m}}{4 \text{ s}}$$

$$= 12.5 \text{ m s}^{-1}$$

Unit 11.2 Activity 4A: Distance versus time graphs

1. Alongside is a speed-time graph of a moving object. Describe the motion of the object.

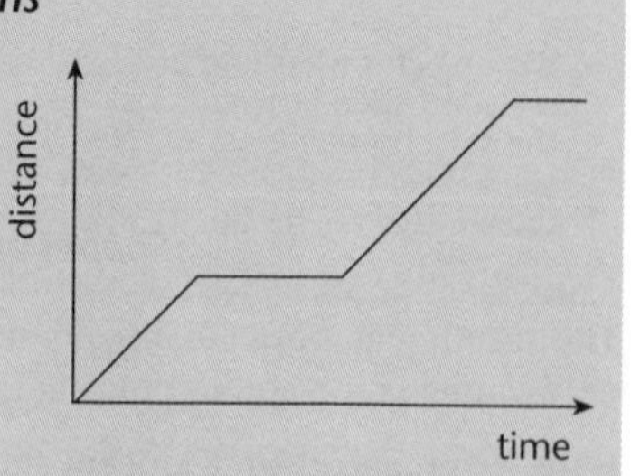

2. The distance-time graph shows the journey of a car. During which section of the graph was the car moving at the greatest speed?

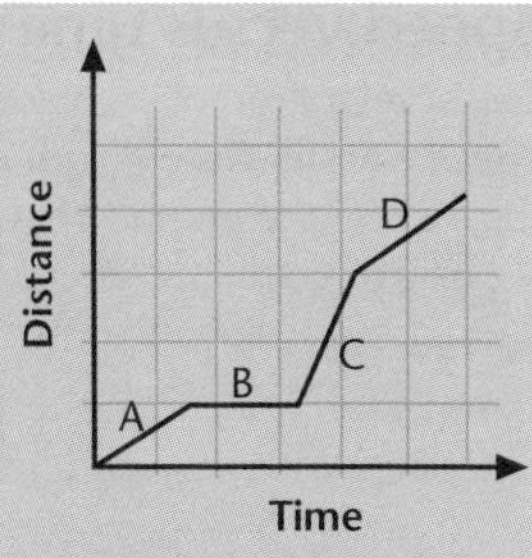

3. Describe the motion in each section of the graph.

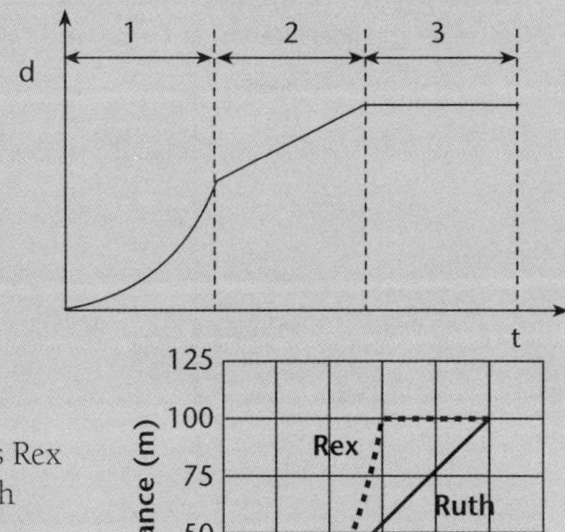

4. Ruth takes her dog Rex out for their usual evening jog. Near the end of the jog she lets Rex off the leash and they have a race. The graph shows the motion during the race.

a. After how many seconds from the start of the race did Rex overtake Ruth?

b. What was Ruth's average speed?

c. What was Rex's speed after 17 seconds?

d. Describe Ruth's speed near the end of the race.

5. A car freewheels from rest down a long ramp and then along a level road.

a. Using the data in the table, plot on a distance-time graph the distance the car has travelled from the top of the ramp.

b. The length of the ramp is 20 metres. How can this be shown from your graph?

c. Use your graph to find:

i. The distance travelled by the car at the end of 5 seconds.

ii. The speed of the car at time 11 seconds.

Time (s)	Distance (m)
2	0.4
4	1.3
6	5.0
8	11.3
10	20
12	30
14	40

Speed versus time graphs

The acceleration of an object at any point in time is equal to the gradient (slope) of the **speed versus time graph** at that point in time. Gradients can be calculated by dividing the 'rise' of the graph ('change in y-direction') by the 'run' ('change in x-direction'). A positive gradient represents acceleration (speeding up). A negative gradient represents deceleration (slowing down).

If the graph is a curve, the acceleration at a particular time can be calculated by drawing a tangent to the curve and finding the gradient of the tangent.

Example B

Calculate the acceleration at $t = 3$ s and $t = 5$ s from the speed versus time graph below.

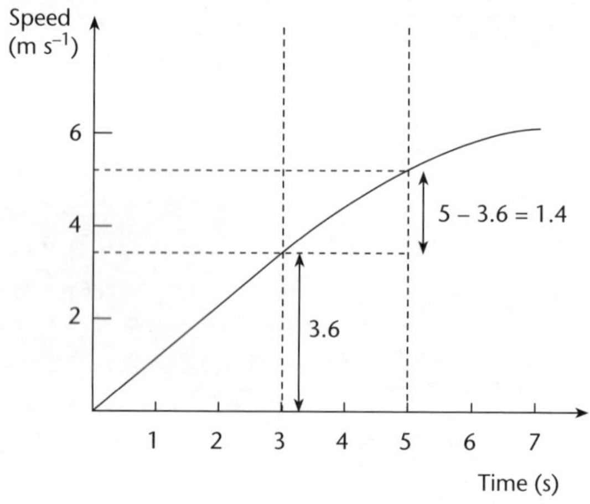

Speed versus time graph for an accelerating object.

Solution:

At $t = 3$ s, the acceleration is equal to the gradient of the tangent to the graph:

$$a = \frac{\Delta v}{\Delta t} = \frac{3.6 \text{ m s}^{-1}}{3 \text{ s}} = 1.2 \text{ m s}^{-2}$$

At $t = 5$ s, the acceleration is equal to the gradient of the graph:

$$a = \frac{\Delta v}{\Delta t} = \frac{2 \text{ m s}^{-1}}{4 \text{ s}} = -0.5 \text{ m s}^{-2}$$

The distance travelled by an object between two points in time is equal to the area under the speed versus time graph between those points in time.

Example C

Calculate the distance travelled by the object in Example B during the first three seconds.

Solution:

distance travelled = area under a v-t graph

$$= \frac{1}{2} \times \text{base} \times \text{height}$$

$$= \frac{1}{2} \times 3 \times 3.6$$

$$= 5.4 \text{ m}$$

Unit 11.2 Activity 4B: Speed versus time graphs

1. A ball is dropped from the top of a building to the ground below. Sketch a speed-time graph for this motion.

2. A cart starts from rest and rolls down a slope. The speed-time graph shows the motion of the cart. What is the acceleration of the cart?

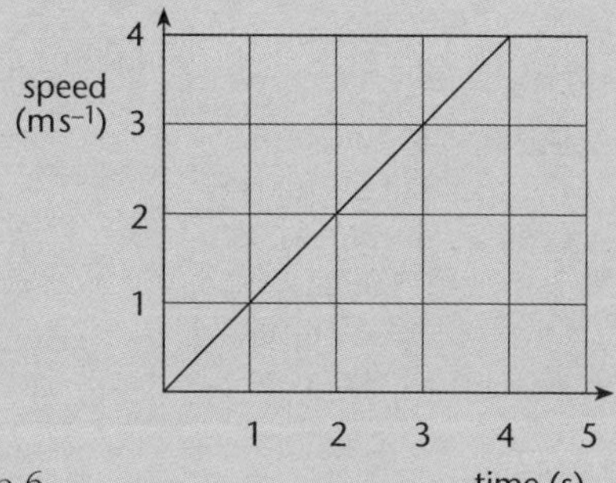

The following information relates to questions 3 to 6.

A car is moving along a straight road. The graph shows how its velocity changes from the start of the journey.

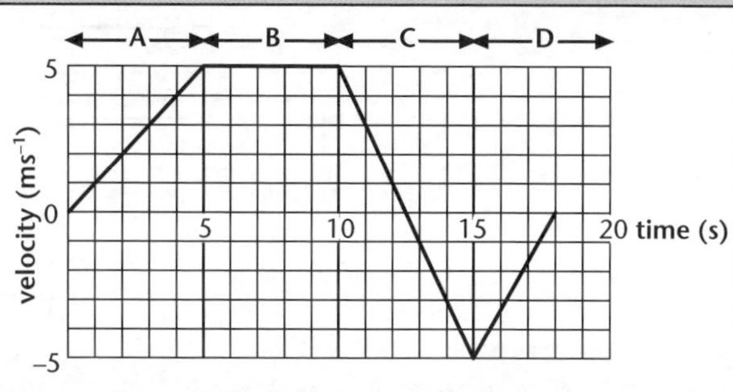

3. What is the total distance travelled by the car?
4. What is the total displacement of the car from its starting point?
5. What is the average speed over the first 12.5 seconds?
6. What is the acceleration of the car over Section C of the graph?
7. A ball is thrown vertically upwards and falls back to the ground. Sketch a velocity-time graph that correctly shows this motion.

8. Tina freewheels down a ramp in her wheelchair and reaches a smooth flat driveway. Sketch a velocity-time graph that describes her motion.

9. The graph shows the velocity of an object during the first 20 seconds of its motion. What is the total distance travelled in 20 seconds?

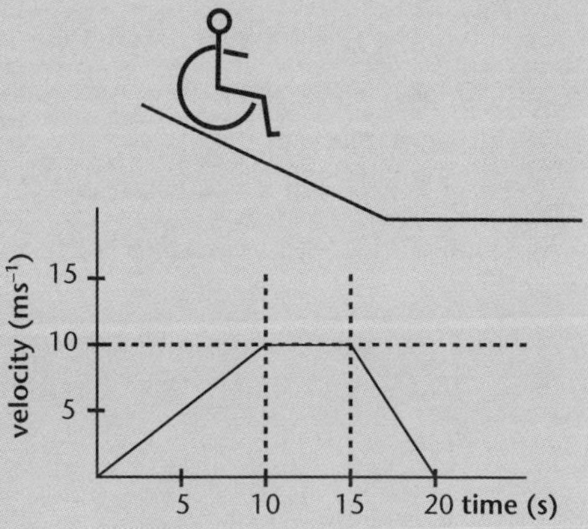

10. The diagram shows a ball at 1-second intervals as it rolls along a horizontal surface.

Sketch a speed-time graph for this motion.

11. The graph shows how the speed of an athlete changes during a work-out. Calculate the average speed for the 24 seconds shown.

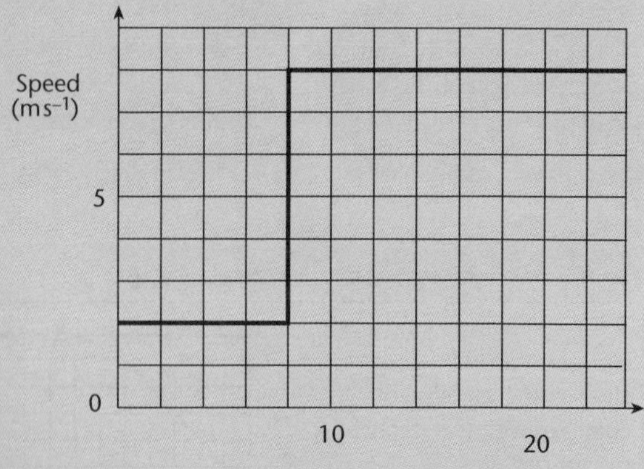

12. John is considering buying a racing car. He takes it for a test drive to test how smoothly it accelerates and whether the brakes work properly.

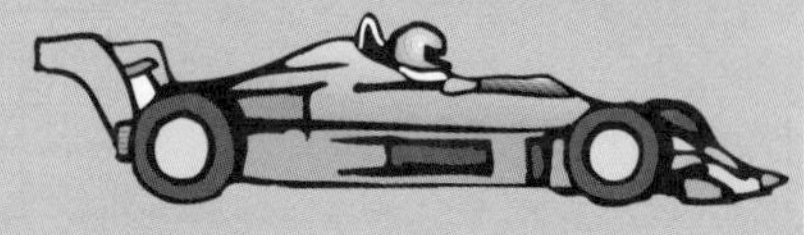

The graph below shows how the speed varied during his acceleration and braking test.

a. Calculate John's acceleration during the first fifteen seconds.

b. Calculate John's acceleration during the last 7.5 seconds.

c. Calculate the total distance travelled during the test drive.

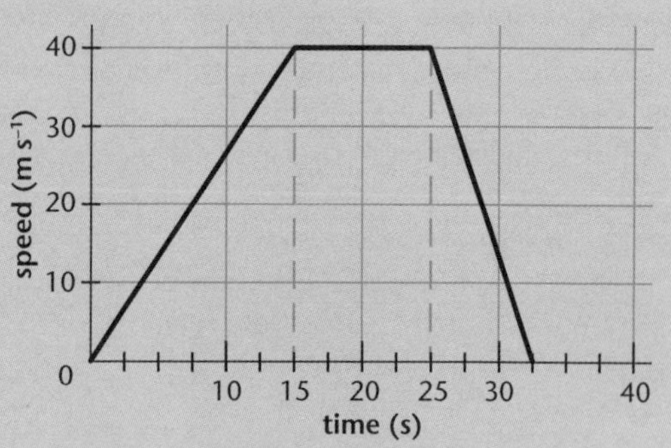

13. Meri slides down the hydroslide at her local swimming pool. You may assume that the effect of friction on her motion is negligible. A graph of her motion on sectors **A**, **B** and **C** of the hydroslide is shown alongside. The hydroslide continues beyond Section **C**.

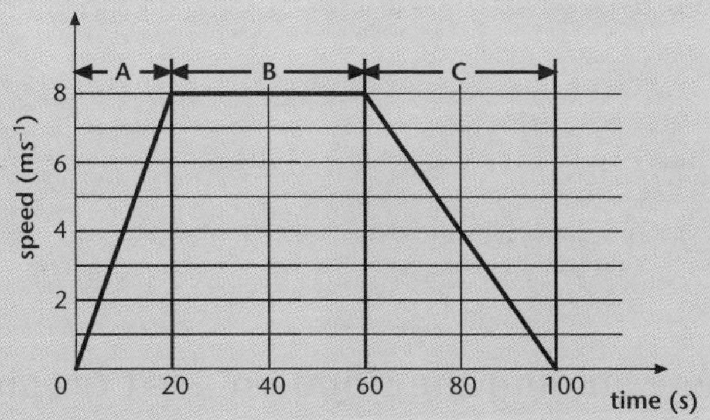

a. Sketch the shape of the hydroslide represented by the graph, using a diagram like that below:

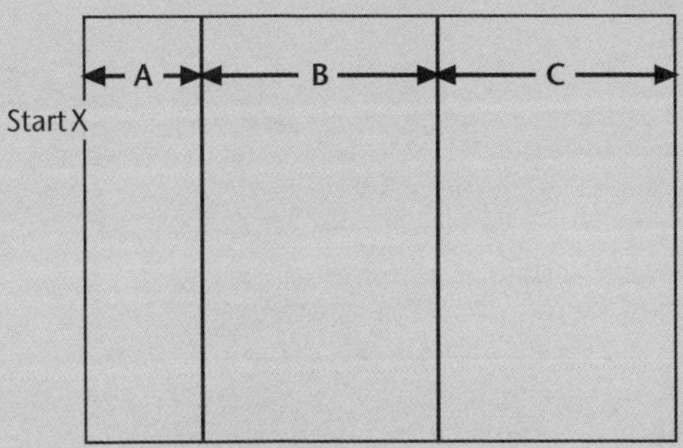

b. Calculate Meri's acceleration on section **A** of the hydroslide.

c. Calculate the total distance travelled by Meri during the time shown on the graph.

d. Calculate Meri's average speed during the time shown on the graph.

14. **a**. Use the graph to estimate the speed of the car at times of 5 and 10 seconds.

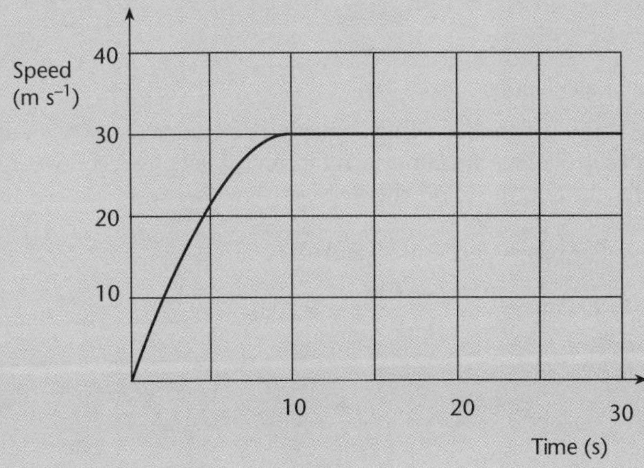

b. Describe the motion of the car shown in the graph.

c. Does the resultant force on the car change increase, stay the same or decrease during the first 10 seconds shown on the graph? Explain how you obtained your answer.

d. How far did the car travel during the time shown on this graph? (Give your reasoning.)

e. When travelling at 30 m s^{-1}, the car brakes steadily and comes to a stop in 90 metres. Graph this part of its motion, making sure you mark the time axis with a suitable scale.

Determining the slope on a v–t graph

The slope of the line on a velocity versus time graph is equal to the *acceleration* of the object.

Consider the velocity versus time graph below.

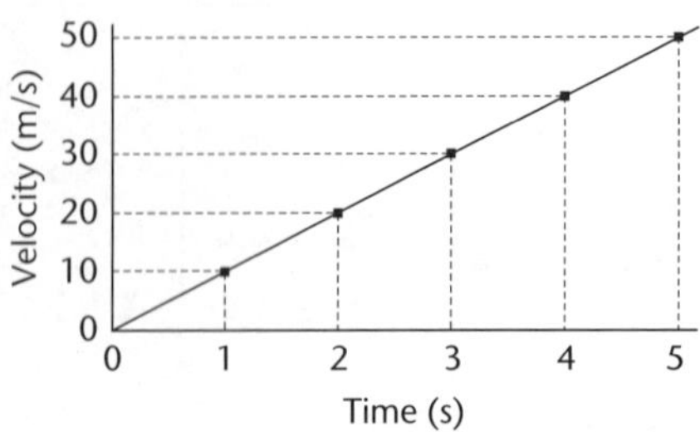

Using the slope equation

$$\text{Slope} = \frac{\Delta y}{\Delta x} = \frac{y_2 - y_1}{x_2 - x_1} = \frac{\text{rise}}{\text{run}}$$

Follow the steps of slope calculation as given earlier.

The calculation below shows this method being applied to determine the slope of the line. Note that three different calculations are performed for three different sets of two points on the line. In each case, the result is the same: the slope is 10 m/s/s.

For points (5 s, 50 m/s) and (0 s, 0 m/s):

$$\text{slope} = \frac{50\,\text{m/s} - 0\,\text{m/s}}{5\,\text{s} - 0\,\text{s}} = 10\ \text{m/s/s}$$

For points (5 s, 50 m/s) and (2 s, 20 m/s):

$$\text{slope} = \frac{50\,\text{m/s} - 20\,\text{m/s}}{5\,\text{s} - 2\,\text{s}} = 10\ \text{m/s/s}$$

For points (4 s, 40 m/s) and (3 s, 30 m/s):

$$\text{slope} = \frac{40\,\text{m/s} - 30\,\text{m/s}}{4\,\text{s} - 3\,\text{s}} = 10\ \text{m/s/s}$$

Unit 11.4 Activity 4C: Velocity–time graph

Consider the velocity–time graph below. Determine the acceleration (ie slope) of the object as portrayed by the graph.

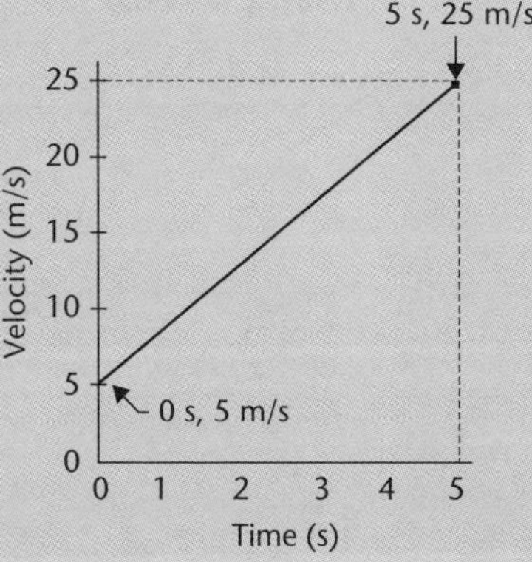

Example D

Examine a more complex case. Consider the motion of a car which first travels with a constant velocity (a = 0 m/s/s) of 2 m/s for four seconds and then accelerates at a rate of +2 m/s/s for four seconds. That is, in the first four seconds, the car is not changing its velocity (the velocity remains at 2 m/s) and then the car increases its velocity by 2 m/s per second over the next four seconds. The velocity–time data and graph are displayed below. Observe the relationship between the slope of the line during each four-second interval and the corresponding acceleration value.

Time (s)	Velocity (m/s)
0	2
1	2
2	2
3	2
4	2
5	4
6	6
7	8
8	10

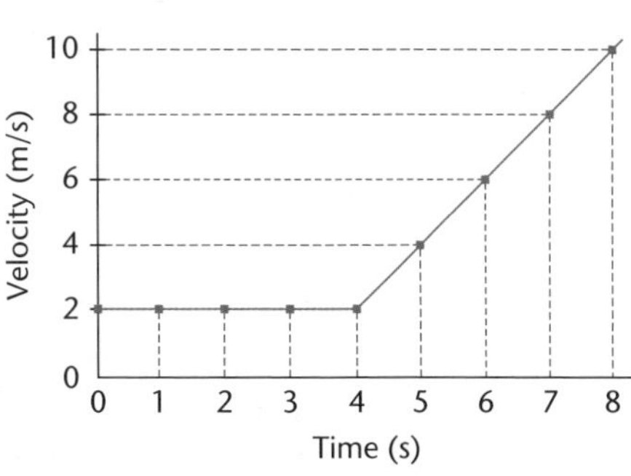

From 0 s to 4 s: slope = 0 m/s/s

From 4 s to 8 s: slope = 2 m/s/s

Unit 11.2 Activity 4D: Two-stage rocket

The velocity–time graph for a two-stage rocket is shown below. Use the graph and your understanding of slope calculations to determine the acceleration of the rocket during the listed time intervals.

a. $t = 0 - 1$ second
b. $t = 1 - 4$ second
c. $t = 4 - 12$ second

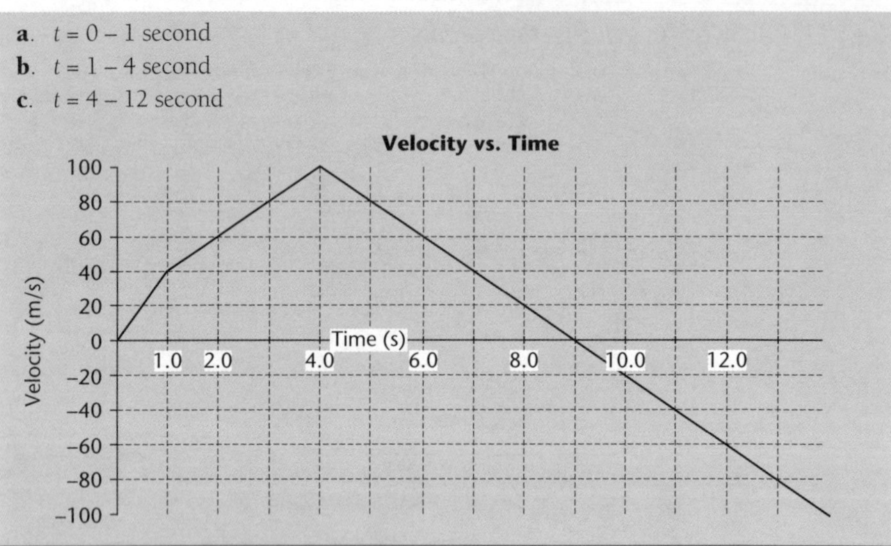

Relating the shape to the motion

The principle can be applied to a variety of types of motion. In each diagram below, a short verbal description of a motion is given (eg 'constant, rightward velocity') and an accompanying *ticker tape* diagram is shown. Finally, the corresponding velocity–time graph is sketched and an explanation is given.

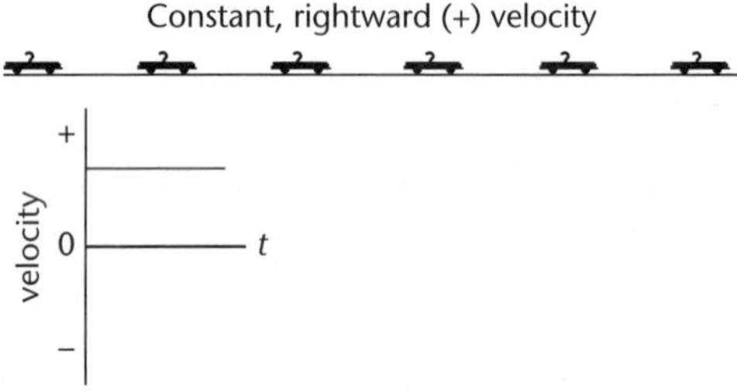

Rightward (+) velocity means + velocity values are plotted; and constant velocity (a = 0 m/s/s) means a zero slope.

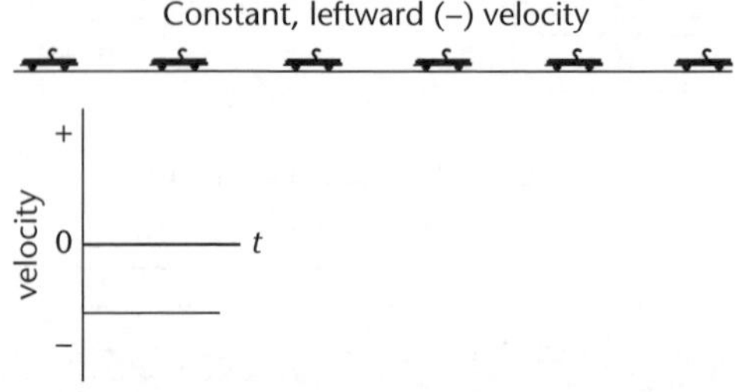

Leftward (–) velocity means – velocity values are plotted; and constant velocity (a = 0 m/s/s) means a zero slope.

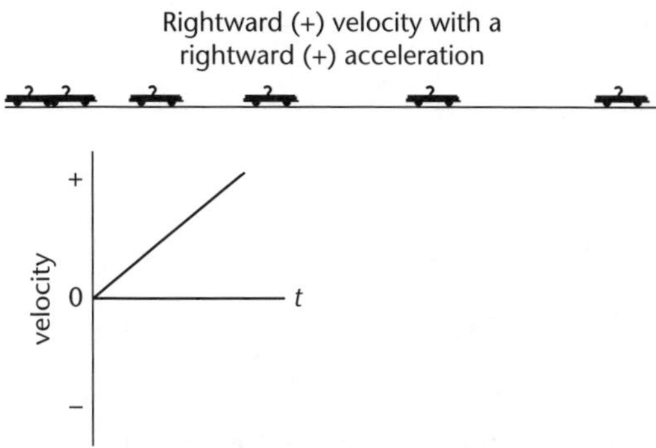

Rightward (+) velocity means + velocity values are plotted; a rightward (+) accleration means a + slope.

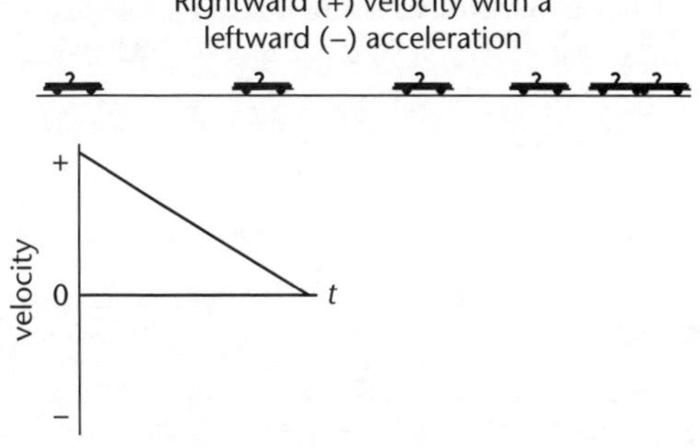

Rightward (+) velocity means + velocity values are plotted; a leftward (–) accleration means a – slope.

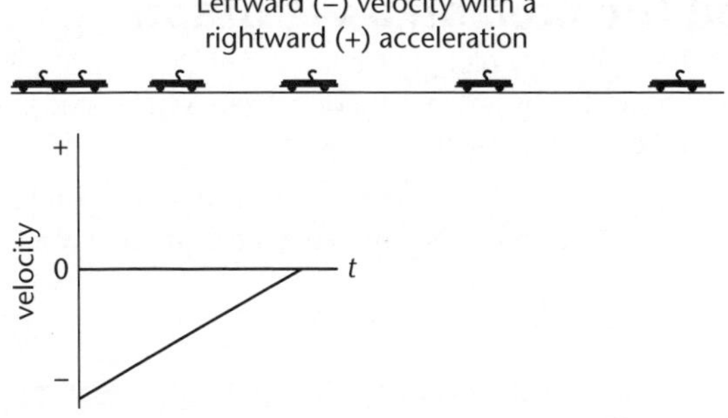

Leftward (–) velocity means – velocity values are plotted; a rightward (+) accleration means a + slope.

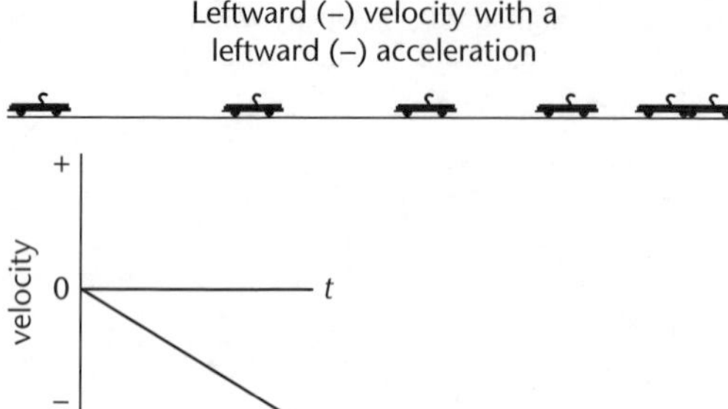

Leftward (–) velocity means – velocity values are plotted; a leftward (–) accleration means a – slope.

Unit 11.2 Activity 4E: Velocity–time graph – motion

Describe the motion depicted by the following velocity–time graphs. In your descriptions, make reference to the direction of motion (+ or – direction), the velocity and acceleration and any changes in speed (speeding up or slowing down) during the various time intervals (eg intervals A, B, and C).

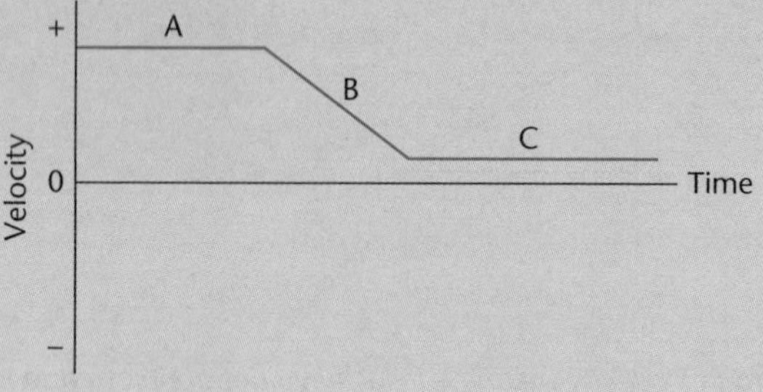

Determining the area on a v–t graph

In this section you will learn how a plot of velocity versus time can also be used to determine the distance travelled by an object. For velocity versus time graphs, the area bound by the line and the axes represents the distance.

The diagram on the next page shows three different velocity–time graphs. The shaded region between the line and the time-axis represents the distance travelled during the stated time interval.

The shaded area is representative of the area travelled during the time from 0 seconds to 6 seconds. This area takes on the shape of a rectangle. It can be calculated using the appropriate equation.

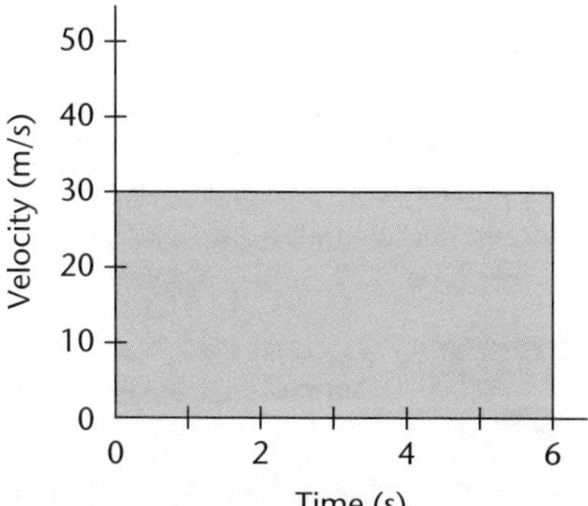

The shaded area is representative of the area travelled during the time from 0 seconds to 4 seconds. This area takes on the shape of a triangle. It can be calculated using the appropriate equation.

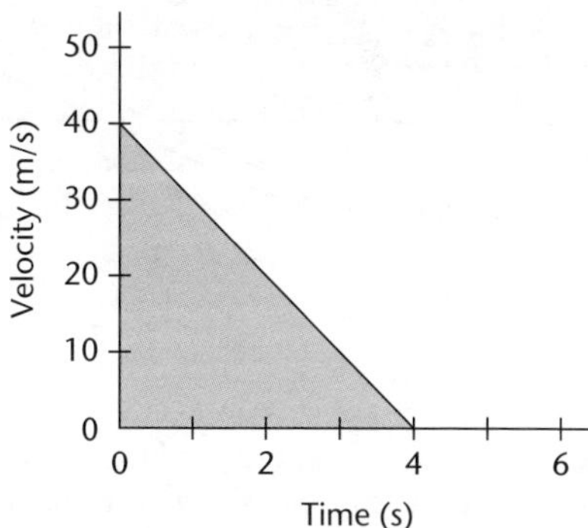

The shaded area is representative of the area travelled during the time from 2 seconds to 5 seconds. This area takes on the shape of a trapezoid. It can be calculated using the appropriate equation.

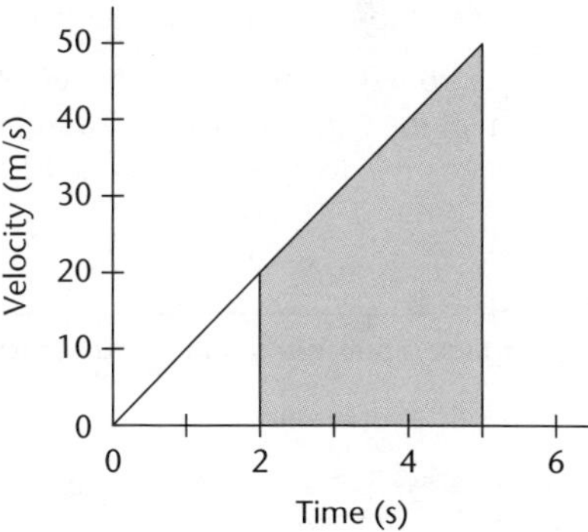

The method used to find the area under a line on a velocity–time graph depends upon whether the section bound by the line and the axes is a rectangle, a triangle or a trapezoid. Area formulae for each shape are given below.

<u>Rectangle</u>

$$\text{Area} = b \times h$$

<u>Triangle</u>

$$\text{Area} = \frac{1}{2} \times b \times h$$

<u>Trapezoid</u>

$$\text{Area} = \frac{1}{2} \times b \times (h_1 + h_2)$$

The ticker timer

In the laboratory, a **ticker timer** is used to record distance and time data for a moving object. A ticker timer is connected to the 50 Hz **alternating current** of the mains supply. It contains an **electromagnet** that switches on and off 50 times each second. The electromagnet attracts a steel strip that vibrates up and down. At intervals of $\frac{1}{50}$ s the strip moves down and a small bump on the strip hits a carbon disc that makes a dot on the paper ticker tape.

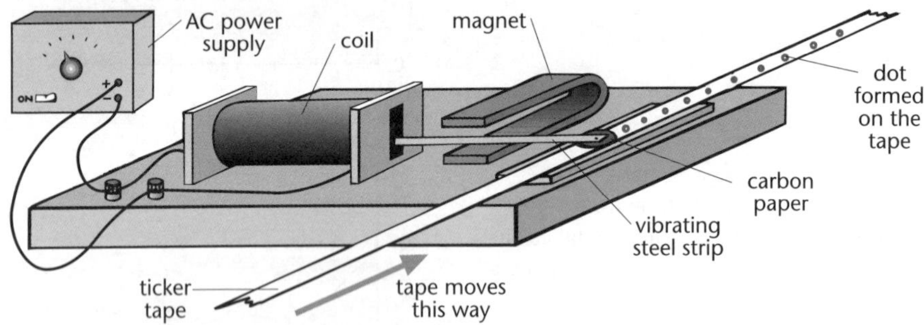

The ticker timer.

The **ticker tape** is connected to a moving object. The length of tape pulled through the ticker timer is an indication of the distance traveled by the object pulling the tape. The number of dots or 'ticks' is a measure of the time. To analyse the motion, divide the tape into sections by drawing a line through the first distinguishable dot on the tape and thereafter every fifth dot (the time interval for 5 dots is $5 \times \frac{1}{50} = 0.1$ s).

The distance from the start of the tape represents the distance travelled. The length of each section of the tape, divided by 0.1, represents the average speed of the object for that section (assume the average speed for a 0.1 s interval is attained at the midpoint of the interval).

Example E

A ticker tape is pulled through a ticker timer by a trolley:

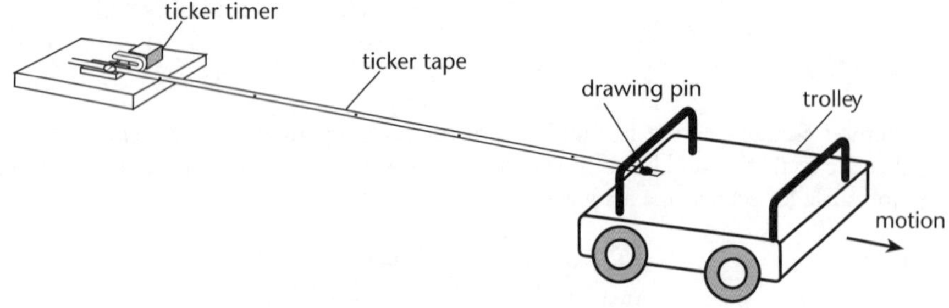

Ticker tape being pulled through a ticker timer by a trolley.

The following tape was obtained:

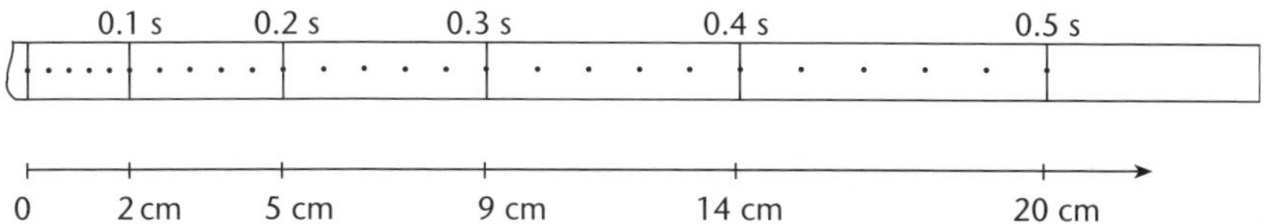

Resultant ticker tape after being pulled through a ticker timer by a trolley.

Total distance travelled (cm)	Time (s)
0	0
2	0.1
5	0.2
9	0.3
14	0.4
20	0.5

Speed (cm s–1)	Time (s)
0	0
20	0.05
30	0.15
40	0.25
50	0.35
60	0.45

Ticker timer tape analysis.

Unit 11.2 Activity 4F: The ticker timer

1. Use the information in the tables above to draw distance-time and speed versus time graphs.
2. The ticker timer tape shown below was produced on a 50 Hz AC power supply (ie 50 dots were made each second).
   Note: The tape is shown in full scale.

   a. What period of time has passed between dot A and dot B?
   b. What distance has the object travelled?
   c. What is the average speed of the object?
3. On the tape shown, the start and the dot corresponding to 0.2 s from the start are shown:

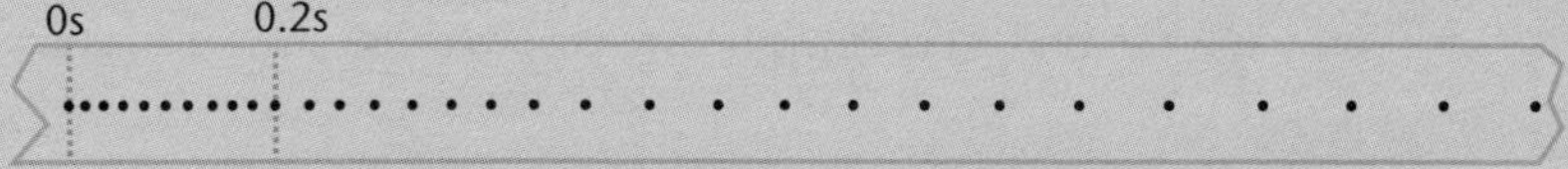

   a. Locate the dots at 0.4 s from the start and 0.6 s from the start, then draw up a table of total distance travelled (cm) against time (s).
   NB: The tape is shown in full scale.
   b. Plot a distance-time graph for the journey.

Free fall and the acceleration of gravity

Introduction

A free-falling object is an object which is falling under the sole influence of gravity. Any object which is moving and being acted upon only by the force of gravity is said to be 'in a state of **free fall**.' This definition of free fall leads to two important characteristics about a free-falling object:

- Free-falling objects do not encounter **air resistance**.
- All free-falling objects (on Earth) accelerate downwards at a rate of approximately 10 m/s/s (to be exact, 9.8 m/s/s).

Acceleration due to gravity

A free-falling object is an object which is falling under the sole influence of gravity; such an object has an acceleration of 9.8 m/s/s downward (on Earth). This numerical value for the acceleration of a free-falling object is known as the **acceleration of gravity (g)**.

The rate at which an object changes its velocity is the ratio of velocity change to time between any two points in an object's path. To accelerate at 10 m/s/s means to change the velocity by 10 m/s each second.

$$a = \frac{\Delta v}{t} = \frac{\sim 10 \text{ m/s}}{1\text{s}}$$

If the velocity and time for a free-falling object being dropped from a position of rest were tabulated, then one would note the following pattern.

Time (s)	Velocity (m/s)*
0	0
1	10
2	20
3	30
4	40
5	50

*(*velocity values are based on the approximated value of 10 m/s/s for g)*

Observe that the velocity–time data above reveal that the object's velocity is changing by 10 m/s each consecutive second. That is, the free-falling object has an acceleration of 10 m/s/s.

Representing free fall by graphs

A position versus time graph for a free-falling object is shown below.

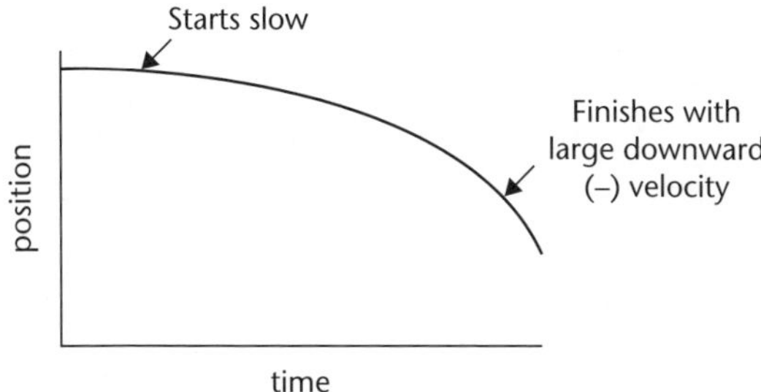

The line on the graph curves. A curved line on a position v. time graph signifies an accelerated motion. Since a free-falling object is undergoing an acceleration ($g = 10$ m/s/s –approximate value), it would be expected that its position–time graph would be curved. A further look at the position–time graph reveals that the object starts with a small velocity (slow) and finishes with a large velocity (fast). Since the slope of any position versus time graph is the velocity of the object, the small initial slope indicates a small initial velocity and the large final slope indicates a large final velocity. Finally, the negative slope of the line indicates a negative (ie downward) velocity.

The velocity versus time graph for a free-falling object is shown below.

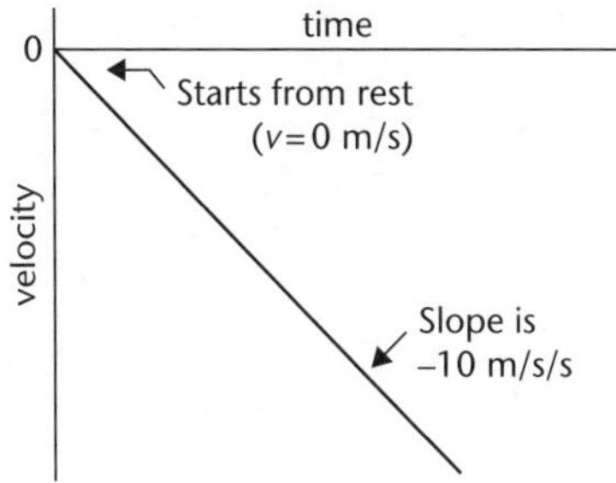

Observe that the line on the graph is a straight, diagonal line. As explained earlier, a diagonal line on a velocity versus time graph signifies an accelerated motion. Since a free-falling object is undergoing an acceleration ($g = 10$ m/s/s), it would be expected that its velocity–time graph would be diagonal. A further look at the velocity–time graph reveals that the object starts with a zero velocity (as read from the graph) and finishes with a large, negative velocity; that is, the object is moving in the negative direction and speeding up. An object which is moving in the negative direction and speeding up is said to have a negative acceleration.

The equation for determining the velocity of a falling object after a time of t seconds is:

$v_f = g \times t$

where g is the acceleration due to gravity.

Unit 11.2 Motion (Kinematics)

Topic 5: Equations of motion

The content to be covered under 'equations of motion' is set out in bullet points on p. 11 of the Syllabus. This Topic aims to cover those equations of:

- Uniform linear motion.
- Uniformly accelerated motions.
- Speed, velocity, average velocity, acceleration, average speed.
- Vertical motion under gravity (free fall).
- Associated word problems.

Introduction

Physics material covered in this topic is for assessed achievement. Understanding of motion in one dimension or straight line will be demonstrated by:

- Examining the relationship among the different studied parameters associated with the motion of an object: distance, displacement, speed, velocity, and acceleration of objects in motion.
- Solving problems related to speed, velocity and acceleration.
- Plotting graphs of the speed, displacement, velocity of the objects in motion.
- Calculating the distance travelled by an object, from the speed time graphs.

For uniformly accelerated motion, we can derive some simple equations that relate displacement (x), time taken (t), initial velocity (v_o), final velocity (v) and acceleration (a). This derivation is done in the following sections and is based on the basic equation for a final velocity $v_f = v_i + at$.

Kinematic equations of motion

Problems involving *uniform acceleration* in a straight line over a time interval can often be quickly solved using a set of formulae called the **kinematic equations of motion**.

If t is the time in seconds, d is the displacement in metres, v_i is the initial velocity in m s^{-1}, v_f is the final velocity in m s^{-1}, and a is the constant acceleration in m s^{-2}, then:

$$v_f = v_i + at \qquad d = \left(\frac{v_i + v_f}{2}\right)t \qquad d = v_i t + \frac{1}{2}at^2 \qquad v_f^2 = v_i^2 + 2ad \qquad d = v_f t - \frac{1}{2}at^2$$

Each equation contains four of the five variables (t, d, v_i, v_f, a). A different variable is missing in each equation.

Note: In the second equation, the expression $\left(\frac{v_i + v_f}{2}\right)t$ is the average velocity, over t seconds. The average velocity is usually given the symbol $\bar{v}$.

So $\bar{v} = \left(\frac{v_i + v_f}{2}\right)$, and the second equation can be written $d = \bar{v}t$

Most of the five kinematic equations are derived from the general velocity/time graph of an object's motion.

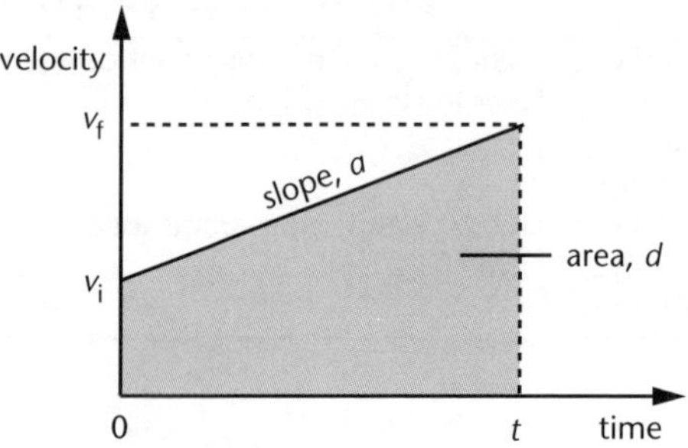

Deriving equations of motion

The graph describes the motion of an object which accelerates with constant acceleration a, from initial velocity v_i, to final velocity v_f. over time t.

Motion graphs are normally drawn for motion along a straight line and sometimes 'speed' is used instead of 'velocity'.

Two important relationships from graphs of motion are:

- The *slope* (gradient) of a velocity/time graph gives the *acceleration*.
- The *area* under a velocity/time graph gives the *displacement*.

Using the first relationship, the slope of the graph above is:

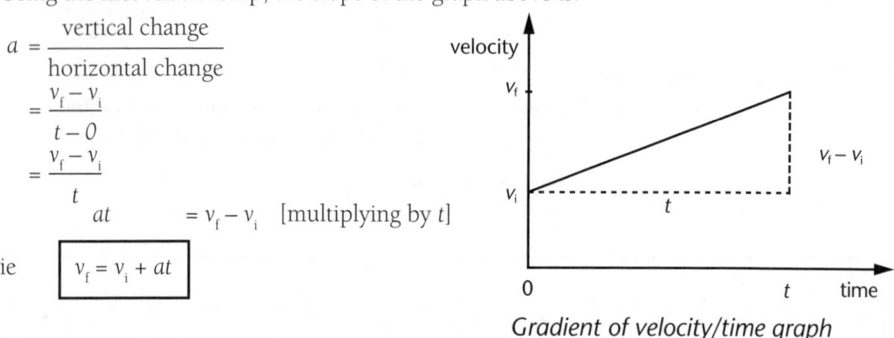

$$a = \frac{\text{vertical change}}{\text{horizontal change}}$$

$$= \frac{v_f - v_i}{t - 0}$$

$$= \frac{v_f - v_i}{t}$$

$$at = v_f - v_i \quad \text{[multiplying by } t\text{]}$$

ie $\boxed{v_f = v_i + at}$

Gradient of velocity/time graph

which is the first kinematic equation.

Using the second relationship, the area under the graph is a trapezium with vertical sides v_i and v_f and base t. The trapezium has area:

$$\boxed{d = \left(\frac{v_i + v_f}{2}\right)t} \quad \text{[area of a trapezium} = \left(\frac{a+b}{2}\right)h\text{]}$$

which is the second kinematic equation.

The third kinematic equation is derived by dividing the area of the graph into a rectangle and a triangle.

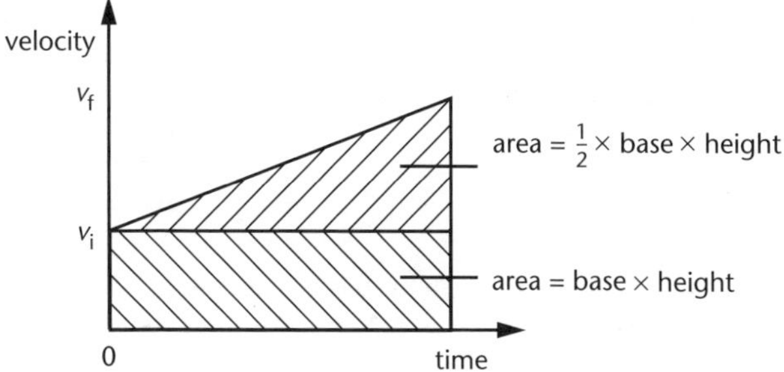

Distance travelled

area = area of rectangle + area of triangle

ie $d = t \times v_i + \frac{1}{2} \times t \times (v_f - v_i)$

but $v_f - v_i = at$ [from the derivation of the first kinematic equation]

so $d = v_i t + \frac{1}{2} \times t \times at$ [rearranging and substituting]

ie $\boxed{d = v_i t + \frac{1}{2} at^2}$ which is the third kinematic equation.

Problem solving using kinematic equations

The kinematic equations are useful for solving problems in which the values of three variables are known and the value of a fourth variable is required. This is done by choosing a kinematic equation which contains the three known variables and the required variable. *The values of the known variables are substituted* and the equation solved to find the required variable.

Example A

A radio-controlled toy car starts from rest and accelerates with a uniform acceleration of 0.50 m s^{-2} in a straight line. What speed does it reach after accelerating for 4.0 seconds?

Solution:

From the problem, $v_i = 0$, $a = 0.50$ m s^{-2}, $t = 4.0$ s and v_f is required.
The equation $v_f = v_i + at$ is used since it contains the three known variables (v_i, a and t) and the required variable (v_f).

$v_f = v_i + at$
$= 0 + (0.50 \times 4.0)$ [substituting $v_i = 0$, $a = 0.50$ m s^{-2}, $t = 4.0$ s]
$= 2.0$ m s^{-1}

If *four* variables are known (and one variable is to be found), two or more kinematic equations could be used. Choose the one involving the simplest rearrangement and arithmetic.

Example B

A ball, initially travelling at 4.0 ms^{-1}, rolls up a slope and slows uniformly to a stop, 16.0 m up the slope.

a. What is the ball's acceleration?

b. How long does the ball take to stop?

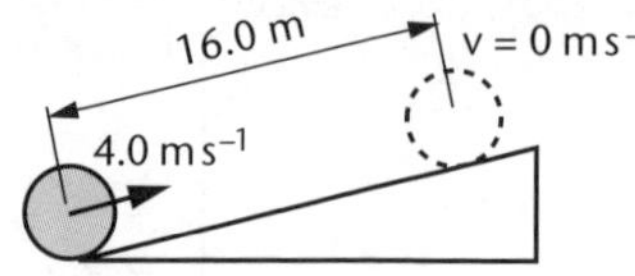

Solution:

a. $v_i = 4.0$ m s^{-1}, $d = 16.0$ m, $v_f = 0$ (when the ball has stopped).
The equation (with t missing) is $v_f^2 = v_i^2 + 2ad$ and substitution in this will give a.

$v_f^2 = v_i^2 + 2ad$

$0^2 = 4.0^2 + 2 \times a \times 16.0$ [substituting v_i, d, v_f]

$-(4.0)^2 = 32a$ [simplifying]

$a = \frac{-16}{32}$ [rearranging to make a the subject]

$= -0.50$ m s^{-2}

The negative acceleration confirms that the object is slowing down.

b. $v_i = 4.0$ m s^{-1}, $v_f = 0$, d = 16.0 m. The equation involving these variables and t is:

$d = \left(\frac{v_i + v_f}{2}\right)t$

$16.0 = \left(\frac{4.0 + 0}{2}\right)t$ [substituting]

$16.0 = 2.0t$ [simplifying]

$t = \frac{16.0}{2.0}$ [rearranging for t]

$= 8.0$ s

Other formulae could be used to solve this since a has been calculated in part (a).

Unit 11.2 Activity 5A: Graphs of motion

1. The graph shows how the velocity of a 50 kg mass varies with time as it travels along a frictionless air track. It is fired from a launching device and at the far end of the track a spring brings the mass to rest and returns it towards the starting point.

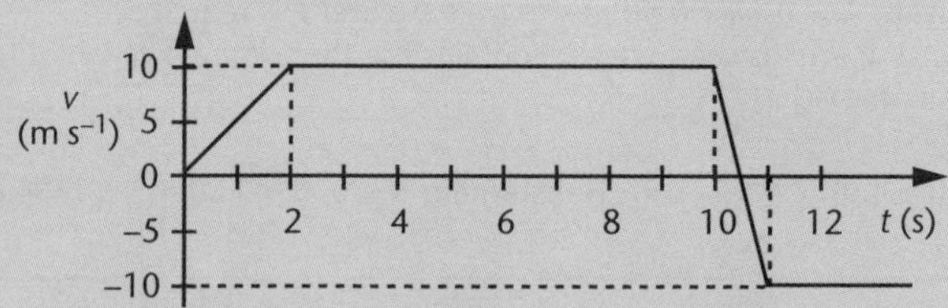

a. What is the acceleration during the first two seconds?

b. How far does the mass travel from the start of the launch until it *first* strikes the spring?

c. What is the *average speed* during the first 10 seconds?

2. A velocity-time graph is shown below for a car and a motorcycle, travelling on a race track.

At $t = 0$ s, the car and the motorcycle are together at the same point.

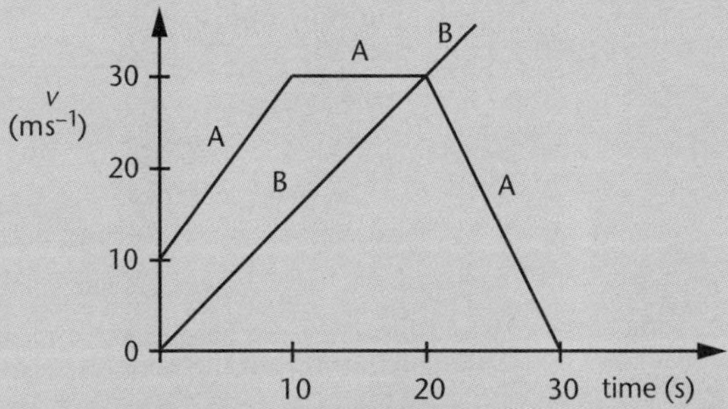

Graph A shows the velocity/time graph of the car.

a. What is the acceleration of the car at time $t = 5$ s?

b. What is the acceleration of the car at time $t = 25$ s?

c. How far does the car travel in 30 seconds?

The straight line (B) represents the velocity/time graph of the motorcyclist travelling in the same direction.

d. At what speed will the motorcyclist be travelling the moment the car comes to a stop?

e. Will the motorcyclist be ahead of the car, behind the car, or at the same place as the car at the moment the car comes to a stop?

3. A velocity/time graph is shown below.

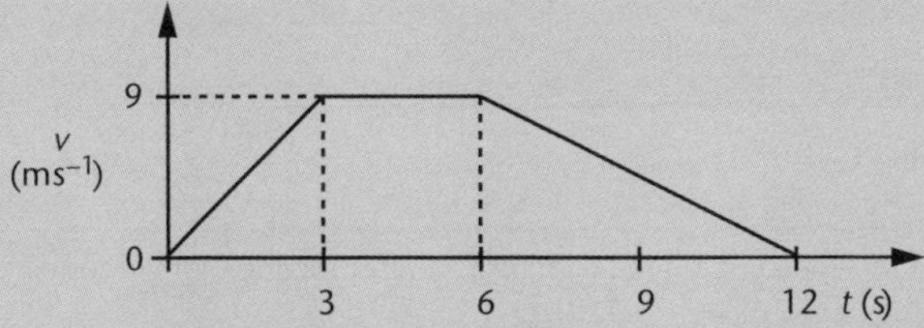

a. Calculate the average velocity over the whole 12 s.

b. Draw the corresponding acceleration/time graph for the motion.

c. If the object has zero displacement at $t = 0$, draw the corresponding displacement/time graph for the motion.

Unit 11.2 Activity 5B: Kinematic equations

1. A skateboarder accelerates uniformly from a standing start at 1.5 m s^{-2} down a steep hill. What is his speed 4.0 seconds after starting?
2. A wind gust accelerates a windsurfer uniformly from a speed of 4.0 m s^{-1} to a speed of 10 m s^{-1} over a period of 50 seconds.
   a. How far does the windsurfer move in that time?
   b. i. What is the windsurfer's instantaneous speed 30 seconds after the wind gust started?
      ii. What is the windsurfer's average speed for the 50 second period?
3. A ball initially moving at 0.50 m s^{-1} rolls down a straight slope, accelerating at 0.25 m s^{-2} for 4.0 s. How far does the ball travel in this time?
4. After stopping at a station, a train accelerates at a steady rate of 0.25 m s^{-2}.
   a. How far does the train travel (at this acceleration) in two minutes?
   b. At what speed is the train moving after it has travelled 1.0 km?
5. What time does it take an aircraft to decelerate uniformly from 360 km h^{-1} (100 m s^{-1}) to a stop if the distance covered along the runway is 1.5 km?
6. A dragster travelling at speed accelerates uniformly at 7.0 m s^{-2} for 8.0 seconds, reaching a new speed of 85 m s^{-1}. What distance did it cover in this 8.0 second period of acceleration?
7. a. Derive the kinematic equation $v_f^2 = v_i^2 + 2ad$ by substituting the first kinematic equation $v_f = v_i + at$ into the second kinematic equation $d = \left(\frac{v_i + v_f}{2}\right)t$ to eliminate time, t. Rearrange to give $v_f^2 = v_i^2 + 2ad$.
   b. Work out a way of deriving the last kinematic equation, $d = v_f t - \frac{1}{2}at^2$. (Hint: The area under a velocity/time graph can be considered to be a larger rectangle minus a triangle.)

Example C

Let us calculate the speed of a free-falling object after six and eight seconds.

Solution:

Using the formula to determine the speed of a falling object as $v_f = g \times t$ where g is the acceleration of gravity, we obtain:

a. for $t = 6$ seconds; $v_f = (10\ m/s^2) \times 6\ s = 60\ m/s$
b. for $t = 8$ seconds; $v_f = (10\ m/s^2) \times 8\ s = 80\ m/s$

Example D

The distance which a free-falling object has fallen from an initial position of rest is also dependent upon the time of fall and can be calculated as the following three cases illustrate using the formula $d = \frac{1}{2}g \times t^2$

a. At $t = 1$ s; $d = (0.5) \times (10\ \text{m/s}^2) \times (1\ \text{s})^2 = 5$ m

b. At $t = 2$ s; $d = (0.5) \times (10\ \text{m/s}^2) \times (2\ \text{s})^2 = 20$ m

c. At $t = 5$ s; $d = (0.5) \times (10\ \text{m/s}^2) \times (5\ \text{s})^2 = 125$ m

Example E

Suppose a falling stone takes 0.2 seconds to fall past a window which is 1 m high from the ground.

From how far above the top of the window was the stone dropped?

Solution:

Using $d = \frac{1}{2}g \times t^2$

$d = \frac{1}{2} \times 10 \times 0.2^2 = (0.5) \times (10) \times (0.04) = 0.20$ m

Thus, the stone was dropped from 0.20 m above the window (1.20 m above the ground level).

Example F

Suppose a stone is thrown vertically upward from the top of a building, and it hits the ground 10 seconds later with a speed of 51 m/s. Find the height of the building.

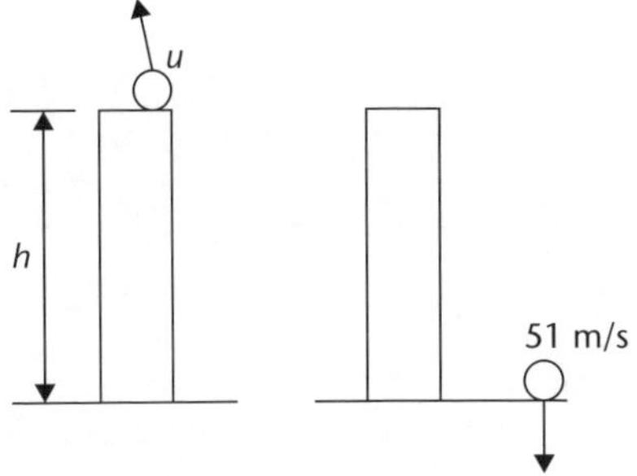

Solution: Taking the downward direction as positive, we have: given: $v = 51$ m/s; $a = g = 9.8\ \text{m/s}^2$, $t = 10$ s; then $h = vt - \frac{1}{2}at^2 = 510 - (50 \times 9.8) = 20$ m

Example G

Suppose you throw a small ball vertically upwards. The ball reaches a maximum height of 10 m, then starts coming down, and finally hits the ground after 8 seconds in the air. Draw a velocity–time graph for this event.

Solution:

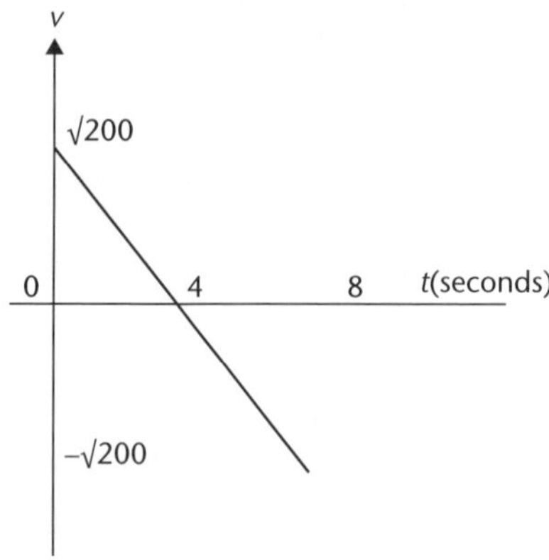

Unit 11.2 Activity 5C: Velocity

1. A particle moving in a straight line with a constant acceleration of 3 m/s^2 has an initial velocity of –1 m/s. Its velocity 2 seconds later is:
   A. 5 m/s
   B. 6 m/s
   C. 4 m/s
   D. 0
   E. –7 m/s
2. An object moves in a straight line and passes through O, a fixed point on the line, with a velocity of 6 m/s. The object moves with a constant deceleration of 2 m/s^2 for 4 seconds and thereafter moves with constant velocity. How long after leaving O does the object return to O?
   A. 3 seconds
   B. 8 seconds
   C. never
   D. 4 seconds
   E. 6 seconds
3. Velocity is the rate of increase of distance. True or False?
4. An object moving under the **action** of its weight only has a constant acceleration g vertically downwards. True or False?
5. If an object moving in a straight line has a negative acceleration then this always means that the speed is decreasing. True or False?

Unit 11.2 Motion (Kinematics)

Topic 6: Projectile motion

The content to be covered under 'projectile motion' is set out in bullet points on p. 11 of the Syllabus. This Topic aims to cover those points:

- Graphs of motion in two dimensions from given experimental data.
- Motion of a projectile and its equations.
- Worked examples on projectile motion.

Introduction

Gravitation is a force in effect at all times in our daily life.

Example A

Babies are aware of the force of gravity at an early stage, even before birth. Later they learn that any toy let go immediately drops out of sight to the ground. The principle that things fall is used to great effect in a high chair at dinner time!

Isaac Newton (1642–1727) developed understanding of gravitation by considering why an object, such as an apple, falls to the ground. He proposed that all objects in the universe attract each other with a force he called *gravitation*. In particular, Newton proposed that gravitation is the force that keeps the planets in their orbits around the sun and keeps the moon orbiting the earth.

Gravitation differs from the other main forces of nature (electric, magnetic and **nuclear forces**) because it is *always attractive*. Each of the other forces can be either attractive or repulsive, depending on certain conditions. If repulsive gravitation could be found, 'antigravity' would be possible. The consequences would be enormous – it would revolutionise transport and eliminate the need for rockets when launching spacecraft.

Vertical motion on the Earth's surface

The **acceleration due to gravity**, g, is fairly constant over the earth's surface and for some kilometres above the surface. The standard value of g is 9.81 m s^{-2} vertically downwards (often rounded to 10 m s^{-2}).

If an object is thrown vertically upwards or dropped downwards, it will *accelerate downwards* at a constant rate of 10 m s^{-2}. The resulting motion can be described using a graph or by kinematic equations (since the acceleration is constant). In this book, air resistance is ignored in these situations to simplify calculations.

Example B

A rock (initially stationary) is dropped from a cliff 50 m above the sea.

a. How far does it fall in 1.0 s?

b. How far does it fall in 2.0 s?

c. How long does it take to fall 50 m?

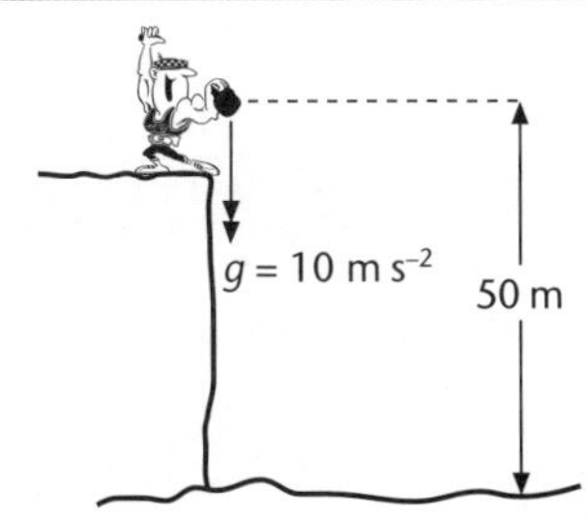

Solution:

For **a.** and **b.**, a velocity/time graph can be used. An acceleration of 10 m s^{-2} downwards means that the stone's velocity increases by 10 m s^{-1} each second.

Velocity/time graph of falling rock

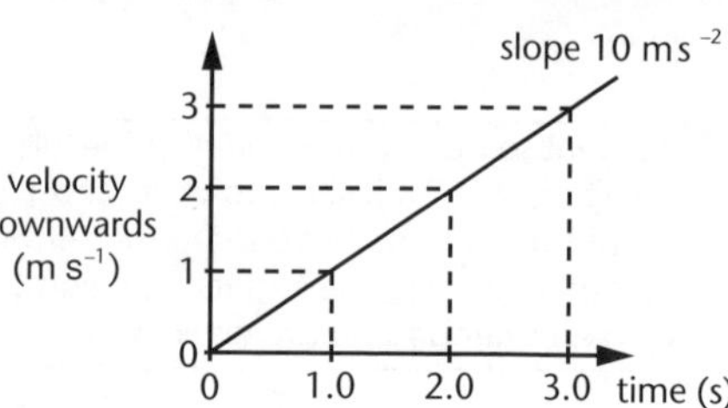

The distance fallen, d, is the 'area under the graph', so:

a. For 1.0 s: $d = \frac{1}{2} \times 1.0 \times 10$ [area = $\frac{1}{2}$ × base × height]

$= 5.0$ m

b. for 2.0 s: $d = \frac{1}{2} \times 2.0 \times 20$

$= 20$ m

c. A kinematic equation can be used to find the time, t. Since the acceleration of the rock is $a = 10$ m s^{-2} and $v_i = 0$ m s^{-1} (the rock is initially stationary), these values can be substituted into a suitable kinematic equation to find t.

Choosing $d = v_i t + \frac{1}{2}at^2$ gives:

$50 = 0 \times t + \frac{1}{2} \times 10 \times t^2$ [substituting]

$t^2 = \frac{50}{5.0}$

$t = \sqrt{10}$

$t = 3.2$ s

A kinematic equation rather than a graph is used in part (c) because it would be difficult to find the answer from the graph.

Example C

A ball is thrown upwards with an initial speed of 20 m s^{-1}.

a. How far does the ball rise in 1.0 s?

b. How long does it take to reach the highest point in its motion?

c. How high above the ground is the highest point in its motion?

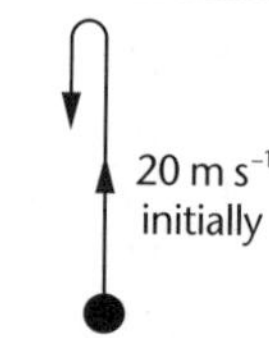

Solution:

The velocity/time graph for the motion is shown alongside.

Velocity/time graph of a ball thrown upwards

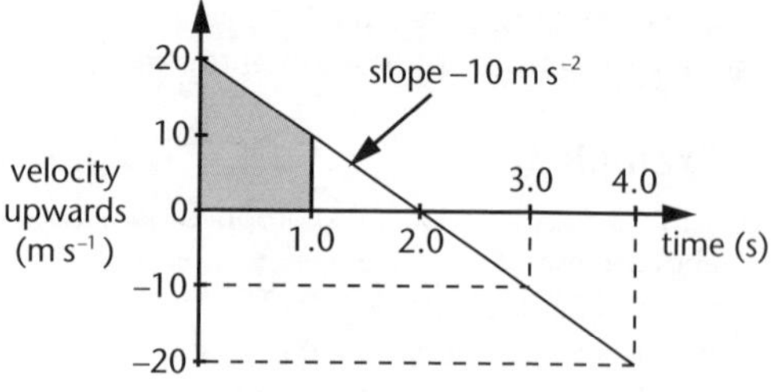

a. Height risen in 1.0 s = area under the graph

area $= \frac{1}{2} \times (20 + 10) \times 1.0$ [the shaded trapezium]

$= 15$ m

b. At the highest point, the ball is momentarily stationary (and is about to move downwards). At this moment, $v = 0$ and this occurs at $t = 2.0$ s.

c. Highest position occurs at $t = 2.0$ s. This height is the area under the graph up to $t = 2.0$ s.

$$\text{area} = \frac{1}{2} \times 20 \times 2.0 \qquad \text{[area of triangle]}$$

$$\therefore \text{height} = 20 \text{ m}$$

In Example C, the upwards direction has been taken as the positive direction. The ball is thrown *upwards* but its acceleration is constantly *downwards*. Thus the slope of the graph (acceleration) is negative. In Example B the *downwards* direction was taken as the positive direction and all velocities and distances were downwards, therefore positive.

Unit 11.2 Activity 6A: Vertical motion under gravity

1. A ball falls vertically from rest from the top edge of a building and strikes the footpath 40 m below. Calculate the velocity of the ball as it first strikes the footpath.

2. A golf ball is thrown vertically upwards with an initial velocity of 25 m s^{-1} and returns to the thrower in exactly 5.0 seconds.

- **a**. Draw a velocity/time graph representing the velocity of the ball over these 5.0 seconds.
- **b**. At what instant during its motion is the ball's velocity zero?
- **c**. Explain why the ball's acceleration is not zero when its velocity is zero.
- **d**. Draw a labelled displacement/time graph representing the displacement of the ball over the 5.0 seconds.

3. A bird drops a snail vertically down from a power line 12 m above the ground. How long does the snail take to fall, if air resistance is ignored?

4. The depth of a dark well is measured by dropping a pebble from rest down the well. If the pebble took 4.0 seconds to hit the bottom of the well, calculate the depth of the well.

Projectile motion

A **projectile** is any object that moves through the air without its own source of power, only under the influence of gravity (eg bullets, shot puts, netballs, water jets and softballs). A rocket or a glider are not projectiles – the rocket propels itself along as it moves, and the motion of air over a glider's wings causes an extra lifting force that can aid the glider's motion.

Another important force that acts on all projectiles is air resistance. This force is not constant because its size changes with the speed of the projectile. Air resistance is ignored in the following Examples because the mathematics describing the motion becomes difficult. It is assumed that once in flight, the only force acting on a projectile is its weight force.

- This means that as the projectile moves, it accelerates *constantly downwards* at 10 m s^{-2}.
- There is no accelerating force acting in the horizontal direction, and so the projectile *moves horizontally at a constant speed.*

The position and acceleration of a projectile at equal time intervals shows that the path travelled by a projectile is a **parabola**.

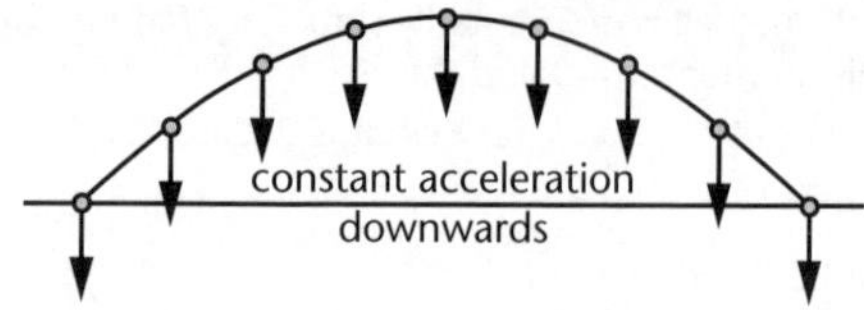

A projectile under the influence of gravity

The **parabolic** path is *symmetric*.

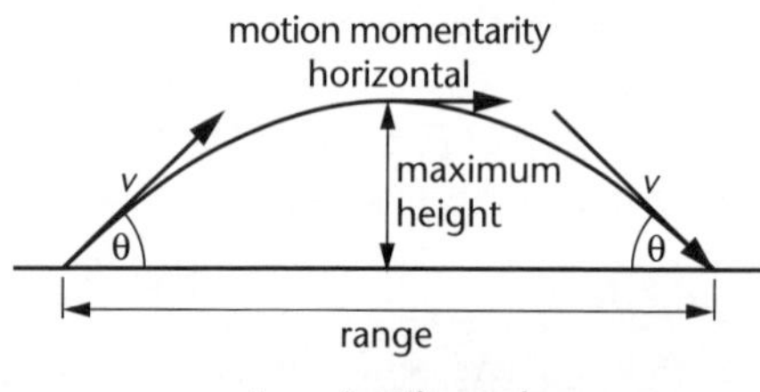

A projectile path

For a projectile moving over horizontal ground:

- The maximum height of the flight is reached when the *vertical component* of the projectile's velocity is zero (momentarily).
- The total time of flight is *twice the time taken to reach the maximum height.*
- The distance travelled horizontally (called the *range*) is the constant horizontal velocity multiplied by the total time of flight.
- The projectile hits the ground at the same speed as it was fired and makes the same angle to the horizontal.

Example D

A soccer player kicks a ball. The initial components of the velocity of the ball are 10 m s^{-1} vertically and 20 m s^{-1} horizontally. Find the range of the ball.

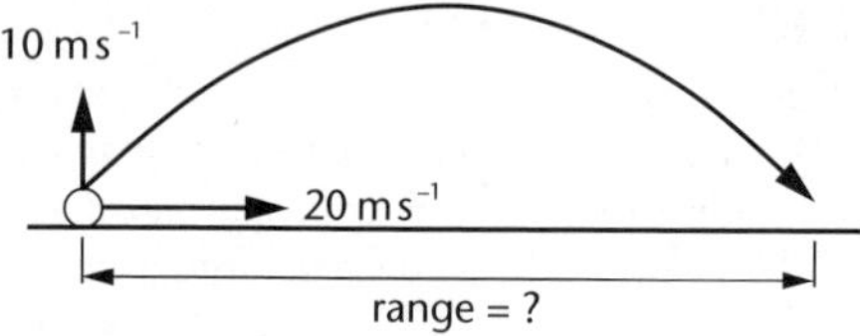

Solution:

At maximum height, the initial vertical speed of 10 m s^{-1} will have reduced to zero. Using

$v_f = v_i + at$, where t is the time taken to reach the maximum height, gives:

$0 = 10 + (-10 \times t)$ [substituting $v_f = 0$, $v_i = 10$ m s^{-1}, a = -10 m s^{-2}]

ie $10t = 10$

$t = 1.0$ s

Time of flight = twice the time to reach maximum height

= 2×1.0

= 2.0 s

The ball travels horizontally at a constant speed of 20 m s^{-1} for 2.0 s. Thus it has a range of:

$d = vt$

$= 20 \times 2.0$

$\therefore$ range $= 40$ m

Often a problem gives the initial velocity of a projectile in the form of a speed, v, and an angle, θ, to the ground.

In such cases, the first step in solving the problem is to calculate the vertical and horizontal components of that initial velocity. This is done by **resolving** the velocity vector into vertical and horizontal components, as in the Figure to the right.

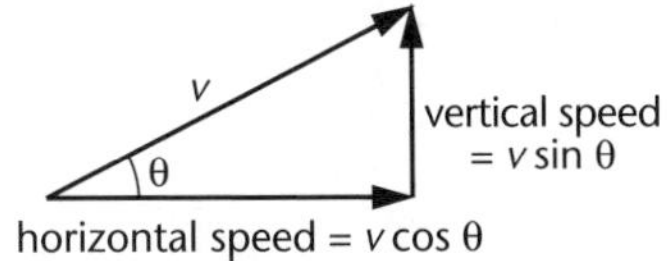

Using components for a projectile

Example E

How high does a softball rise if it is hit with an initial speed of 40 m s^{-1} at an angle of 40° to the ground?

Solution:

The initial vertical speed is:

$v \sin\theta = 40 \sin 40°$

$= 25.7$ m s^{-1}.

At the maximum height, the vertical speed will have reduced from 25.7 m s^{-1} to zero.

The relevant kinematic equation that does *not* involve time is:

$$v_f^2 = v_i^2 + 2ad$$

ie $0 = 25.7^2 + (2 \times -10 \times d)$ [substituting $v_f = 0$, $v_i = 25.7$ m s^{-1}, $a = -10$ m s^{-2} to find height, d]

$d = \dfrac{-(25.7)^2}{-20}$ [rearranging]

$= 33$ m high [2 sig figs]

Applications in sport

Several field events involve projectile motion, especially the shot put and the hammer throw.

For a throw at a particular speed, what is the best angle of delivery that will make a shot put or javelin travel the furthest? How could you show that your chosen angle is the best angle?

In the game of badminton, a shuttlecock is hit using a racquet. Air resistance is very significant in the shuttlecock's motion and its motion forms a 'squashed' parabola as shown in the figure at right.

The time for a shuttlecock to rise to maximum height does not equal the time to fall. Which of these times is the longer?

A badminton hit

Unit 11.2 Activity 6B: Projectile motion

Ignore air resistance in all questions.

1. A tennis ball is thrown into the air. A graph of height against horizontal distance, d, is shown.

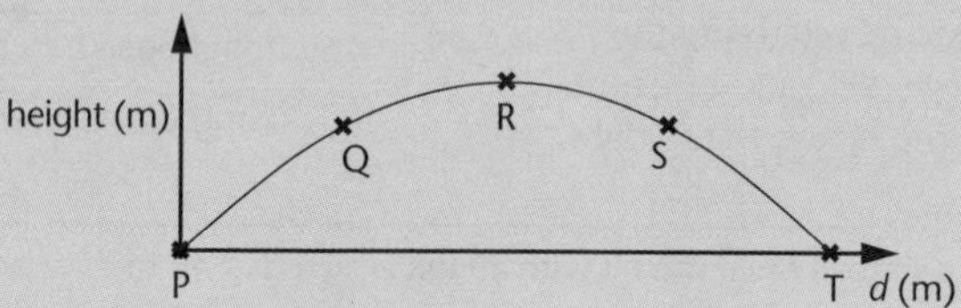

Points P, Q, R, S and T are at equal intervals in time.

a. Describe how the height of the tennis ball varies with time during its motion.

b. Describe how the speed of the tennis ball varies with time during its motion.

c. At which other point or points would the tennis ball have the same acceleration as at point Q?

d. Somehow, the gravity is doubled and the ball is thrown again in exactly the same way as before. Sketch a new graph of height against distance, showing how the ball's motion would be different from the original motion.

2. A ball is launched horizontally from P, the top of a cliff. It hits the sea sometime later, at point T.

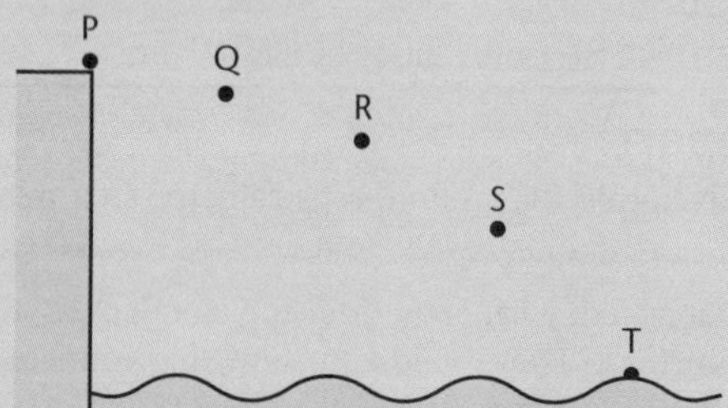

P, Q, R and S are positions of the ball at equal time intervals.

a. Draw a set of vectors that represent the acceleration at Q, R and S.

b. Draw a set of vectors that represents the horizontal velocity at Q, R and S.

c. Draw a set of vectors that represents the vertical velocity at Q, R and S.

3. The diagram shows an object that has been fired horizontally above the earth's surface. A, B, and C show its position at $t = 0$ s, $t = 1.0$ s and $t = 2.0$ s, respectively.

   Copy the diagram and indicate the object's position at $t = 3.0$ s, and by use of an arrow, show the *direction* of the resultant force on the object at this instant.

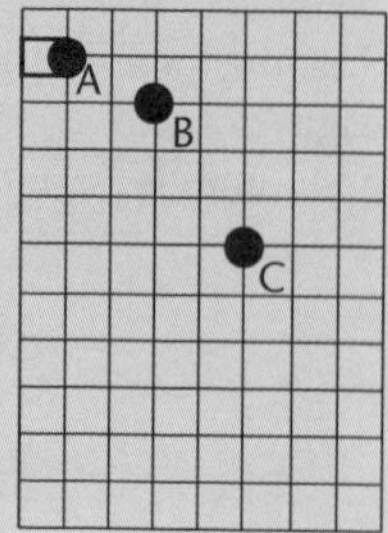

4. The diagram shows a toy engine which travels in a straight line from left to right across a level floor. A device inside the engine fires a small ball vertically upwards through the engine's funnel, while the engine is moving.

   a. Describe the motion of the engine in order that the ball lands back in the funnel.
   b. Draw a diagram that shows the path of the ball through the air as seen by a stationary observer standing to the side of the track, when the ball is on its way down from its highest point.
5. A hockey ball is 'flicked' off the ground with initial velocities of 2.0 m s^{-1} upwards and 10 m s^{-1} horizontally. Calculate the distance from the point where the flick takes place to the point where the ball hits the ground again.
6. A javelin is thrown with a speed of 20 m s^{-1} at an angle of 30° to the horizontal.
   a. Calculate the initial vertical and horizontal speeds of the javelin.
   b. Calculate the javelin's time of flight.
   c. Calculate the distance from the throwing point that the javelin hits the ground (the range).
7. A rugby ball, 50 m in front of a goal post, is 'place kicked' with initial velocities of 15 m s^{-1} upwards and 20 m s^{-1} horizontally.

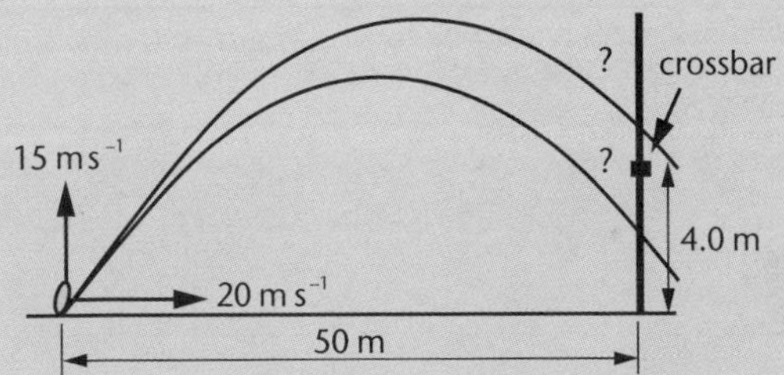

   If the ball is kicked accurately so that it passes between the posts, determine whether or not the ball goes over the crossbar (which is 4.0 m above the ground).
8. Some students may see a mathematical way to handle projectile questions using a vector equation such as:

$$\frac{v_x}{v_y} = \frac{20}{30-10t}$$

   a. Find the initial horizontal and vertical components of motion by setting $t = 0$.
   b. Find the time of flight.
   c. Find the velocity with which this projectile hits horizontal ground.
   d. Find the speed after 1.5 s.
9. A cricket ball is hit to the boundary. During its flight, and ignoring the effects of the air, what happens to:
   a. the horizontal and vertical components of the velocity?
   b. the horizontal and vertical components of its acceleration during its ascent and its descent, and at the topmost point of its flight?

10. At what point in the path of a projectile is the speed:

a. a maximum?

b. a minimum?

11. Suppose you toss a guava directly upward while riding in a moving car. Does the guava tend to land behind you, in front of you, or back in your hands, if the car is:

a. travelling at a constant speed

b. increasing in speed

c. decreasing in speed

12. The following figure shows three paths for a kicked football. Ignoring the effects of air on the flight, rank the paths according to:

a. time of flight

b. initial vertical velocity component

c. initial horizontal velocity component

d. initial speed

Place the greatest first in each case.

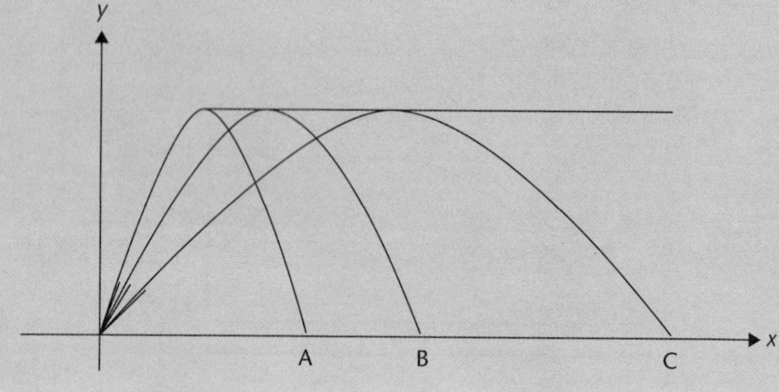

Unit 11.2 Motion (Kinematics)

Topic 7: Circular motion

The content to be covered under 'circular motion' is set out in bullet points on p. 11 of the Syllabus. This Topic aims to cover:

- Circular motion, its quantities and their units and provide worked examples on circular motion.

Note that 'equations of circular motion and the relationship between linear and angular quantities' is covered in Unit 11.3 Topic 4: 'Rotational motion and equilibrium'.

Introduction

In everyday life there are many examples of motion in a circle. The blades in a food processor move rapidly in a circular motion; the slow orbiting of the moon around the earth is almost circular.

In this chapter, only objects that move at a *constant speed* in a circle are considered.

Circular motion

A **rotation** (or **revolution**) means motion once around a circle. The time taken to go once around a circle is called the **period** of rotation (symbol T).

The SI unit for period is the second, s.

Example A

A ball tied to the end of a length of string is whirled around at a constant speed in a horizontal circle.

The distance that the ball moves in one rotation is the circle's circumference: $d = 2\pi r$, where r is the circle's radius and π is a constant ($\pi = 3.142$ approximately).

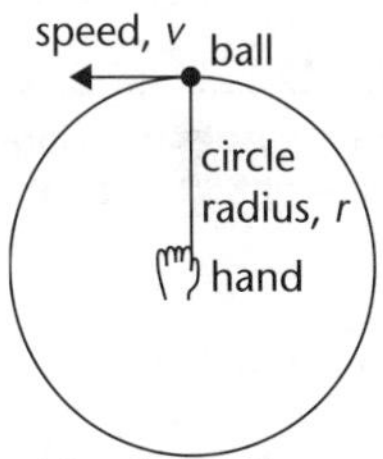

Circular motion

The speed, v, of an object travelling in a circular path can be calculated from the distance travelled and the time taken.

Example B

Clothes are flung to the sides of a round tub during the spin-cycle of a washing machine. If the tub's period of rotation is 0.20 s and the tub's radius is 0.30 m, how fast are the clothes moving?

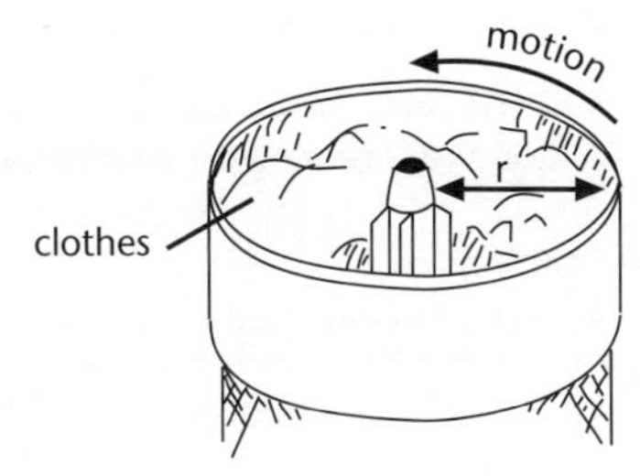

Solution:

The distance the clothes move in one rotation is:

$$\begin{aligned} d &= 2\pi r \\ &= 2\pi \times 0.30 \qquad [r = 0.30\text{ m}] \\ &= 1.88\text{ m} \end{aligned}$$

Since the time taken to go through one revolution is 0.20 s, the speed of the clothes is:

$$v = \frac{d}{t}$$
$$= \frac{1.88}{0.20}$$
$$= 9.4 \text{ m s}^{-1} \qquad [2 \text{ sf}]$$

In general, the speed of a rotating object is found by dividing the distance travelled in one rotation by the time taken for one rotation. If r is the radius (in metres) and T is the period (in s), then the speed v (in m s^{-1}) is:

$$v = \frac{2\pi r}{T}$$

The **frequency** of rotation, f, is the number of rotations made per second.
The SI unit for frequency is **hertz** (symbol Hz).

Example C

In Example B, a rotation period of 0.20 s means that 5 rotations are made per second, ie $f = 5.0$ Hz.

Period, T, and frequency f, are reciprocals of one another:

$$T = \frac{1}{f} \quad \text{or} \quad f = \frac{1}{T}$$

The relationship $f = \frac{1}{T}$ shows that 1 Hz = 1 s^{-1}; ie hertz means *per second*.

Example D

The propeller blades of a single-engined aircraft rotate with a frequency of 40 Hz. If the blades have a radius of 85 cm, how fast do the tips of the propeller blades move?

Solution:

The period of rotation for the blades is: $T = \frac{1}{40}$ $\qquad [T = \frac{1}{f}]$

$= 0.025$ s

The speed of the blades' tips is: $v = \frac{2\pi r}{T}$

$= \frac{2\pi \times 0.85}{0.025}$ $\qquad$ [85 cm = 0.85 m]

$= 210 \text{ m s}^{-1}$ $\qquad$ [2 sf]

The frequency of rotation is sometimes expressed in a non-SI unit called **revolutions per minute** (**rpm**). The frequency in rpm is 60 times greater than the frequency in hertz, because there are 60 seconds in a minute.

Example E

A racing car's rev counter indicates an engine speed of 4000 rpm. Calculate this in Hz.

Solution:

4000 revolutions in 60 seconds is a frequency of $\frac{4000}{60} = 67$ Hz $\qquad$ [2 sf]

Force and acceleration

When a ball on a string is whirling around in a circle, various forces are acting. There is **tension** in the string which pulls the ball *inward*. The direction of this force is at right angles to the ball's motion. This force *changes the ball's direction*, but because the force is always at right angles to the direction of the ball's motion, *the speed of the ball does not change*.

The force on the ball provided by the tension in the string is called the **centripetal force**. It causes the ball to accelerate, by changing the ball's direction, but not its speed.

The ball is continually accelerating in the direction of the centripetal force, ie inwards towards the centre of the circle. The acceleration is called the **centripetal acceleration**.

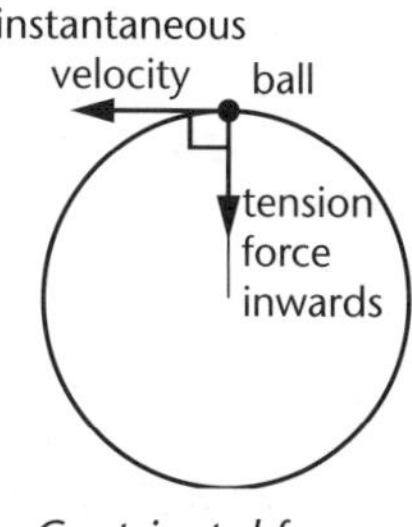

Centripetal force

If v is the speed of the ball (in m s^{-1}) and r is the radius of the circle (in m), then the centripetal acceleration a (in m s^{-2}) is:

$$a = \frac{v^2}{r}$$

The centripetal force is related to the centripetal acceleration using Newton's Second Law:

$$\begin{aligned} F &= ma \\ &= m\left(\frac{v^2}{r}\right) \qquad \text{[substituting } a = \frac{v^2}{r}\text{]} \\ &= \frac{mv^2}{r} \end{aligned}$$

$$F = \frac{mv^2}{r}$$

Example F

During a hammer throw, a 7.0 kg steel ball is swung horizontally with a speed of 10 m s^{-1} in a circle of radius 2.0 m.

The force required to keep the ball moving in a circle is:

$$\begin{aligned} F &= \frac{mv^2}{r} \\ &= \frac{7.0 \times 10^2}{2.0} \\ &= 350 \text{ N inwards} \end{aligned}$$

When people accelerate, *apparent forces* are experienced.

Example G

When a lift accelerates upwards, a person in the lift feels heavier, ie a greater force is felt acting downward, which is in the opposite direction to the person's acceleration. In the same way, a person moving in a circle experiences a force towards the *outside* of the circle.

The *apparent* force experienced towards the outside of a circle is called the **centrifugal force** and is due to the mass of the object resisting the inward centripetal acceleration that the object is experiencing. A stationary observer outside the circular motion observes the centripetal force only. No centrifugal force is observed.

Example H

A centrifuge uses centripetal forces to separate liquids with different densities, such as cream and milk, or blood serum and plasma.

A typical centrifuge uses a wheel that rotates in a horizontal plane. Buckets are attached to the wheel which are vertical when the wheel is stationary and horizontal when the wheel is moving around.

In the horizontal (spinning) position, the liquids separate, with the denser liquids moving further out from the centre of the circle. When the spinning stops, the denser materials are found in the bottom of the separating containers.

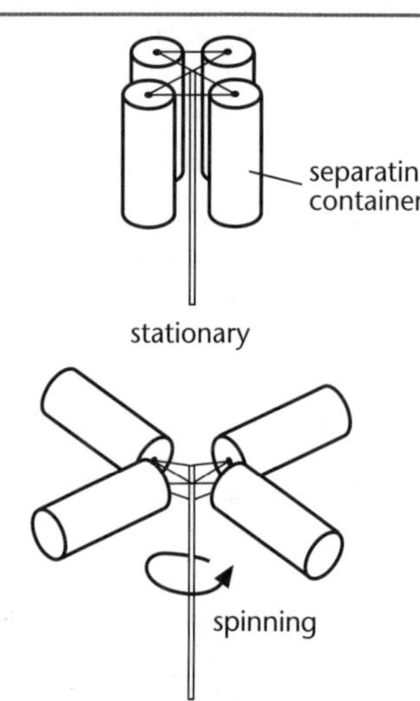

Often, it is easier to measure the period of rotation, T, rather than the speed of rotation, v. The centripetal acceleration, a, can be calculated for T and r using the formula developed below:

$$a = \frac{v^2}{r}$$

$$= \frac{\left(\frac{2\pi r}{T}\right)^2}{r} \qquad \text{[substituting } v = \frac{2\pi r}{T}\text{]}$$

$$= \frac{4\pi^2 r^2}{T^2} \times \frac{1}{r} \qquad \text{[expanding and rearranging]}$$

$$= \frac{4\pi^2 r}{T^2} \qquad \text{[cancelling]}$$

$$\boxed{a = \frac{4\pi^2 r}{T^2}}$$

The centripetal force is: $F = ma$

$$= \frac{4\pi^2 mr}{T^2}$$

$$\boxed{F = \frac{4\pi^2 mr}{T^2}}$$

Example I

Friction prevents a 1.5 kg mass from falling off a rotating turntable, which rotates at 30 rpm. If the mass is 10 cm from the centre of rotation, what is the friction force preventing the mass moving outwards, and falling off?

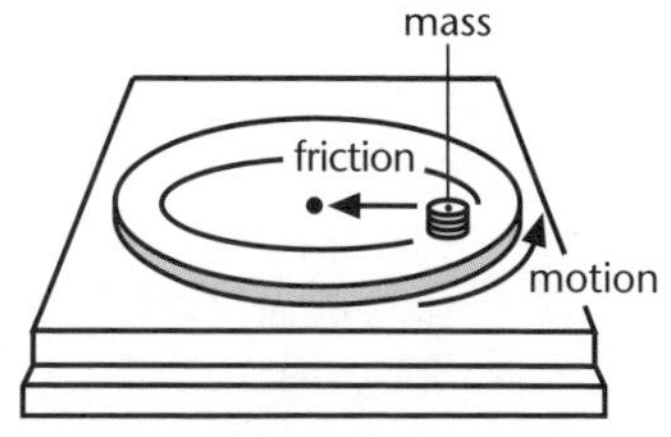

Solution:

Friction provides the centripetal force.

The period of revolution is: $T = 2.0$ s [30 revolutions in 60 s]

$$\text{centripetal force} = \frac{4\pi^2 mr}{T^2}$$

$$= \frac{4\pi^2 \times 1.5 \times 0.10}{2.0^2} \quad \text{[10 cm = 0.10 m]}$$

$$= 1.5 \text{ N inwards}$$

$$\text{Friction} = 1.5 \text{ N inwards}$$

Unit 11.2 Activity 7A: Circular motion

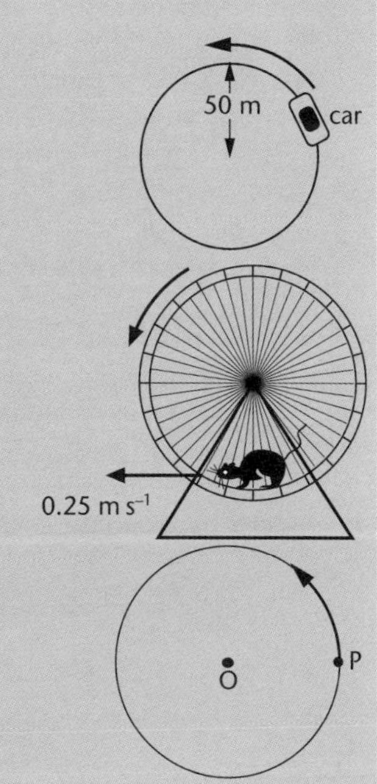

1. Calculate the speed of a car which completes a full circular lap of radius 50 m in a time of 20 s.

2. A mouse runs at a speed of 0.25 m s^{-1} in a circular 'mouse exerciser' of radius 0.050 m (5.0 cm).

a. How long does the exerciser take to revolve once?

b. How many revolutions per second is this?

c. How many rpm is this?

3. The diagram shows a particle moving anticlockwise in a circle at constant speed. Indicate the direction of its velocity, v, and its acceleration, a, when it is at point P.

4. The position of an object travelling in a circle at constant speed was recorded every 0.20 s (as shown in the diagram by the points W, X, Y and Z). The mass of the object was 2.0 kg and the radius of the path was 0.80 m.

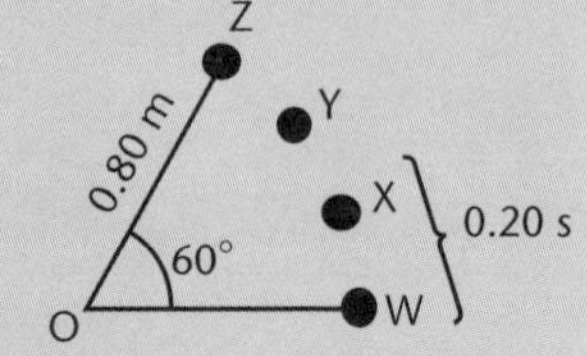

   a. Comment on the correctness of each of the following statements:
      i. The instantaneous speed of the object is the same at X as it is at Z.
      ii. The direction of the velocity at Z is at right angles to the line OZ.
      iii. The magnitude of the acceleration at X is the same as the magnitude of the acceleration at Z.
      iv. The accelerating force acting on the object at W is directed at right angles to the line OW.
      v. The direction of the acceleration at X and Z is directed towards the centre of the circle.
   b. Determine the frequency of revolution for the object.
   c. Calculate the instantaneous speed at point X.

5. The position of an object travelling in a circle at constant speed is shown every 0.20 s as it travels in the arc from A to B.

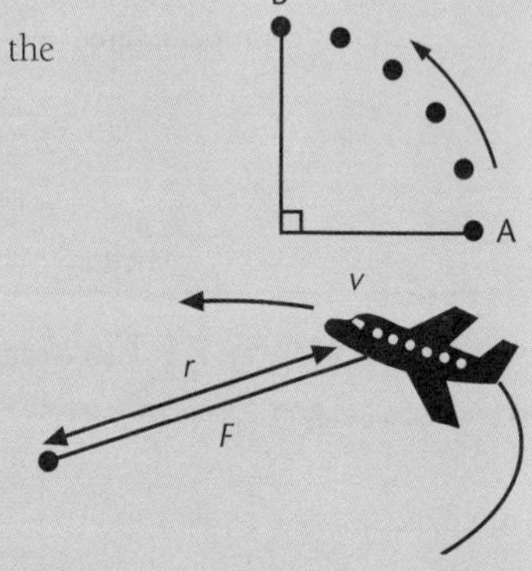

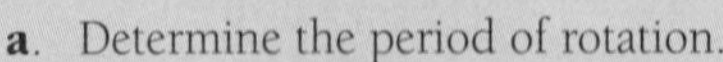

   a. Determine the period of rotation.
   b. If the radius of the circle is 50 cm, calculate the speed of the object around the circle.

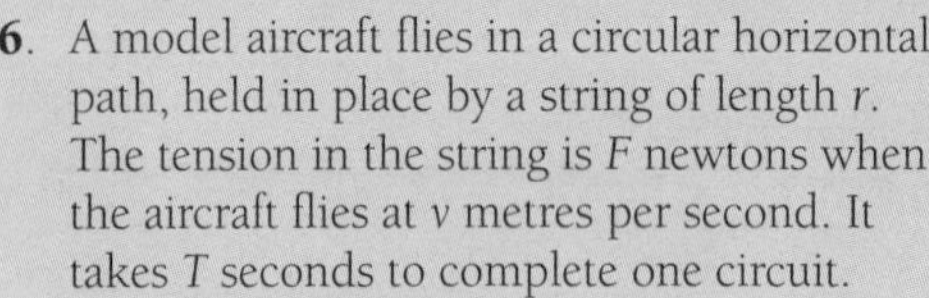

6. A model aircraft flies in a circular horizontal path, held in place by a string of length r. The tension in the string is F newtons when the aircraft flies at v metres per second. It takes T seconds to complete one circuit.

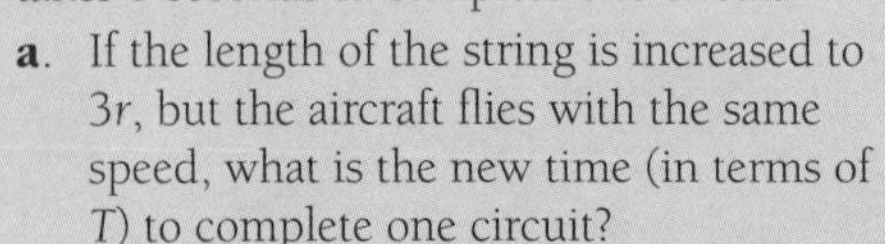

   a. If the length of the string is increased to $3r$, but the aircraft flies with the same speed, what is the new time (in terms of T) to complete one circuit?
   b. If the length of the string is maintained at $3r$, and the aircraft flies (again) with speed v, then what is the new tension in the string?
   c. A new engine is fitted to the aircraft, so that the total mass remains unchanged, but the airspeed increases to $3v$. At radius r, what is the tension in the string now?
   d. Draw a diagram (looking from above) of the aircraft travelling in its circular path. Include vectors showing the directions of the aircraft's acceleration and the tension force acting on the aircraft.

7. A student performing a centripetal force experiment whirls a rubber bung of mass 0.050 kg at a constant speed of 6.0 m s^{-1} in a horizontal circle of radius 2.0 m, as shown below.

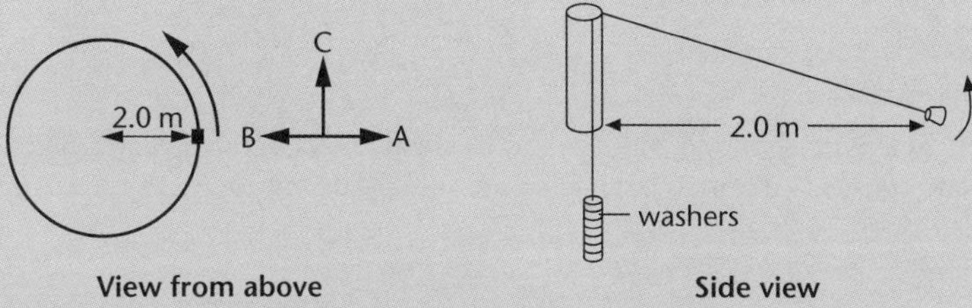

   a. Calculate the time in seconds for one revolution.
   b. Determine the magnitude of the bung's acceleration.
   c. State what would happen to the bung if the washers providing the centripetal force were to fall off at the point shown in the diagram.

8. A car travelling at a constant speed of 5.0 m s^{-1} turns a corner in a circle of radius 10 m. Calculate the acceleration of the car during the turn.

9. A 0.40 kg mass *P* on a frictionless horizontal table is attached to a weight of mass 0.60 kg by a string passing through a smooth hole in the centre of the table. The mass *P* is moving in a circle about the hole with a uniform speed of 3.0 m s^{-1}. Calculate the radius of the circular path.

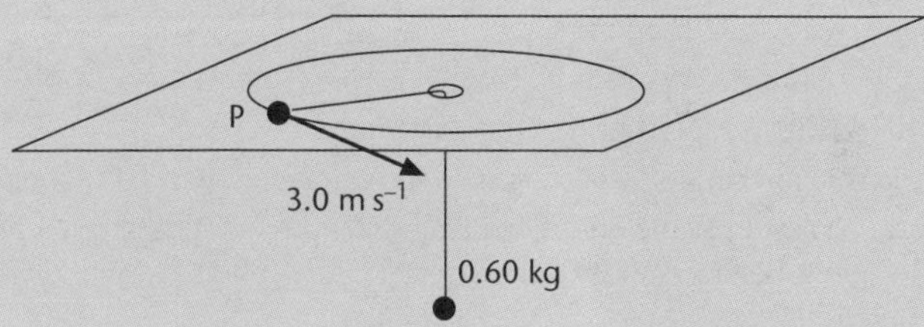

10. A very tall man whirls a 0.50 kg stone on the end of a 2.0 m length of fishing line which has a breaking strain of 25 N. He whirls the stone 2.45 metres above the ground, faster and faster. Eventually, the fishing line breaks (at the position shown) and the stone flies off horizontally as a projectile. Find the following:
   a. The speed with which the stone flies off.
   b. The time of flight of the stone.
   c. How far from his feet the stone lands.

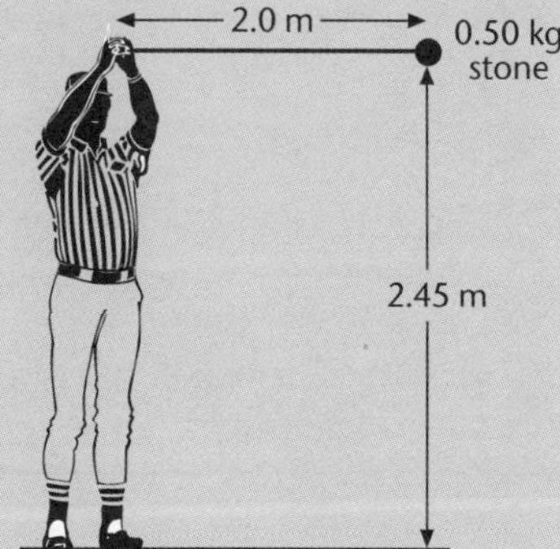

11. A small object of mass m is travelling at a constant speed v round a circle of radius r. Its acceleration is:

A. rv^2

B. mv^2/r

C. v^2/r

D. mrv^2

12. A string of length ℓ has one end fixed and a small object of mass m attached to the other end travels in a horizontal circle of radius r. The tension in the string is:

A. mg

B. $\dfrac{m\ell g}{\sqrt{\ell^2 - r^2}}$

C. mgr/ℓ

D. $mg\ell/r$

13. A bead is threaded on to a circular wire fixed in a vertical plane. The bead travels round the wire. The acceleration of the bead is:

A. Towards the centre and constant.

B. Towards the centre and varies.

C. Made up of two components one radial and one tangential.

D. Away from the centre and varies.

14. A vehicle can travel round a curve at a higher speed when the road is banked than when the road is level. This is because:

A. Banking increases the friction.

B. Banking increases the radius.

C. The normal reaction has a horizontal component.

D. When the track is banked the weight of the car acts down the incline.

15. A small bead is hanging at the end of a string of length a and is given a horizontal velocity V so that it begins to travel in a vertical circle. The bead will describe a complete circle if:

A. $V \geq \sqrt{4ga}$

B. $V < \sqrt{5ga}$

C. $V > \sqrt{2ga}$

D. $V \geq \sqrt{5ga}$

16. A small object travelling in a circle of radius r has an acceleration of constant magnitude v^2/r towards the centre of the circle. Therefore, the object has a constant velocity. True or False?

17. Every object in motion and describing a circle has a constant acceleration towards the centre. True or false?

Unit 11.3 Force and Motion (Dynamics)

Topic 1: Force

In line with the Syllabus, this Unit combines motion and force – but force is introduced first, on its own. This Topic covers:

- Definition of force and its units.
- Types of forces.
- Calculation of net force.
- Resolving force into its vertical and horizontal components
- Introduction to the work of Isaac Newton.

A **force** is a push or a pull. A force can:

- Cause an object which is at rest to move.
- Cause an object that is travelling at constant speed to accelerate or decelerate.
- Change the shape of an object.

There are two types of forces – **contact forces** and **field forces**.

Contact forces

A contact force needs to touch an object to produce an effect, eg hitting or kicking a ball, striking a match, **friction**.

Friction exists when there is movement between two surfaces that are touching. Friction occurs because there is no such thing as a perfectly smooth surface. All surfaces have microscopic 'roughnesses' that produce friction and **heat** when surfaces move over each other. A friction force is always in the opposite direction to the direction of motion.

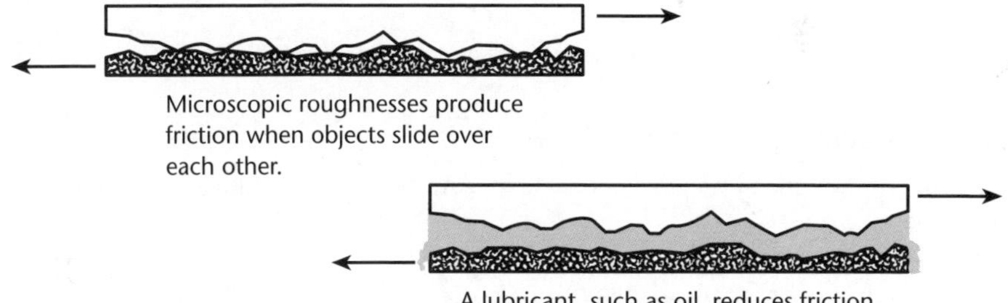

Two surfaces – one with, and one without lubrication.

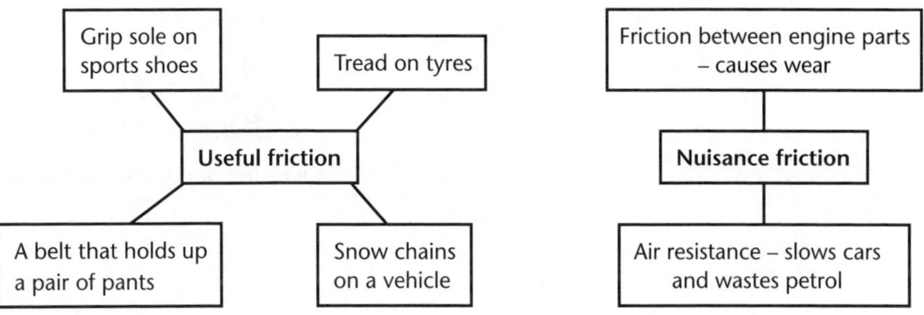

Using **lubricants** such as grease and oil can reduce friction.

Field forces

Field forces can act at a distance through a force field.

There are three types of field forces – electrostatic, magnetic and gravitational forces.

Electrostatic forces

Charged objects such as the dome of a **van der Graaf generator** are surrounded by an electric field.

Whenever a charged particle, such as an **electron**, enters an **electric field**, it will have an electrostatic force exerted on it. The direction of the force is given by the rule:

Like charges repel, unlike charges attract.

Magnetic forces

The poles of a **magnet** and current-carrying wires are surrounded by a **magnetic field**. Whenever a magnetic pole enters the field, it will have a magnetic force exerted on it. The direction of the force is given by the rule:

Like poles repel, unlike poles attract.

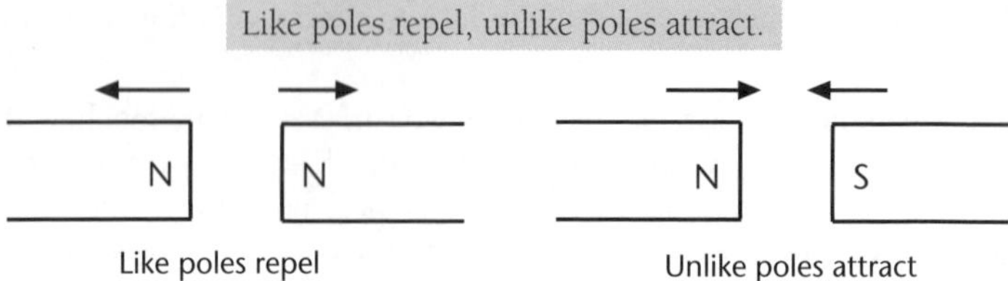

Magnetic poles interacting.

Gravitational forces

The Earth is surrounded by a gravitational field. Any mass in the field has a weight force exerted on it.

Sir Isaac Newton was an English scientist who was born in 1642 and died in 1727. He was the first scientist to explain the laws of gravity. Newton was inspired to come up with the theory of gravity after watching an apple falling from a tree. He wondered why the apple always fell down towards the Earth's surface, rather than sideways or upwards. He decided that the reason for this must be that there is a drawing power in the centre of the Earth.

Newton and the falling apple.

Tension

Tension is the force in connecting strings and ropes. Tension tends to pull in both directions along the string or rope.

Example A

A stationary 0.5 kg mass hangs on the end of a string from a ceiling. Tension is caused in the string by the weight force (5 N) of the hanging mass. Tension acts *downward on the ceiling* due to the weight force of the mass.

The tension also acts *upward on the hanging mass* to balance the weight force acting down.

The effect on the string is that tension acts to pull the string apart.

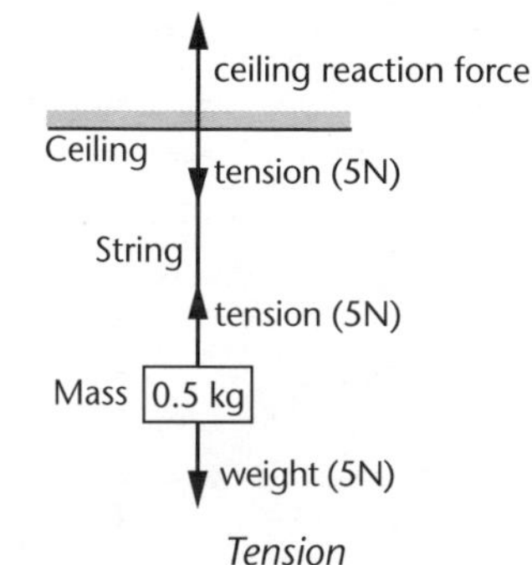

Tension

Example B

Two masses of 1.0 kg and 2.0 kg are suspended by strings. Calculate the size of the tensions in the strings shown in the diagram alonside.

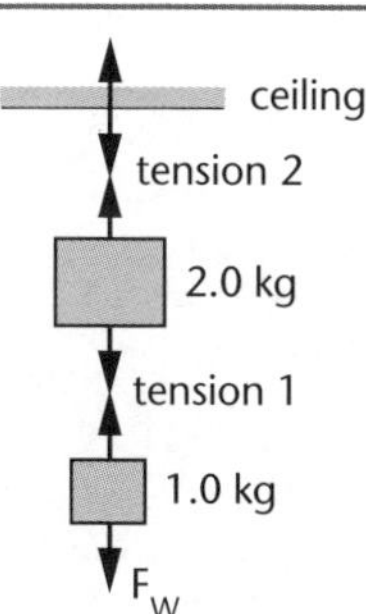

Solution:

Free body force diagrams can be drawn to show forces acting on each mass.

The forces acting on the 1.0 kg mass are shown:

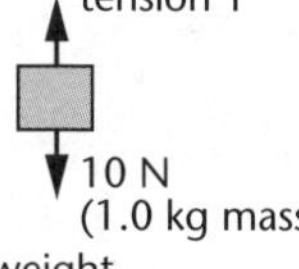

The weight force of the 1.0 kg mass is:

$F_w = 1.0 \times 10$

$= 10$ N

Thus tension 1 also must be 10 N to balance the weight.

The forces acting on the 2.0 kg mass are shown alongside – the weight force of the 2.0 kg mass is 20 N (2.0×10) and so the total force acting downwards is $20 + 10 = 30$ N. Thus tension 2 must be 30 N to balance the forces downwards.

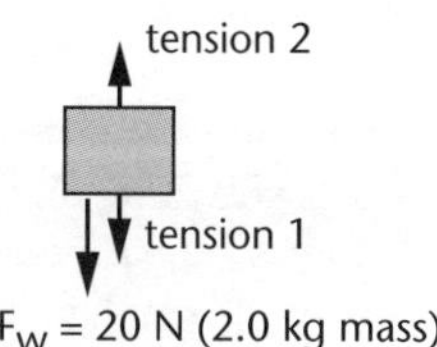

Unit 11.3 Activity 1A: Using Newton's Laws

1. Four forces are applied to an object, as shown.
   a. What is the magnitude of the resultant force?
   b. What is the direction of the object's acceleration?

2.0 N
6.0 N
4.0 N
4.0 N

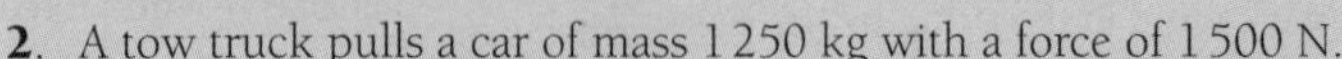
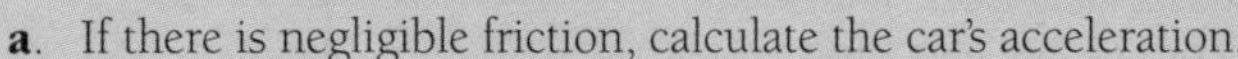

2. A tow truck pulls a car of mass 1 250 kg with a force of 1 500 N.
   a. If there is negligible friction, calculate the car's acceleration.
   b. If there was friction totalling 600 N, calculate the car's acceleration.
3. Two students set up the apparatus shown. The object has a mass of 4.0 kg and is connected by a light cord passing over a pulley to a cart of mass 6.0 kg. The students angle the table slightly so that the tilt compensates for the friction in the apparatus. Take the value of g to be 10 m s^{-2}.

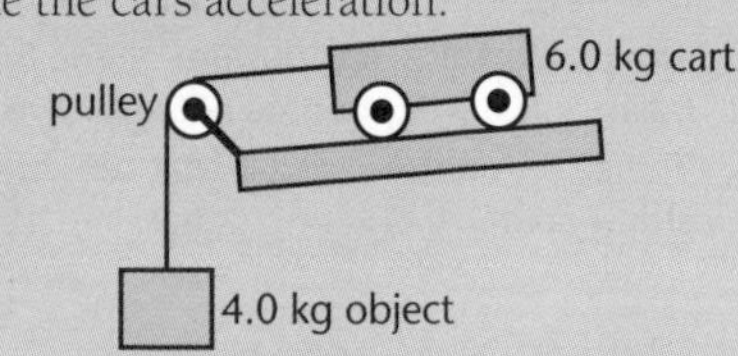

   a. What is the force of gravity on the object, in newtons?
   b. What is the combined mass of the cart and object?
   c. What is the acceleration of the cart and 4.0 kg object?
   d. What is the speed of the cart after it has moved forward 50 cm from rest?
   e. What is the tension in the light cord? (Hint: the tension is the force that acts to accelerate the cart.)
4. A group of students investigate the acceleration of a trolley using various forces applied by a spring balance over a pulley as shown. Their results are recorded in the table.

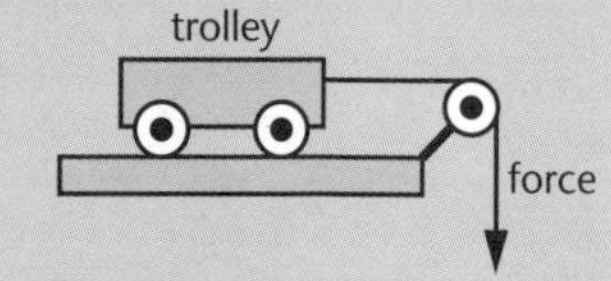

Acceleration (m s^{-2})	0	0	1.0	2.0	3.0	8.0
Force (N)	1.0	2.0	3.0	4.0	5.0	10.0

   a. Graph these results (put acceleration on the horizontal axis).
   b. Calculate the gradient of this graph.
   c. What is the mass of the trolley?
   d. Give a reason why there is no acceleration when the force is 2.0 N or less.
   e. The experiment is repeated with a mass of 1.0 kg on the trolley. What effect will this have on the gradient of the graph?
5. A parachutist of mass 100 kg (including parachute) jumps from a height of 2 000 m. She immediately accelerates downwards at 10 m s^{-2}. Air resistance increases on her as she falls faster and faster until she reaches a terminal velocity (maximum speed when forces are balanced) of 50 m s^{-1} (approx 190 kph).
   a. What is the parachutist's total weight?
   b. What is the force of air resistance on her at terminal velocity?

The parachutist now pulls her ripcord and as her canopy unfurls she slows to a steady speed of 5.0 m s^{-1} in 2.0 seconds.

c. Calculate her average deceleration over the 2.0 seconds.

d. Calculate the average resultant force acting on her over the 2.0 seconds.

e. What is the force of air resistance on her and her parachute now that she is travelling at the steady speed of 5.0 m s^{-1}?

6. A small object of 5 N is attached to the end of a vertical string (remember that weight is the force of gravity). Determine the tension force in the string when the object is moving upwards with a uniform velocity.

7. A vehicle of mass 1000 kg is brought to rest from a speed of 50 m/s in a distance of 100 m. Determine the breaking force of the vehicle assuming that is constant and that there is a constant resistance to motion (friction) of 100 N.

8. An object of mass 5 kg is pulled along a smooth horizontal surface using a horizontal string. Find the tension in the string when the acceleration of the object is 3 m/s^2.

9. An object of mass 8 kg is pulled along a smooth horizontal surface by a string inclined at 30° to the horizontal. Determine the acceleration of the object when the measured tension in the string is 10 N.

10. A small object of mass 2 kg has an acceleration of 5 m/s^2. What is the magnitude of the resultant force acting on the object?

11. A small car has a mass of 300 kg and it is brought to rest in 4 seconds from a speed of 20 m/s. If there is no resistance to motion, find the force exerted by the brakes, assuming it to be constant.

12. A book lies on a horizontal table. The book and the table are both in an elevator (a lift), moving upward at a constant speed.

a. Is the magnitude of the normal force N greater than, less than or equal to the magnitude of its weight, mg?

b. Answer the same question but consider this time that the elevator is increasing its speed. Explain your answer.

13. Explain how birds fly. Is this an application of Newton's second or third law?

14. Explain why an automobile moves forward.

15. Explain why/how rockets accelerate.

16. Name three main sources of forces that can act on an object.

Unit 11.3 Force and Motion (Dynamics)

Topic 2: Friction

Topic 2 discusses the nature and properties of friction (Syllabus p. 13):

- Static and kinetic friction.
- Coefficient of friction.

Friction

Friction is the force produced when:

- Two surfaces (eg a shoe and the floor) come in contact and 'grip' together without slipping past each other.
- One surface slides over another (eg when sandpaper is rubbed over wood).

Friction forces act in many situations and always *oppose the relative motion of the two surfaces involved*. **Air resistance** is friction caused when an object moves through air.

Friction must be allowed for when it is involved in a situation. Usually the friction force is subtracted from the applied force to give a resultant force.

Nature and properties of friction

The harder two surfaces press together, the stronger the grip between them, ie the stronger the frictional force being developed.

The same type of molecular forces are at work in the case of two surfaces interacting, as in 'springs' (ie two kinds of forces operate on the molecules in any material: an attractive force that pulls molecules together, and a repelling force that pushes them apart). Normally these forces balance so the molecules stay a certain distance apart; therefore, the forces between the molecules in the surfaces pull the surfaces together. The closer the molecules get, the stronger the frictional forces being developed.

Friction can be considered a property of contact between objects. So, if two surfaces are able to move, one across the other, without encountering any resistance to that motion, you consider them in frictionless contact, or smooth contact.

On the other hand, if there is resistance to the motion, you know there is friction between the objects, so their contact is rough. The results of many experiments conducted with objects in close contact have shown that the behaviour of frictional forces confirms the following facts that are useful for you to remember/consider when you analyse objects in close interaction and or in motion:

i. Friction always opposes the movement of an object across the surface of another object if they are considered in rough contact.

ii. You can consider the direction of the frictional force to be always opposite to the potential direction of the motion.

iii. The magnitude of the frictional force is considered only just sufficient to prevent movement of an object, and increases as the tendency to move increases, up to a limiting value. Once that limiting value is reached, the frictional force cannot increase any further and, therefore, motion will be about to begin. (This is considered as a limiting equilibrium.) When the frictional force reaches its limit, its value is related to the normal (**reaction**) force N as follows:

$$F_{fr} = \mu N$$

The constant μ is called the coefficient of friction and each pair of surfaces has its own value for this constant.

In limiting equilibrium: $F_{fr} = \mu N$ (kinetic friction)

In general case: $F_{fr} \leq \mu N$ (static friction)

It takes a certain minimum force to start an object moving over a surface. Once the sliding has started, the force needed to keep it moving at constant velocity is lower than this 'minimum' force. If an object does not move when a force is applied to it (F_{app}) there is a frictional force opposing the applied force (F_s). This opposing force is called **static friction**.

Static friction will increase as the applied force increases and will balance this applied force until a certain limiting value is reached, ie when the object starts to slide. The maximum static frictional force can be determined if the coefficient of static frictions (μ_s) is known.

$$F_{s(max)} = \mu_s N$$

Once the object starts to slide it is subject to a lowered frictional force called kinetic friction (F_k) or sliding friction. The coefficient of kinetic friction (μ_k) is different to that of static friction.

$$F_k = \mu_k N$$

Example A

The following figure shows a 20 kg object that you are dragging along the floor applying a force of 50 Newton (*N*). Assume you are applying the force at an angle of 30°. You now wish to determine the magnitude of the normal force (upward force), exerted by the floor on the object and the acceleration of the object, assuming a coefficient of kinetic friction equal to 0.20.

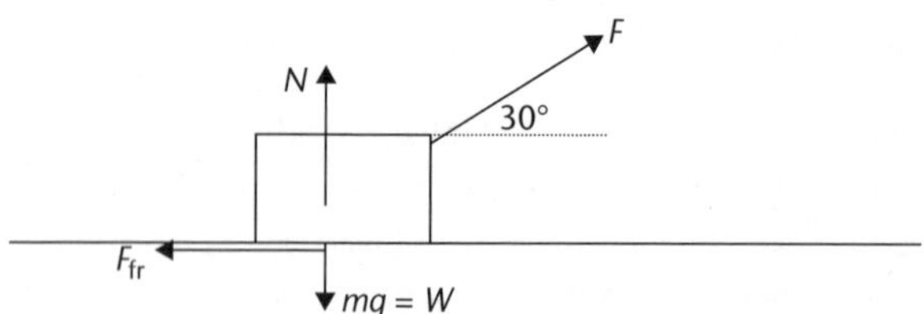

Solution:

The normal force acts in the vertical (upward) direction so, in this case, you will have the addition of forces in the vertical direction:

$$ma_y = N - W + F_y$$

where F_y is the resolved component of F in the vertical direction.

Since mg = (20Kg)(9.8 m/s^2) = 196 Newton and Fy = (50 N)(Sin 30°) = 25 N. and since ay = 0 (no motion in the vertical direction), then:

$$0 = N - 196 + 25$$

Therefore you find the upward force (N) exerted by the floor is: N = 171 Newton.

Now you can determine the acceleration by knowing the fact that the frictional force always opposes the direction of motion and is parallel to the surface of contact, as illustrated in the diagram above. Since the motion is in the x direction, you only need to consider the addition of forces in the x direction to find the net force in that direction and apply Newton's second law of motion. Therefore,

$$ma_x = F_x - \mu_k N$$

and since $F_{fr} = \mu_k N = (0.20)(171) = 34.2\ N$, and $F_x = (50)(\text{Cos}30°) = 43.30\ N$; then

$$a_x = (F_x - \mu_k N)/20\text{Kg} = 9.1/20 = 0.455\ \text{m/s}^2$$

Example B

The **thrust** from the outboard motor of a powerboat is 1 000 N. If the boat has a mass of 500 kg and the friction force opposing the motion of the boat through the water is 200 N, what is the power boat's acceleration?

Solution:

The resultant force on the powerboat is:

F_{res} = 1 000 – 200

= 800 N (forward)

Using $F_{res} = ma$, the acceleration is:

$a = \frac{F_{res}}{m}$

$= \frac{800}{500}$

= 1.6 m s^{-2}

1000 N

800 N

200 N

Using Law 2

Unit 11.3 Activity 2A: Force

Consider the following figure, where a plastic box is lying on the floor, and the box does not slide when a force F_1 = 12 Newton is applied to it as shown:

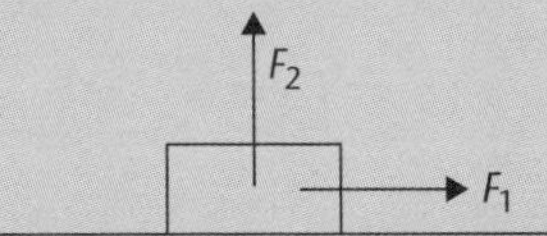

If you start increasing the magnitude of a vertical force F_2 from zero, but before the plastic box begins to slide, what can you conclude about the following quantities?

a. the magnitude of the normal force on the box from the floor

b. the magnitude of the frictional force on the box

c. the maximum value of the static frictional force on the box?

Express your answers by saying if those quantities increase, decrease, or stay the same.

Unit 11.3 Force and Motion (Dynamics)

Topic 3: Newton's Laws

Following on from the introduction to the work of Isaac Newton in Topic 1, we now look at Newton's Laws of Motion in Topic 3 (Syllabus p. 14) with worked examples and some applications.

Introduction

Forces are a fundamental concept in all areas of physics.

- **Force** is a vector, and so has size (magnitude) *and* direction.
- A **resultant** (or **net**) force is produced when two or more forces are added or subtracted (vectorially).
- When a resultant force acts on an object, the object will accelerate in the direction of the resultant force. The relationship between the resultant force, F_{res}, the object's mass, m, and its acceleration, a, is:

$$F_{\text{res}} = ma$$

  This relationship is the second of Newton's Laws of Motion.
- Gravitational attraction between an object and Earth causes the object to have a weight force. The relationship between the weight force, F_w, the mass of the object, m, and the acceleration due to gravity, g, is:

$$F_{\text{w}} = mg$$

g is usually rounded to a value of 10 m s^{-2} to make calculations easier.

Newton's Laws of Motion

Isaac Newton discovered several relationships between forces and their effects on different objects.

Law 1

If the resultant force on an object is zero, the acceleration of the object will be zero. An object with an acceleration of zero will either have a speed of zero (ie it will be stationary) or it will have a constant velocity (constant speed in a straight line).

Example A

Combinations of forces with a zero resultant.

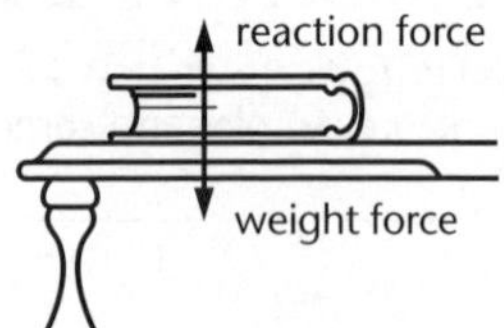

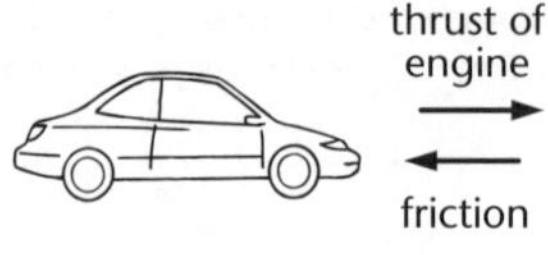

5 N
12 N
13 N

A book on a table. The weight force downward is balanced by an equal sized reaction force acting upward.

A car travelling at constant speed on the motorway. The force caused by the motor is balanced by the friction force.

Three forces acting on an object. The forces are balanced because the resultant force is zero.

No resultant force

Law 2

The relationship between the resultant force, F_{res}, that acts on an object of mass m to produce the acceleration, a, of the object is:

$$F_{res} = ma$$

F_{res} is measured in N, m in kg and a in m s^{-2}.

The direction of the object's acceleration is the same as the direction of the resultant force on the object.

Usually the 'res' subscript is omitted and Law 2 is written simply as $F = ma$.

Example B

A 1.0 kg mass is connected by a string to a trolley which has a mass of 3.0 kg. What will the trolley's acceleration be? (Take $g = 10$ m s^{-2} and neglect friction in the axles of the trolley or pulley.)

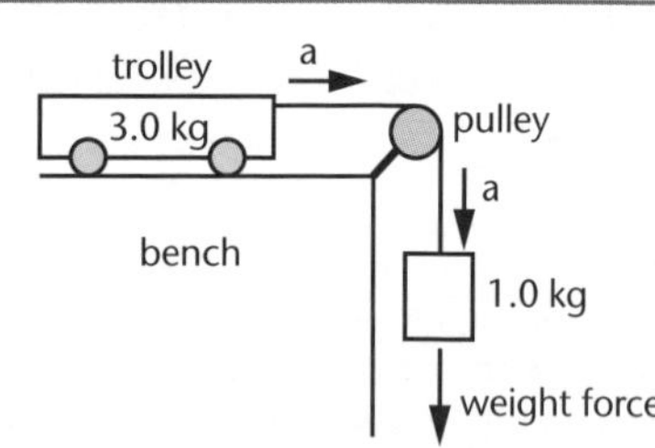

Accelerating masses

Solution:

The objects which accelerate are the 3.0 kg trolley *and* the 1.0 kg mass (they are tied together and therefore have the same acceleration).

Total mass of accelerating objects $= 3.0 + 1.0$
$= 4.0$ kg

The force causing the acceleration of the two objects is the weight force on the 1.0 kg mass. The weight force is:

$$\begin{aligned} F_w &= mg \\ &= 1.0 \times 10 \\ &= 10 \text{ N} \end{aligned}$$

The acceleration of the trolley is the same as the trolley and mass together:

$$\begin{aligned} a &= \frac{F}{m} \\ &= \frac{10}{4.0} \\ &= 2.5 \text{ m s}^{-2} \end{aligned}$$

The weight force of the 3.0 kg trolley has no effect since it is *perpendicular* to the trolley's direction of acceleration and is balanced by the reaction force from the bench.

Newton's second law

Newton's second law states that if an unbalanced force acts on an object it will cause the object to accelerate. The size of the acceleration depends on the size of the force and the mass of the object.

$F = m\,a$ $\qquad$ F = force, m = mass, a = acceleration

The SI unit for force is the **newton**. An unbalanced force of 1 N acting on a 1 kg object will cause an acceleration of 1 m s^{-2}.

Example C

John (mass 65 kg) is wearing skates and standing still on an ice rink. The ice is an almost frictionless surface. His friend pushed him from behind with a force of 600 N. Calculate John's acceleration.

Solution:

$$a = \frac{F}{m} = \frac{600}{65} = 9.2 \text{ m s}^{-2}$$

Ice skating – a near-frictionless surface.

Example D

Meri is riding a bicycle. The combined mass of Meri and her bicycle is 85 kg. The forward force acting on Meri and the bike is 500 N; the friction force acting is 150 N. Calculate how fast Meri accelerates on her bike.

Solution:

$Fnet = 500 - 150 = 350$ N

$$a = = \frac{F}{m} = 4.1 \text{ m s}^{-2}$$

Mass and weight

Mass is the amount of matter in an object, measured in kg. The mass of an object is the same regardless of where it is located in the universe. An astronaut of mass 85 kg will have the same mass on Earth, on the Moon and during space travel. Mass is a scalar quantity.

The **weight** of an object is the force of gravity that acts on the object. The weight of an object depends on where it is located in the universe. The weight of an object on the Moon is approximately $\frac{1}{6}$ of the weight of the object on Earth. In outer space objects are weightless.

The weight (W) of an object can be calculated using the formula:

$W = m\,g$ $\qquad$ where g = 10 m s^{-2}

Example E

Calculate the weight of an astronaut of mass 85 kg:

a. On Earth. b. On the Moon. c. In outer space.

Solutions

a. The weight of the astronaut on Earth is $W = mg = 85 \times 10 = 850$ N.

b. The weight of the astronaut on the Moon is $\times 850 = 142$ N.

c. The weight of the astronaut in outer space is 0 N.

Unit 11.3 Activity 3A: Forces

1. Two low-friction trolleys are connected by a stretched rubber band and then released. They collide at point **C**.

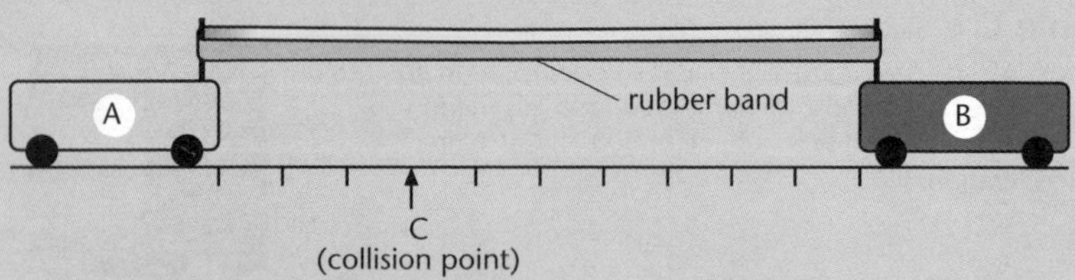

Students make the following comments about the experiment:

I. The rubber band exerts equal forces on the trolleys.

II. Trolley A has more mass than trolley B.

III. Trolley A is travelling faster than trolley B.

Which of these comments are correct?

2. Five forces are acting on the moving car shown. Two of the forces are labelled. What are the correct labels for forces 1, 2, and 3?

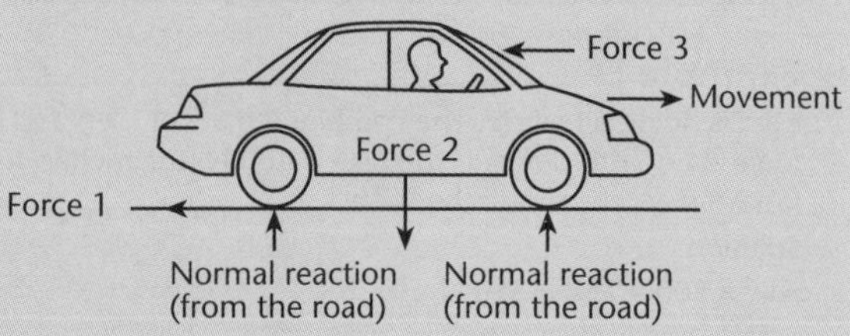

3. Two forces F_1 and F_2, acting at right angles to each other, are applied to a stationary 5.0 kg mass.

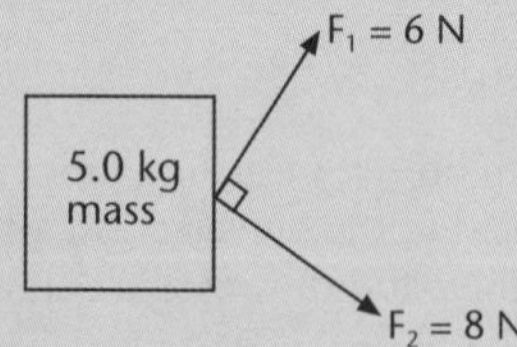

Draw a vector diagram to determine the total force acting on the 5.0 kg mass.

4. A space ship is moving through empty space far from any stars when its rockets are turned off. Describe how this will affect the motion of the space ship.

5. The diagram below shows a 'physics mobile' of signs made from light bamboo sticks, signs set against a background grid.

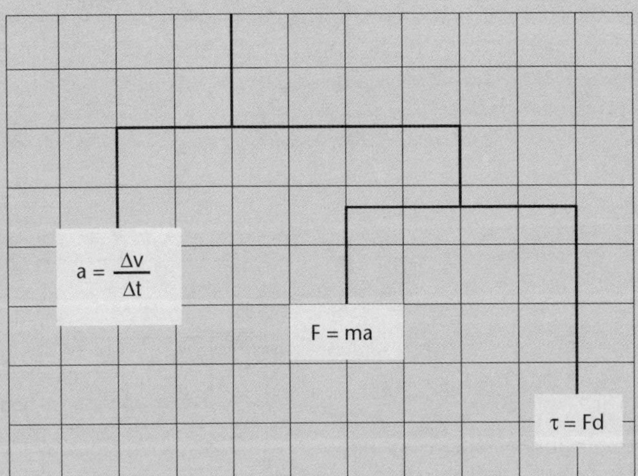

If the τ sign has a mass of 15 grams, calculate the mass of the other two signs.

6. A front-wheel drive car is travelling at constant velocity. The various forces acting on the car are shown in the diagram.

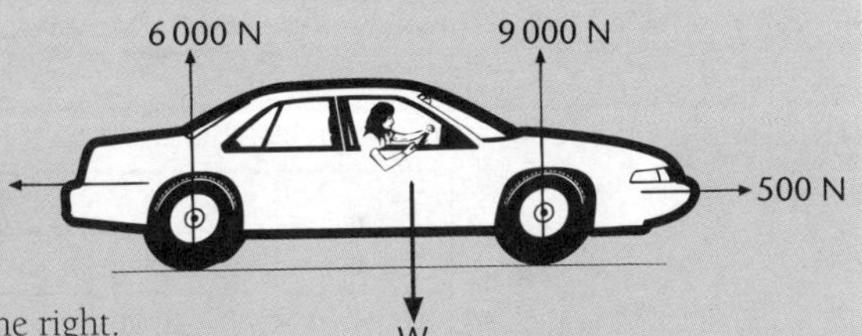

   a. Name the 500 N force on the right.
   b. What is the mass of the car and its passengers?
   c. What is the value of the force F on the left?

   The force shown on the right is now doubled. The other forces remain constant.

   d. Calculate the new resultant force on the car.
   e. Calculate the acceleration of the car.

7. A skydiver jumps out of an aeroplane and begins to fall vertically down to earth.

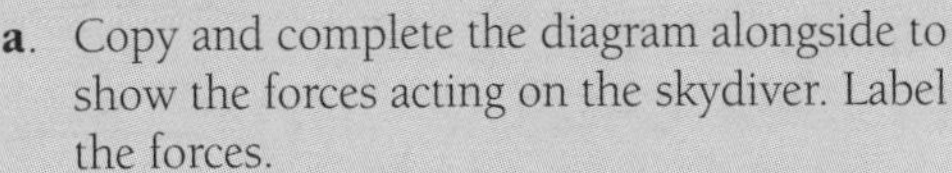

   a. Copy and complete the diagram alongside to show the forces acting on the skydiver. Label the forces.

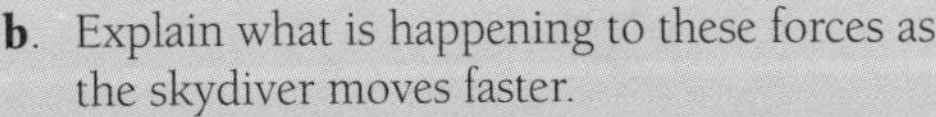

   b. Explain what is happening to these forces as the skydiver moves faster.
   c. What happens to the skydiver's speed after free-falling for a long time? Explain why this happens.

d. Using axes like that shown, sketch a speed-time graph for the skydiver's descent prior to opening his parachute.

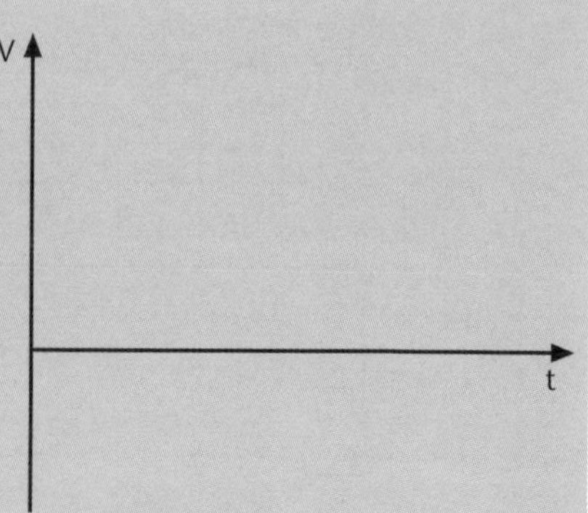

8. A 2 kg trolley is pulled with a force of 16 N and accelerates at 5 m s^{-2}. Calculate the frictional force acting on the trolley.

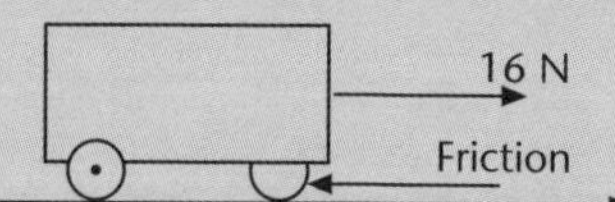

9. The diagram shows three horizontal forces acting on a cart. What single force will have the same effect?

10. Complete the following table by calculating **a** to **h**.

Force	Mass	Acceleration
a	2 kg	8 m s^{-2}
b	50 kg	0.05 m s^{-2}
c	25 000 kg	800 m s^{-2}
10 N	20 kg	d
15 000 N	0.6 kg	e
60 N	1 500 g	f
800 N	g	2.5 m s^{-2}
1 000 N	h	0.05 m s^{-2}

11. What is the reading on the scale of the force meter?

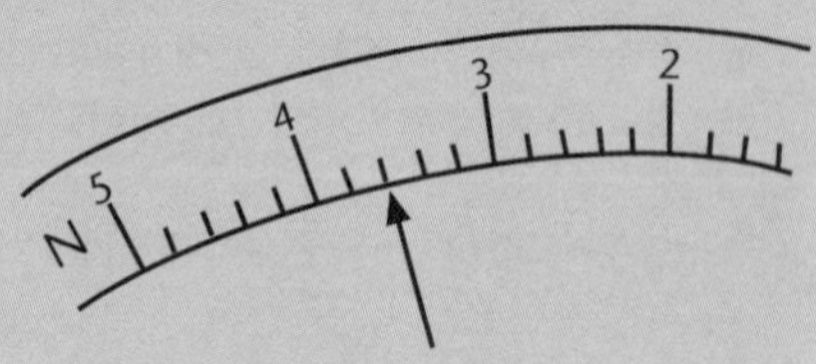

12. A Moon rock is brought to Earth. How does its mass and weight on Earth compare with its mass and weight on the Moon?

13. In an experiment to investigate the relationship between the force and acceleration of a trolley, the graph shown alongside was obtained.

a. What is the acceleration of the trolley?

b. If the mass of the trolley was 4 kg, what is the size of the force accelerating the trolley?

c. If the force is now doubled, what will the new acceleration be?

d. Sketch the speed-time graph already shown and on it sketch the new speed-time line which you would expect when the force was doubled.

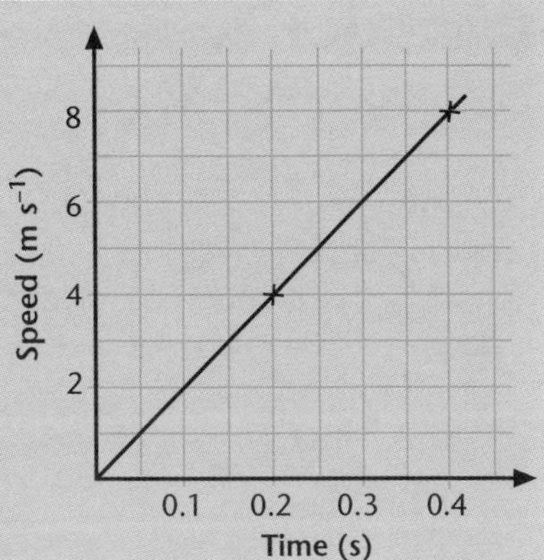

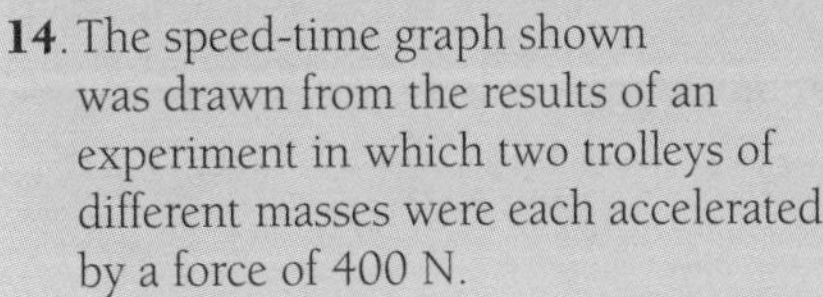

14. The speed-time graph shown was drawn from the results of an experiment in which two trolleys of different masses were each accelerated by a force of 400 N.

a. Which trolley had the greatest mass?

b. Calculate the acceleration of trolley A.

c. Find the mass of trolley A using your result from (b).

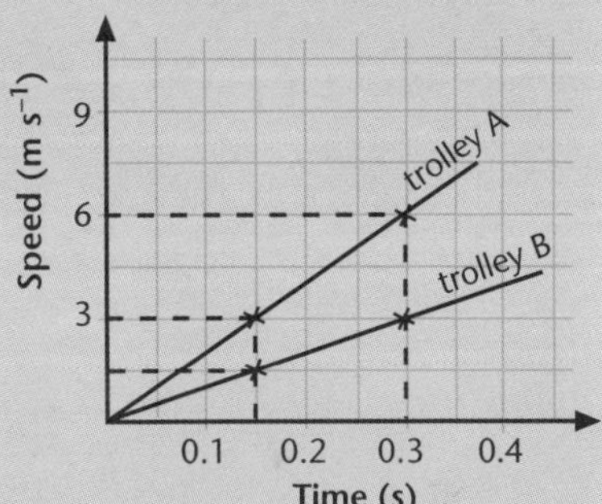

15. Which of the identical trolleys below will have the greatest acceleration?

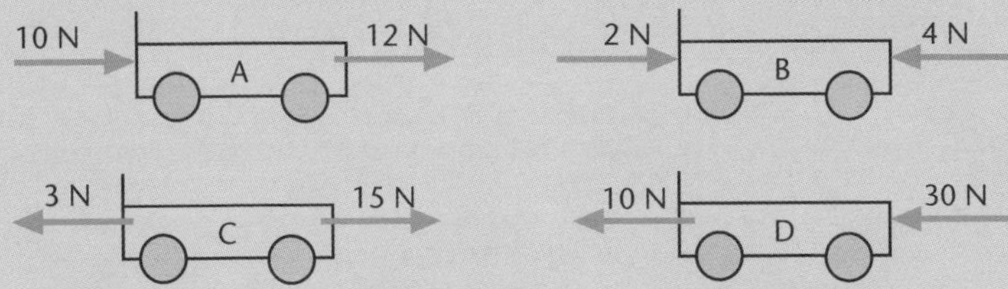

16. The wind pushes a sand yacht along on dry sand. When the unbalanced force on the yacht is 280 N its acceleration is 2 m s^{-2}.

a. Calculate the mass of the yacht.

b. The yacht is blown onto wet sand where the friction force has the same size as the wind force. What happens to the speed of the yacht?

c. When the yacht moves on to soft sand the friction force is larger than the wind force. What happens to the speed of the yacht?

17. Two tug boats are pulling a container ship into port as shown in the diagram below.

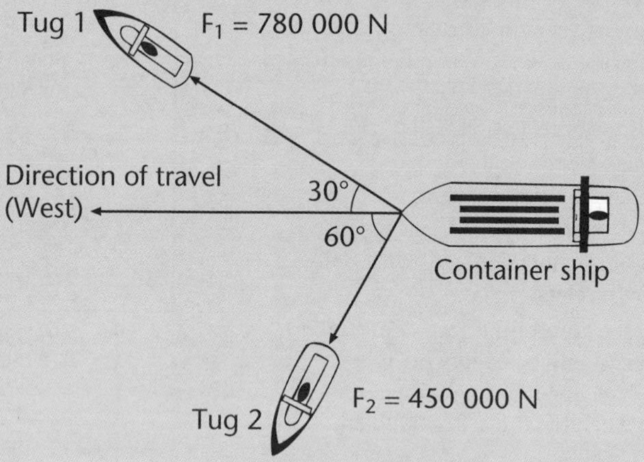

The container ship moves at a steady speed in the direction shown.

a. On the grid below, draw a scale diagram to find the resultant force exerted by the two tugs on the container ship. Use a scale of 1 cm = 100 000 N.

b. Calculate the resultant force of the two tugs on the container ship.

c. What is the size and direction of the frictional force on the container ship?

(Source: NCEA Examination paper)

18. The diagram below shows an astronaut standing on the Moon.

a. On the free-body diagrams below, draw in and label the forces acting on the astronaut and the moon.

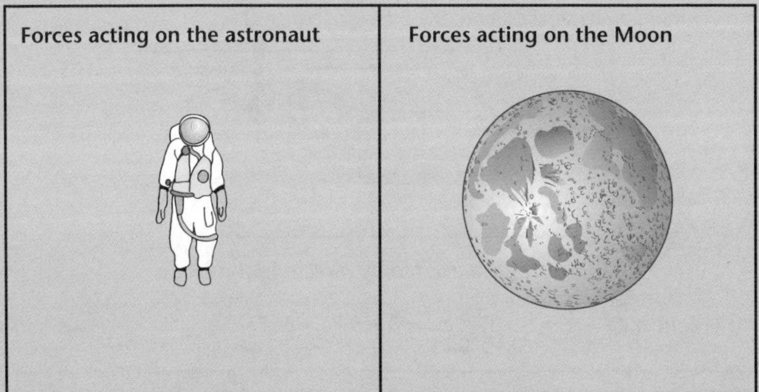

b. The astronaut is in equilibrium. What can be said about the forces acting on the astronaut?

(Source: NCEA Examination paper)

Pressure

The effect of a force acting at a right angle to a surface depends on the size of the force and the area over which the force acts.

$$P = \frac{F}{A}$$

Pressure = Force ÷ Area

The SI unit for **pressure** is the **pascal** (Pa). One Pa = 1 N m^{-2}.

Example F

A rectangular block 10 cm × 5 cm × 2 cm weighs 40 N. Calculate the largest and smallest pressure it can exert on a table.

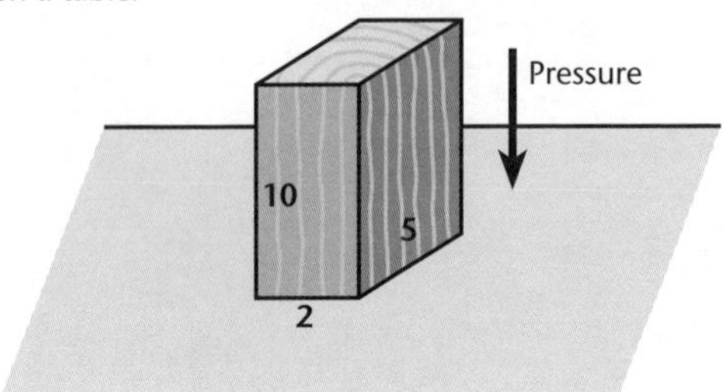

Largest pressure exerted by a rectangular block.

Answer:

The largest pressure $= \frac{F}{A} = \frac{40}{0.05 \times 0.02} = 40\,000$ Pa = 40 kPa

(ie block standing on its smallest end, the 5 cm × 2 cm end)

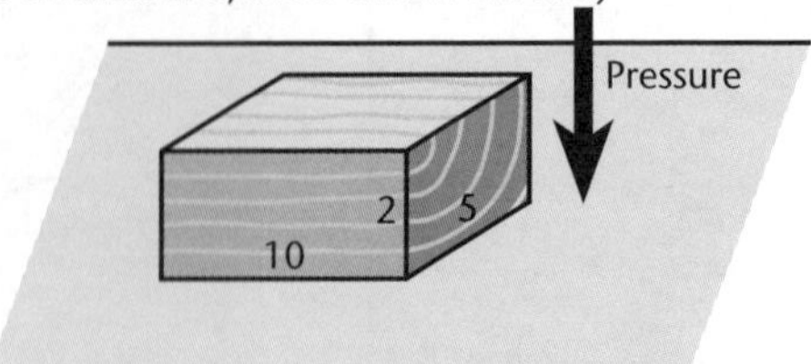

Least pressure exerted by a rectangular block.

The smallest pressure $= \frac{F}{A} = \frac{40}{0.1 \times 0.05} = 8\,000$ Pa = 8 kPa

Example G

Compare the pressure exerted on the ground by a 6 tonne (6 000 kg) elephant (foot area = 0.32 m^2) and a 65 kg woman wearing stiletto-heeled shoes (sole area of one shoe = 0.006 m^2).

Comparing pressures.

Answer:

The weight of the elephant = $mg = 6\,000 \times 10 = 60\,000$ N

The force exerted on the ground by one foot $= \frac{60\,000}{4} = 15\,000$ N

The pressure exerted by the elephant foot $= \frac{F}{A} = \frac{15\,000}{0.32} = 46\,875$ Pa

The weight of the woman = $mg = 65 \times 10 = 650$ N

The force exerted on the ground by one foot $= \frac{650}{2} = 325$ N

The pressure exerted by the woman's shoe $= \frac{F}{A} = \frac{325}{0.006} = 54\,167$ Pa

The pressure exerted by the woman's shoe is greater than the pressure exerted by the elephant's foot!

Unit 11.3 Activity 3B: Pressure and density

1. An empty bus is parked in a car park. The diagram below shows the part of a front tyre which is resting on the road. Only 60% of the area shown below is in contact with the ground, due to the grooved surface of the tyre. The force exerted by this front tyre on the road is 6048 N.

   Calculate the pressure exerted by this front tyre on the road. Give the correct unit with your answer.

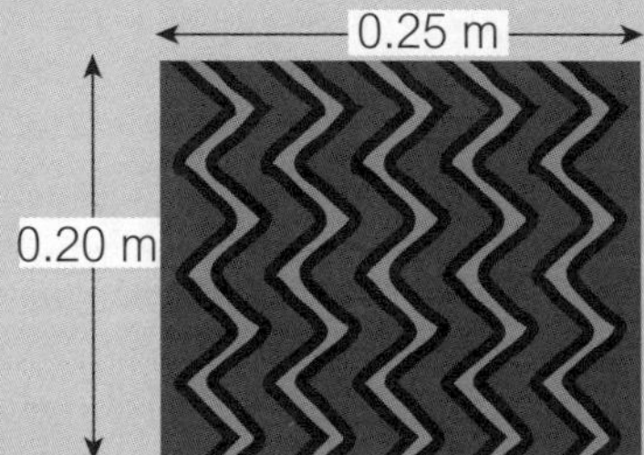

(Source: NCEA Examination paper)

2. Tom is wearing smooth-soled shoes and Dipa is wearing shoes with moulded soles that have ridges on them. The diagram below shows the pattern of the soles of their shoes.

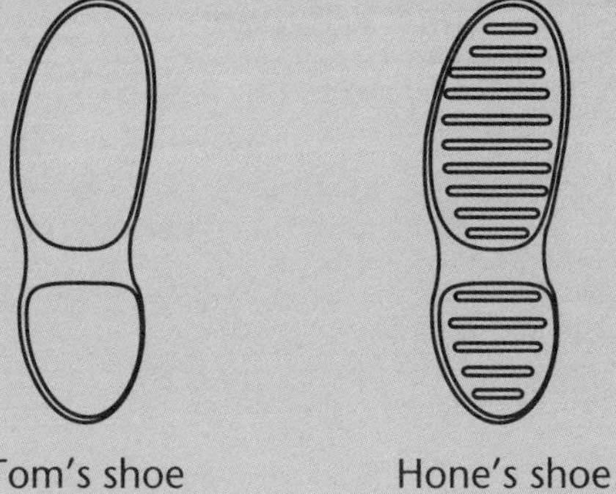

   Each boy has a mass of 60 kg and their shoes are the same size.

   a. Carefully explain why Dipa exerts more pressure on the hard surface of the field where they are playing.

   b. Calculate Dipa's weight. Give the correct unit with your answer.

   c. The total area of the ridges on Dipa's shoes is 16 cm^2 (1.6×10^{-3} m^2). Calculate the pressure that Dipa exerts on the ground. Give the correct unit with your answer.

(Source: NCEA Examination paper)

3. After a kick during a training session, a ball of mass 0.44 kg hit the ground. The average force acting on the ball was 75 N.

   At the instant the ball hit the ground, the surface area of the ball in contact with the ground was 70 cm^2 (0.007 m^2).

   Select a formula and calculate the pressure exerted by the ball on the ground. State the appropriate SI unit with your answer.

(Source: NCEA Examination paper)

4. A hollow steel buoy in the form of a sphere of radius 0.5 m is fixed by a wire to the base of a river. If the mass of the buoy is 20 kg, calculate the tension in the cable. (Take density of water = 1000 kgm^{-3}).

5. A hot air balloon has a volume of 600 m^3 and hovers just clear of the ground when the air in it is heated so that its density is 0.80 kgm^{-3}. If the density of the air outside the balloon is 1.25 kgm^{-3}, calculate (take g = 10 ms^{-2}):

a. the total mass of the balloon including the air inside it

b. the mass of the envelope of the balloon and its load.

6. Calculate the distance below the surface of the Earth at which a piece of iron (density 7870 kgm^{-3}) will float in air if the density of air at sea level is 1.3 kgm^{-3}.

7. The volume of a deep sea diver enclosed in his diving suit is 0.15 m^3 and his total mass is 95 kg. Find the mass of lead that must be attached to his feet so that the diver will just sink in sea water of density 1030 kgm^{-3}. (Take density of lead = 11400 kgm^{-3}).

8. The wind exerts a pressure of 100 Pa on a wall of a house. If the area of the wall is 6 m^2, what is the force on the wall?

9. If a metal object has a mass of 2600 kg and dimensions of 0.5 m by 1.0 m by 2.0 m, what is the maximum pressure it exerts on the ground when resting on it?

10. If the density of water is 1000 kgm^{-3}, what is the pressure at the bottom of a swimming pool 2 m deep when it is full of water?

Unit 11.3 Force and Motion (Dynamics)

Topic 4: Rotational motion and equilibrium

The content of Topic 4 relates to motion and equilibrium (Syllabus p. 14), in particular:

- Angular velocity and angular acceleration.
- Rotational motion and kinematic equations.
- Circular motion, its quantities and their units, with worked examples on circular motion.

When studying this Topic it is recommended that students refer back to Unit 11.2, especially Topic 7 'Circular motion'.

Angular velocity and angular acceleration

To explain angular velocity, it is helpful to understand the relationship between linear velocity and angular velocity. Linear velocity, as the name suggests, deals with speed in a straight line; the units are often km/h or m/s or mph (miles per hour). Linear velocity (v) = change in distance/change in time:

$$v = d/t$$

Angular velocity, as the name suggests, deals with speed through an angle, or in a circular motion; the units are often radians per second or degrees per second. Angular velocity (w) = change in rotation/change in time:

$$w = \theta/t$$

To relate the two types of velocity to each other, we can use the following relationship:

$$v = r \times w$$

where r = radius.

You can also explain angular velocity in terms of its similarity to linear velocity by using the concept/definition of velocity you studied in Unit 11.2 Topic 1 (page 46): instead of the change of linear displacement travelled, you can use the change of angular displacement $\Delta\theta$. By doing this, you can define the average angular velocity as:

$$\bar{\omega} = \theta/t$$

where θ represents the angle through which an object has rotated during a time frame t. Following this line of reasoning, you can also define instantaneous angular velocity as a small angle $\Delta\theta$, through which an object turns in a very short interval of time Δt, as:

$$\omega = \Delta\theta/\Delta t \qquad [\Delta t \text{ approaching zero}].$$

You can measure angular velocity in radians per second. You can relate the angular velocity of an object to its linear speed v and its distance r from the axis of rotation by considering that v = distance/time, hence, using the equation for $\theta = d/r$:

$$v = r\theta/t$$

or, since $\theta/t = \omega$ then:

$$v = r\omega$$

This equation is valid for the instantaneous and the average case.

Angular acceleration is similar to the concept of ordinary linear acceleration, as defined on page 47. Therefore, angular acceleration is defined as the change in angular velocity divided by the time interval required to make that change; thus:

$$\overline{\alpha} = (\omega - \omega_0)/t$$

This equation is a definition of the average angular acceleration over a time interval t, where ω_0 is the initial angular velocity and ω represents the angular velocity after the time t has elapsed.

You can also define an instantaneous angular acceleration in the usual way, as:

$$\alpha = (\Delta\omega)/\Delta t$$

In most cases you will be dealing mainly with cases where the angular acceleration is constant; therefore, you can consider it equal to the average angular acceleration. Since you usually measure ω in rad/s and t in seconds, you can express α in rad/s^2; and when radian measured is used, α can be related to the linear acceleration of the object as:

$$a_T = r\alpha$$

In this equation, r is the radius of the circle in which the object is moving, and the T in aT tells you that you are dealing with a 'tangential' acceleration, because in this case, the acceleration is along the circle on which the object is moving (ie tangential to the circle).

This prompts the question: how often is the object rotating or going around the circle?

This feature of the circular motion is called frequency of rotation, f, and it is defined as the number of complete revolutions (rev) made in one second. Therefore, since one revolution represents an angle of 2π radians, then 1 rev/s = 2π rad/s. The general relationship between frequency f and angular velocity ω can be written as:

$$f = \omega/2\pi \text{ or } \omega = 2\pi f.$$

Similar to linear motion, angular velocity and angular acceleration are vector quantities; therefore, they have a magnitude and a direction associated with them. Generally speaking, it can be concluded that the angular velocity ω for a rotating object is usually represented by a vector pointing along the axis of rotation.

Example A

Calculate the linear speed of a point on the edge of an old 33-rpm (revolutions per minute) vinyl record whose diameter is 30 cm.

Solution:

First you need to find the angular velocity in radians per second, so you are required to convert 33 rpm to rev/second, as follows:

f = 33 rpm = 33 rev/60s = 0.55 rev/s.

Therefore, $\omega = 2\pi f$ = 3.5 rad/s. Since the radius of the record is r = 0.15 m, then:

$v = r\omega$ = (0.15 m)(3.5 rad/s) = 0.52 m/s.

Example B

A rotating object (centrifuge rotor: motor) is accelerated from rest to 20,000 rpm in 5.0 minutes. Determine its angular acceleration.

Solution:

For this case you first find ω as:

ω = (20,000 rpm)(2π rad/rev)/(60 s/min) = 2100 rad/s;

therefore, since $\omega_0 = 0$;

α = (2100 rad/s - 0)/300 s = 7.0 rad/s^2.

Uniformly accelerated rotational motion: kinematic equations

During your study of linear motion of objects in Unit 11.2, Topic 2 (pp.45–46) you derived important equations that gave you relationships between acceleration, velocity, and distance for the situation and conditions of uniform acceleration.

You derived those equations based on the definitions of linear velocity and acceleration, assuming constant acceleration. For rotational motion, the definitions of angular velocity and angular acceleration are the same as for their linear counterparts, except that in this case, θ replaces distance d, ω replaces v, and α replaces a. Therefore, the angular equations for the constant angular acceleration case are similar to the linear equations you previously developed and applied, with d replaced by θ, v replaced by ω, and a replaced by α; and they are summarised here, in the following table:

Angular	Linear case
$\omega = \omega_0 + \alpha t$	$v = v_0 + at$
$\theta = \omega_0 t + 1/2\alpha t^2$	$d = v_0 t + 1/2at^2$
$\omega^2 = \omega_0^2 + 2\alpha\theta$	$v^2 = v_0^2 + 2ad$
$\varpi = (\omega + \omega_0)/2$	$v = (v + v_0)/2$

Example C

Consider the rotating object in Example B (p. 128). Determine how many turns the object (centrifuge rotor) made during its acceleration period of 5 minutes.

Solution:

The object was accelerated from rest, $\omega_0 = 0$, and $\omega = 2100$ rad/s, $\alpha = 7.0$ rad/s^2, and $t = 300$ s. Since you need to find the number of turns and a turn is, by definition, a complete cycle or 360° around a circle, then you first need to determine the angle θ (since an angle is the equivalent to distance travelled, in the rotational case), using one of the above equations for the rotational motion, as follows:

$= \omega_0 t + 1/2\alpha t^2 = 0 + ½(7.0 \text{ rad/s}^2)(300 \text{ s})^2 = 3.2 \times 10^5$ rad

Now, to find the number of revolutions (turns), you only need to divide by 2π (since a complete circle is equivalent to an angle of 2π radians) and obtain the total number of turns; thus: 3.2 x 105 rad/2π rad = 5.0×10^4 revolutions/turns.

Circular motion of an object: centripetal acceleration and uniform circular motion

Let us consider a very important and special case of an object moving in a circle of radius r with constant angular velocity ω. This special case is known as 'uniform circular motion'. Since this is a case of zero angular acceleration ($\alpha = 0$), there is no reason to apply the equations you developed in the previous section of this Unit, on page 131.

Although the angular acceleration is zero, the linear acceleration of the object is not zero. This is true because the magnitude of the linear velocity of the object, v, is constant and equal to rω; but as you should have realised by now, the direction of its velocity is continuously changing since it is moving around a circular path, as shown in the following figure.

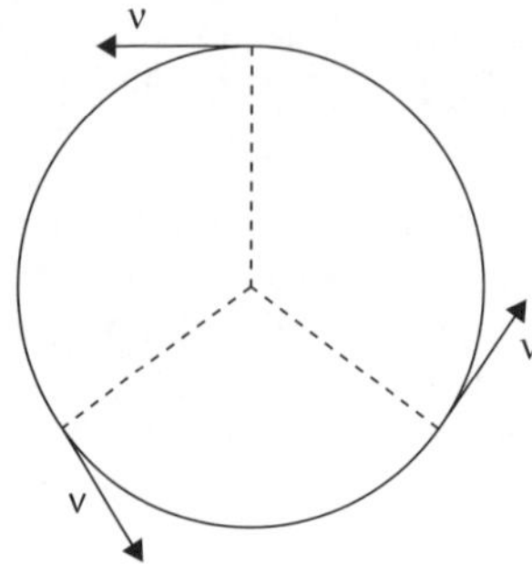

Based on the general definition for acceleration as 'the rate of change of velocity', a change in direction of velocity must also constitute an acceleration, just as a change in magnitude of the velocity would have. Therefore, you will now investigate this acceleration in a quantitative form.

You previously defined acceleration as: $a = (v - v_0)/\Delta t = \Delta v/\Delta t$

where Δv is the change in velocity during a short interval of time Δt. You also learnt that by making Δt approach zero you can determine the instantaneous acceleration, but for the purpose of illustrating the concept you can consider a small non-zero time interval as in the following figure:

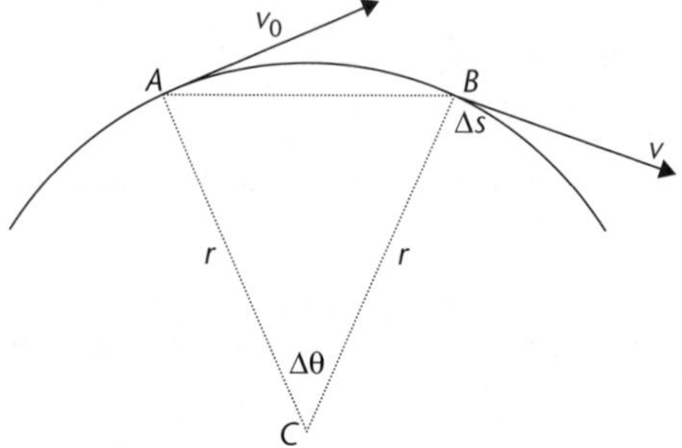

By analysing the figure you understand that during a time Δt, the object moves from point A to point B, covering the distance Δs which subtends an angle $\Delta\theta$, and that the change in the velocity vector is given by $v - v0 = \Delta v$. By rearranging this equation you find $v = v0 + \Delta v$. Since velocity is a vector quantity, this equation is representing an addition of vectors, so Δv must be a vector, and you can illustrate this relationship as in the following figure:

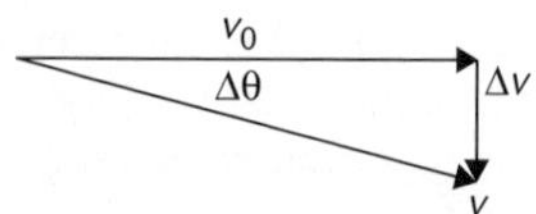

From this vector diagram you notice that when Δt is approaching zero, and therefore Δs and $\Delta\theta$ are also very small, v ($v2$ in the diagram below) must be almost parallel to v_0 ($v1$ in the diagram below) and Δv is essentially perpendicular to them. Based on this analysis you conclude that Δv points toward the centre of the circular path.

This can be all graphically shown in the diagram below.

Since, by definition, acceleration is in the same direction as change in velocity Δv, it must also point toward the centre of the circle. Therefore, this acceleration is called centripetal acceleration, and is denoted by the symbol a_c.

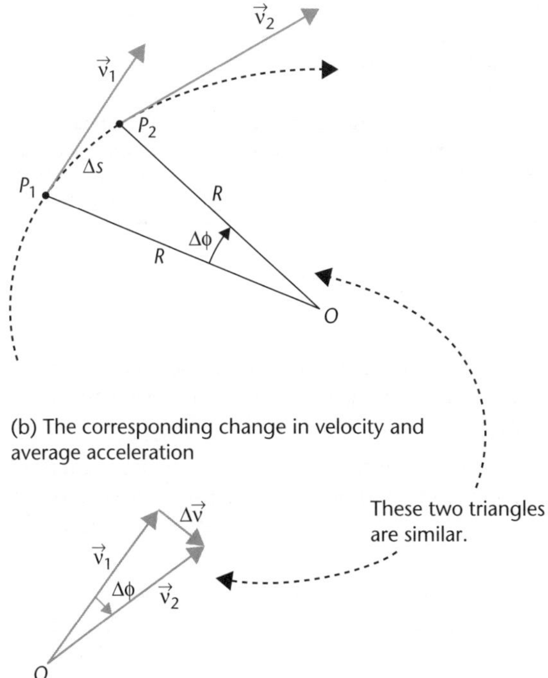

Following an analysis of the figures above and by the use of geometry you can find out that the relationship between the centripetal acceleration and the linear velocity that changes to give rise to it is usually expressed as:

$$a_c = v^2/r$$

where r is the radius of the circle.

In the previous section of this Unit (page 129) you developed the relationship between the linear velocity and the angular velocity, and expressed it as $v = r\omega$; so, using this relationship, you can express the centripetal acceleration in terms of the angular velocity, ω, as:

$$ac = \omega^2 r.$$

From all of this analysis, you are now able to conclude the following:

Any object moving in a circle (circular path), such as a ball on the end of a string, must have a force applied to it to keep it in that path; that is, a force necessary to give it a centripetal acceleration, and the magnitude of that required force is found by applying Newton's second law of motion ($F = ma$), where centripetal acceleration is $a_c = v^2/r$, giving you the total force as: $F = ma_c = m(v^2/r)$.

Since ac is directed toward the centre of the circle, this force too must be directed toward the centre of the circle. So, you can see intuitively that a force is necessary in this case because, if no force were applied on the object, it would not move in a circle but in a straight line, as Newton's first law tells you.

Example D

Calculate the magnitude of the acceleration of a speck of dust on the edge of an old 33-rpm vinyl record whose diameter is 30 cm.

Solution:

In the previous section you analysed and understood an example concerning this old 33-rpm record and found out that the angular velocity ω is 3.5 rad/s and the linear velocity v at the edge of the record was 0.52 m/s. Therefore, since the radius is 0.15 m, you find that $ac = \omega^2 r = 1.8$ m/s²; and using equation $a_c = v^2/r$, you will get the same result.

Non-uniform circular motion

You should now have a clear understanding that a necessary condition for a case of circular motion at a constant speed is to have a force on an object, directed toward the centre of a circular path. If the force is not directed towards the centre of the circle, but at an angle as for the situation, illustrated in the following figure, then you realise the force can be considered as having two components: F_T and F_c.

The component of the force directed toward the centre of the circle, F_c, is responsible for the centripetal acceleration of the object, and keeps the object moving around the circle, while the component F_T, tangent to the circular path, is responsible for the increase (or decrease) of the speed.

Therefore, you can conclude there must be a component of the acceleration tangential to the circle, a_T, ($a_T = r\alpha$), so when an F_T is acting on the object, the speed of the object is changed.

Conclusion: a_T is related to the change in the magnitude of the velocity of the object, whereas a_c is responsible for the change in direction of the velocity.

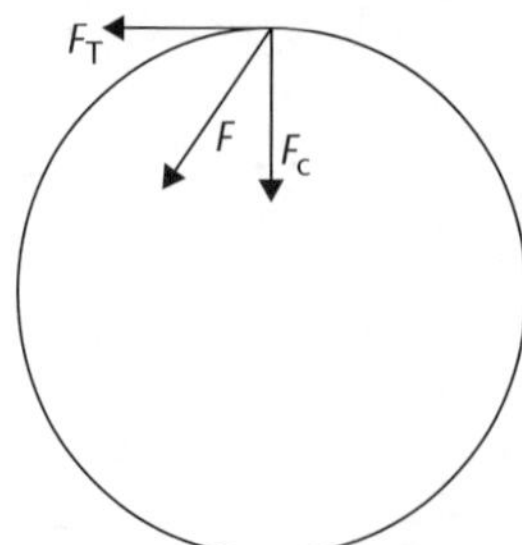

Unit 11.3 Activity 4A: Force and motion

1. Consider a 1000 kg car rounding a curve on a flat road of radius 50 m at a speed of 50 km/h (14 m/s).
   a. Will the car make the turn if the pavement is dry and the coefficient of static friction is 0.60?
   b. Will the car make the turn if the pavement is icy and $\mu = 0.20$?
2. A particle moves along a circular path of radius 3 m with an angular velocity of 20 rad/s. Calculate the following:
   a. the linear speed of the particle
   b. the angular velocity in revolutions per second
   c. the time for one revolution
   d. the centripetal acceleration.

(from Roger Muncaster *A Level Physics 4th edition* Nelson Thornes)

Torque

Torque, τ, refers to the **turning effect** about a **pivot**. Torque (or **moment** or **leverage**) is involved in situations such as the turning of the steering wheel of a car, the turning on/off of a tap or the lifting of a cap on a bottle.

The 'moment of a force' is the torque that the force produces. The size of a torque depends on the size of the force *and* the perpendicular distance from the pivot to the point where the force is applied.

Example E

A small force, F acting at a small distance, $d_\perp$ from a pivot produces a small torque, τ.

The same force acting at a larger distance from a pivot produces a larger torque.

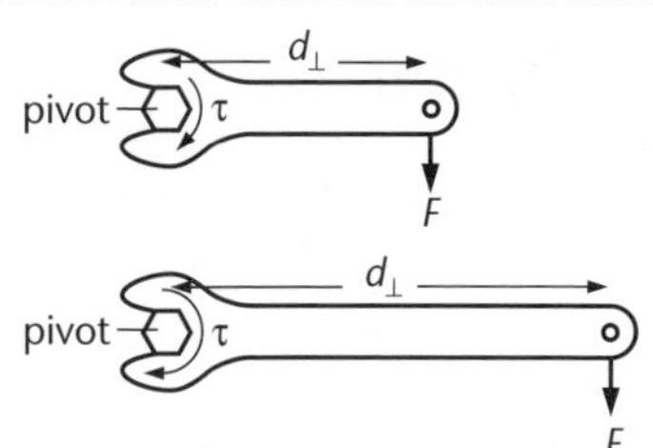

Torque depends upon distance

The torque, τ, that a force produces about a pivot is the product of the force, F, and the *perpendicular* distance, $d_\perp$, of the force's line of action from the point.

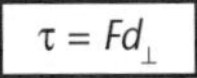

$$\tau = Fd_\perp$$

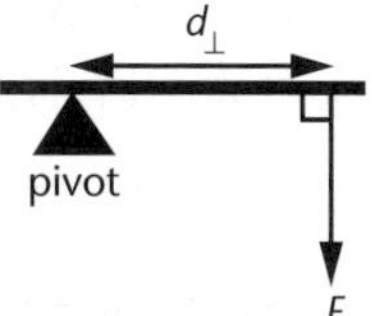

Torque diagram

The unit for torque is newtons × metres (symbol N m).

Torque can act in a clockwise or anticlockwise **sense**.

Example F

A force of 10 N which acts at 8.0 cm from the pivot is needed to lift the cap off a bottle of soft drink.

The torque applied is:

$\tau = Fd_{\perp}$

$= 10 \times 0.080$ [8.0 cm = 0.080 m]

$= 0.80$ N m anticlockwise

This torque is the same as a force of 40 N acting 2.0 cm from the pivot.

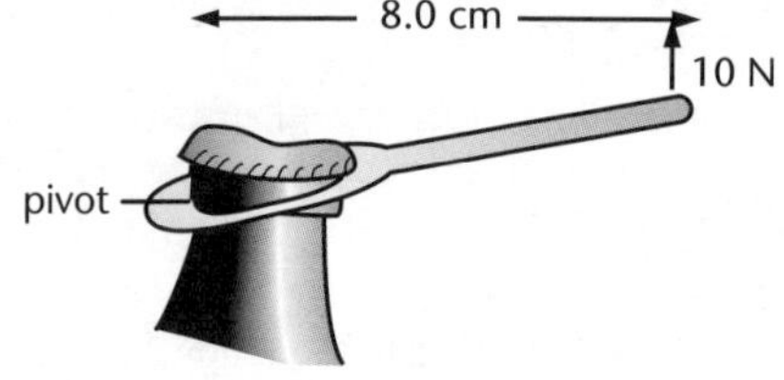

A lever in action

A **couple** occurs where *two equal and oppositely directed forces* act at a distance apart. A couple causes rotation only.

Example G

The diagram shows the situation of a couple applied to a tap handle.

The torque produced by the couple is caused by two equally sized forces.

The size of the torque is:

$$\tau = \left(F \times \frac{d}{2}\right) + \left(F \times \frac{d}{2}\right)$$

$$= F \times d$$

ie $\tau = Fd$

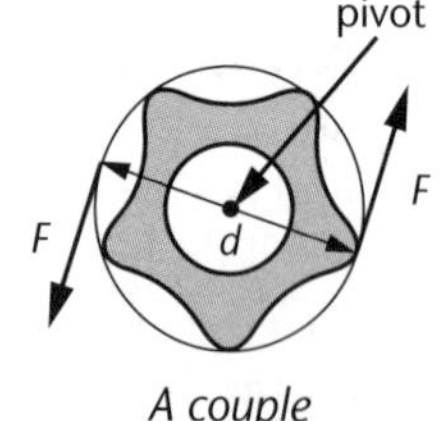

A couple

Equilibrium

Equilibrium occurs when an object is *at rest* or *moving uniformly*, as described in Newton's First Law. An object is described as being in equilibrium when *both* the resultant force is zero *and* the sum of all the torques acting on the object is zero.

Equilibrium

The resultant force acting on the object is zero; ie the vector sum of the forces acting is zero.	The sum of all the torques acting on the object is zero.
mathematically: $\Sigma \underset{\sim}{F} = 0$	mathematically: $\Sigma\tau = 0$
The acceleration is zero so the object will either be stationary or have a uniform motion.	As a result, the object will not twist or rotate. The clockwise moments equal the anticlockwise moments about any point on the object.

Many everyday situations involve objects in equilibrium.

Example H

A painter organising trestles and planks, or an engineer working on the design of a bridge, are both working with equilibrium situations.

Example I

The three forces shown act on the small object P, so that object P is in equilibrium. On a vector diagram, show how the forces add together and find the angle θ to the horizontal in which the 13.0 N force must act.

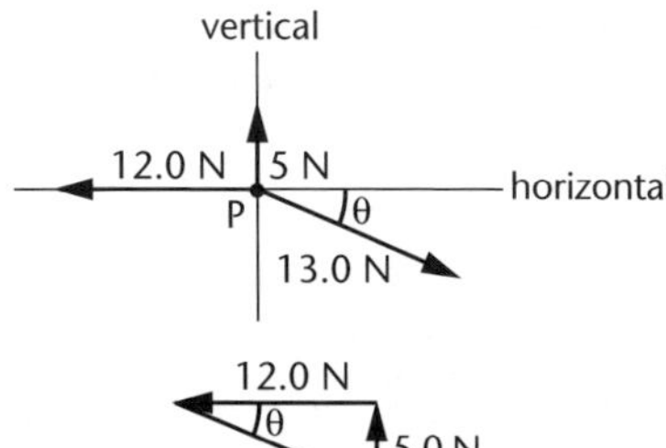

Solution:

The resultant of the three forces is zero; ie the three forces are balanced and add together to give a closed triangle.

12.0 N
θ
5.0 N
13.0 N

The vector diagram is a right-angled triangle. The angle required, θ, is:

$$\tan\theta = \frac{5.0}{12.0}$$

$$= 0.4167$$

$$\theta = 23^\circ$$

The second condition for equilibrium is also met in this example, since each of the three forces act through P, producing no torque on P.

Example J

A metre ruler of negligible mass is pivoted at P, at the 20 cm mark. A mass weighing 15 N is suspended from the 80 cm mark.

What weight, F_W, must be suspended at the 10 cm mark so that the metre ruler is balanced?

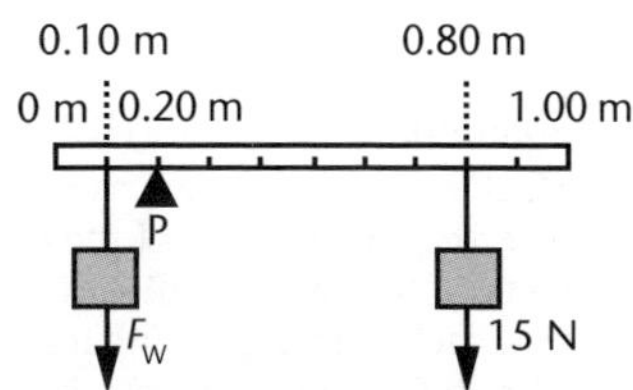

Solution:

For the ruler to be balanced, the sum of clockwise moments must equal the sum of anticlockwise moments. Taking moments about pivot point, P:

clockwise moments	$= Fd_\perp$	
	$= 15 \times (0.80 - 0.20)$	[changing cm to m]
	$= 15 \times 0.60$	[the 15 N force acts 0.60 m from P]
	$= 9.0$ N m	
anticlockwise moments	$= Fd_\perp$	
	$= F_w(0.20 - 0.10)$	[substituting $d_\perp = 0.20$ m $- 0.10$ m]
	$= 0.10 \times F_w$ N m	

since sum of clockwise moments = sum of anticlockwise moments:

$$9.0 = 0.10\ F_w$$

$$F_w = \frac{9.0}{0.10}$$

$$= 90\ \text{N}$$

The two weights add to give a total force of 105 N (90 + 15) acting downward on the pivot. There must also be a reaction force of 105 N upward on the metre ruler at the pivot. Overall, the forces acting on the metre ruler are 105 N upward and 105 N downward so that the metre ruler is in equilibrium.

The weight of an object acts down through the object's **centre of mass**. This is the point at which, if an object is suspended or pivoted, the object will balance.

Example K

The centre of mass of a uniform metre ruler is the point in the middle of the ruler.

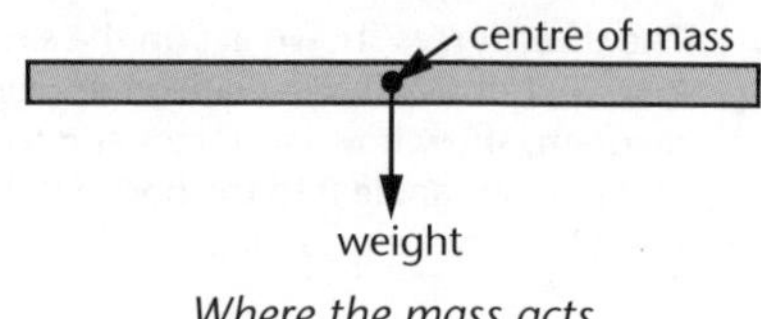

Where the mass acts

Example L

A house painter uses a uniform wooden plank 2.00 m long and of weight 300 N. It is supported at both ends by trestles A and B. The painter (of weight 800 N) stands 0.50 m from trestle A.

Calculate the size of the forces with which trestles A and B hold up the plank and painter.

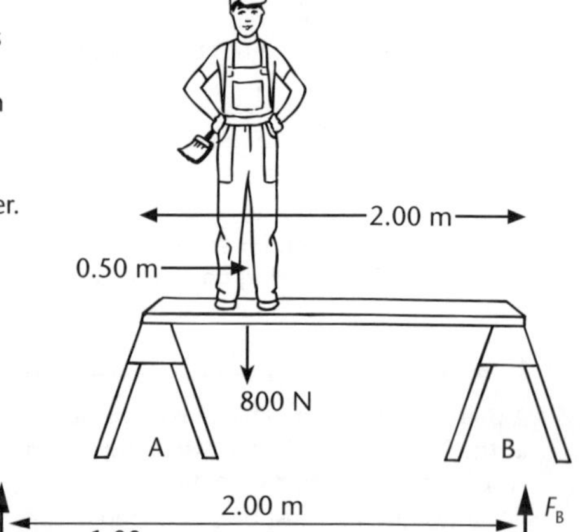

Solution:

The diagram is redrawn alongside with the weight force of the plank (acting through its centre of mass), the reaction forces on the trestles (F_A and F_B), and the painter's weight force.

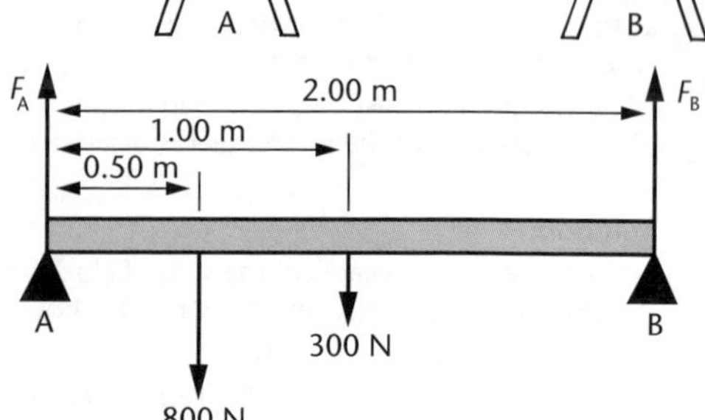

To find the size of the force on pivot B (F_B), moments are taken about pivot A. At this point, F_A produces no torque, since it acts through A.

$$
\begin{aligned}
\text{clockwise moments} &= (800 \times 0.50) + (300 \times 1.00) \\
&= 400 + 300 \\
&= 700 \text{ N m} \\
\text{anticlockwise moments} &= F_B \times 2.00 \\
\text{clockwise moments} &= \text{anticlockwise moments} \qquad \text{[plank is in equilibrium]} \\
700 &= 2.00\, F_B \\
F_B &= \frac{700}{2.00} \\
&= 350 \text{ N}
\end{aligned}
$$

F_A can be calculated by considering the other condition for equilibrium; ie the vector sum of forces = 0.

$$
\begin{aligned}
\text{sum of the forces upward} &= \text{sum of the forces downward} \\
F_A + F_B &= 800 + 300 \qquad \text{[equilibrium of forces]}
\end{aligned}
$$

$$F_A + 350 = 1100 \quad \text{[substituting } F_B = 350 \text{ N]}$$
$$F_A = 750 \text{ N}$$

The calculation could be checked by taking moments about pivot B.

It would not help to take moments about the centre of the plank or the painter, since both F_A and F_B would be present together in the one equation of moments.

Example M

Determine the mass of a retort stand using a 500 g mass, a ruler, and a laboratory stool as a pivot.

Solution:

Step One: Turn the laboratory stool upside down and slide the retort stand shaft along one of the metal braces until the retort stand balances.

The point along the retort stand shaft above the pivot is the centre of mass for the retort stand. Mark its position carefully.

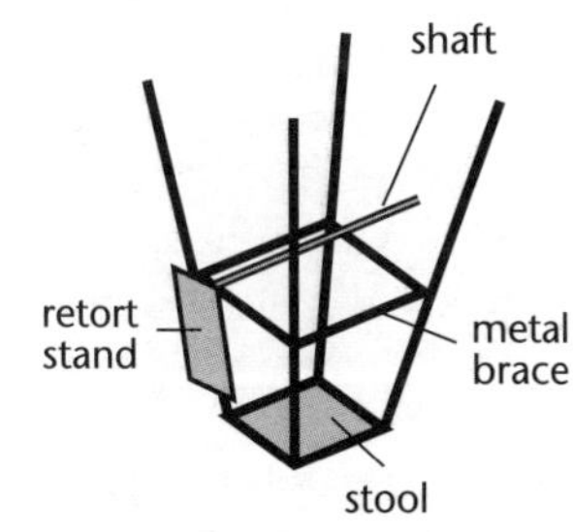

Step Two: Hang the 500 g at the top of the retort stand shaft and rebalance the retort stand. Measure the distance of the centre of mass to the new pivot position and the distance from the 500 g mass to the new pivot position.

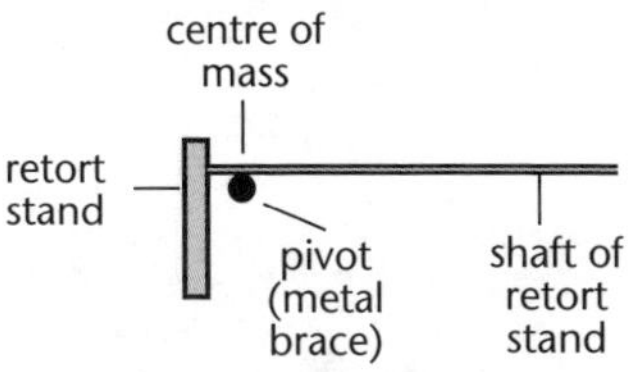

Step Three: Calculate the weight of the retort stand from the measurements taken using equilibrium of moments.

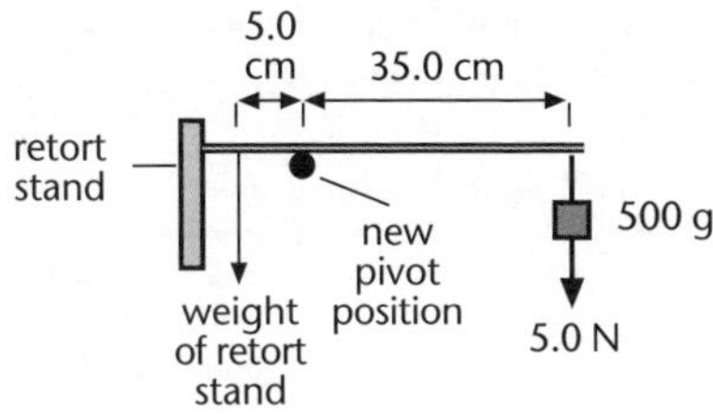

anticlockwise moment = clockwise moment

$$\text{Weight of retort stand} \times 5.0 = 5.0 \times 35.0$$

$$\text{Weight of retort stand} = \frac{5.0 \times 35.0}{5.0}$$

$$= 35.0 \text{ N}$$

Finally, the mass of the retort stand can be determined from W = mg.

Substituting gives $35.0 = m \times 10$

$$m = \frac{35.0}{10}$$

$$= 3.5 \text{ kg}$$

Distances in cm are used rather than converting them to metres. This works since the units cancel in the calculations.

Unit 11.3 Activity 4B: Force, motion and equilibrium

1. Three forces act on an object so that there is *no* resultant force on it.
   a. Draw a vector diagram showing how these three forces add to give a zero resultant.
   b. Use your diagram, or otherwise, to determine the size of θ (to the nearest degree).

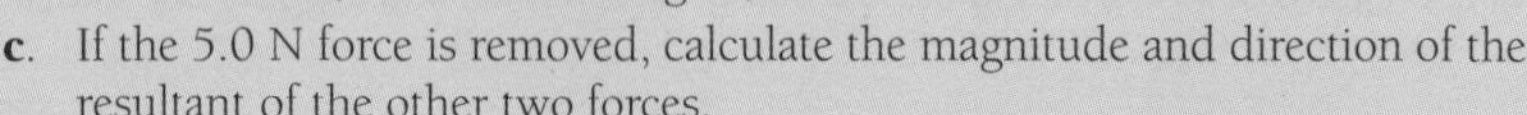

   c. If the 5.0 N force is removed, calculate the magnitude and direction of the resultant of the other two forces.
2. An object of mass 3.0 kg is held by two strings attached to a roof at P and Q. The strings are each inclined to the vertical at 60°.
   a. What is the force of gravity on the 3.0 kg mass? (g = 10 m s^{-2})

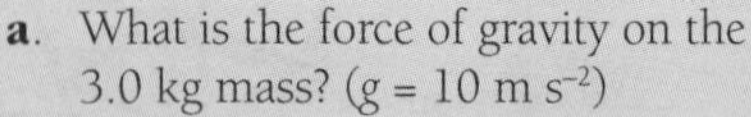

   b. Draw a labelled vector diagram to show how the forces add to give *zero* resultant force.

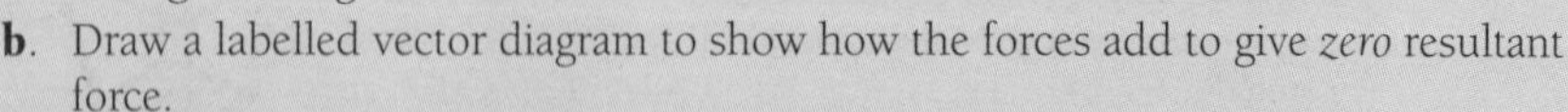

   c. Calculate the value of T, the tension each the string that keeps the object stationary.
3. A 3.0 kg mass is suspended from a ceiling by a light cord. It is pulled horizontally by a 40 N force as shown in the diagram and equilibrium is gained when θ = 53°.
   a. Copy the diagram and draw in all the forces that act on the 3.0 kg mass.
   b. Draw a vector diagram of the forces acting on the 3.0 kg mass.
   c. Calculate the weight force, F_W, of the mass.
   d. Calculate the tension, T, in the light cord.
4. The diagram shows a stationary upright bicycle, with a weight force W newtons hanging from a pedal. A spring balance measures the forward force F newtons which that weight produces.

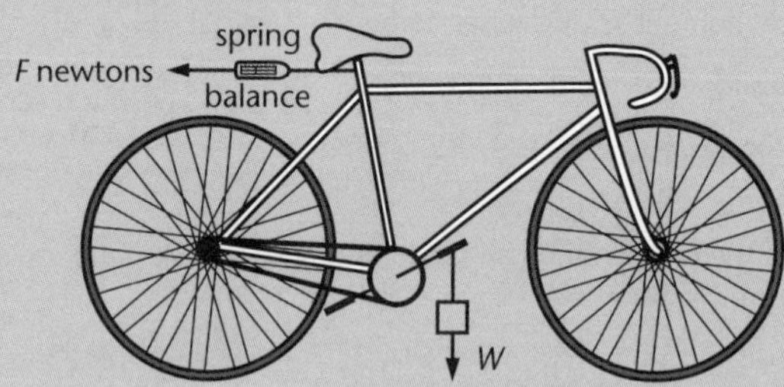

Draw a sketch diagram that shows the position of the pedal for which there will be a maximum measured value of F, for a given W.

5. A cyclist exerts a force of 300 N on pedal 1. How much force should the cyclist exert to produce the same turning effect on pedal 2?

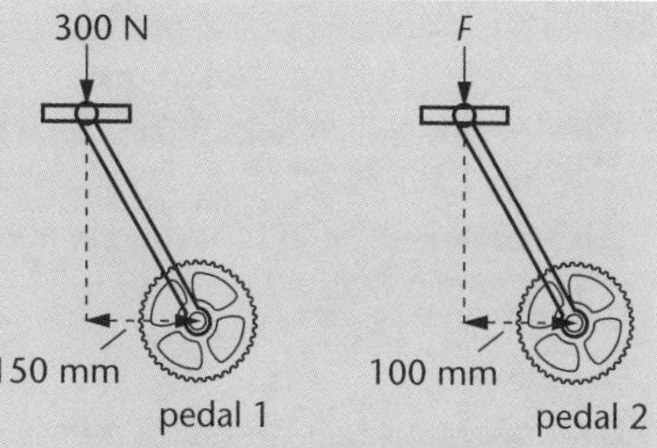

6. A worker wants to lift a crate 40 cm upwards using a 3.0 metre beam (the mass of which can be ignored). The pivot is placed 50 cm from the centre of the crate.
   a. Calculate the force, *F*, needed to lift the crate.
   b. Give one practical reason why this method would not be suitable to lift the crate 40 cm upwards.

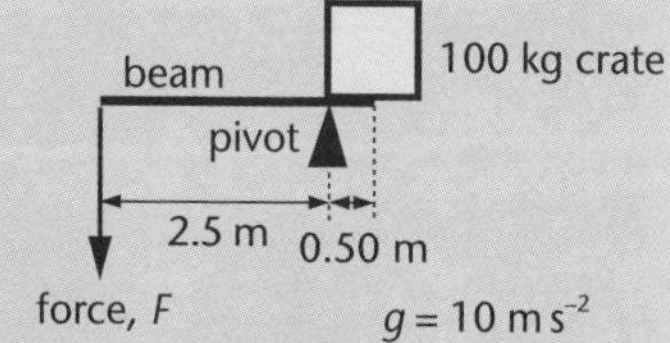

7. AE is a light rod 1.0 metre long which is free to move and rotate. A, B, C, D, E are five points equally spaced along the rod. There is a 5 N force at A and a 10 N force at E. The mass of the rod may be ignored. State where a pivot should be placed to keep the rod stationary.

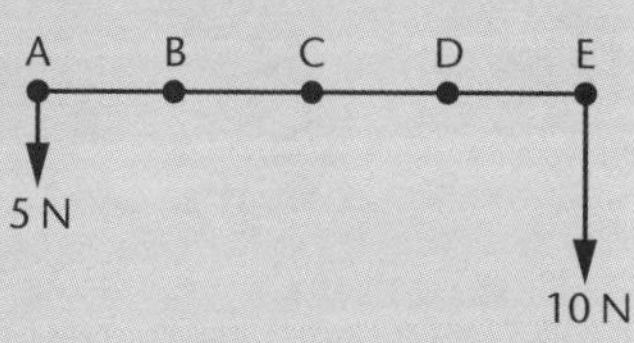

8. Four parallel forces are applied to a light rod at points X (two forces), Y and Z. Determine:
   a. The resultant of the four forces.
   b. Their total torque about X.
   c. Their total torque about Y.
   d. Their total torque about Z.
   e. Whether the rod is in equilibrium under their action.

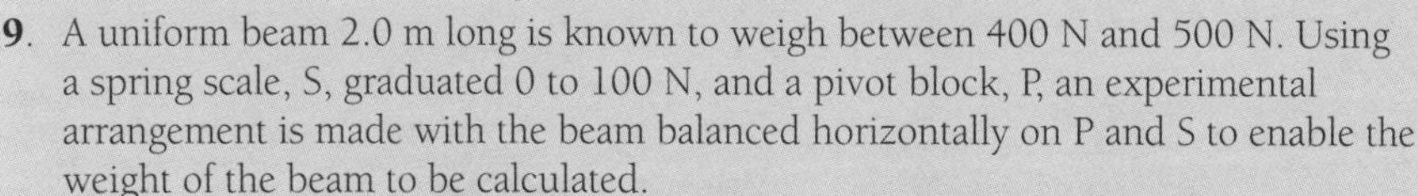

9. A uniform beam 2.0 m long is known to weigh between 400 N and 500 N. Using a spring scale, S, graduated 0 to 100 N, and a pivot block, P, an experimental arrangement is made with the beam balanced horizontally on P and S to enable the weight of the beam to be calculated.
   a. Draw a diagram of the arrangement of the beam, S, and P, that would make this possible.
   b. If in the correct arrangement of the equipment, the spring scale S reads 85 N when P is 0.75 m from one end, determine the correct weight of the beam.

10. PQ represents a uniform beam of mass 2.0 kg supported at ends P and Q. An object of mass 4.0 kg is placed so that it is 0.5 metres from Q. The beam is 2.0 m long.

a. By how much does the force on Q exceed the force on P?

b. Calculate the total anticlockwise torque (moment) about P.

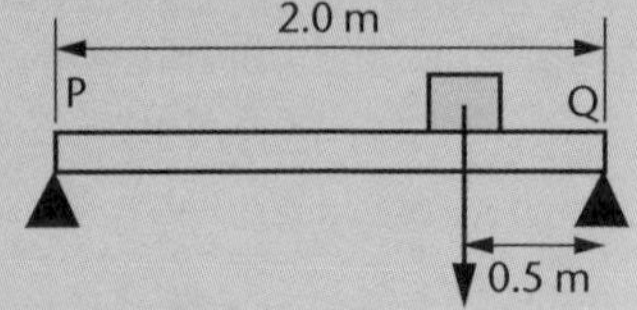

11. A uniform beam in equilibrium is suspended by a cord at X which is 2.0 m from one end of the beam and 3.0 m from the other end. Masses provide forces on the ends of the beam of 28 N and 10 N.

Determine the weight force of the beam.

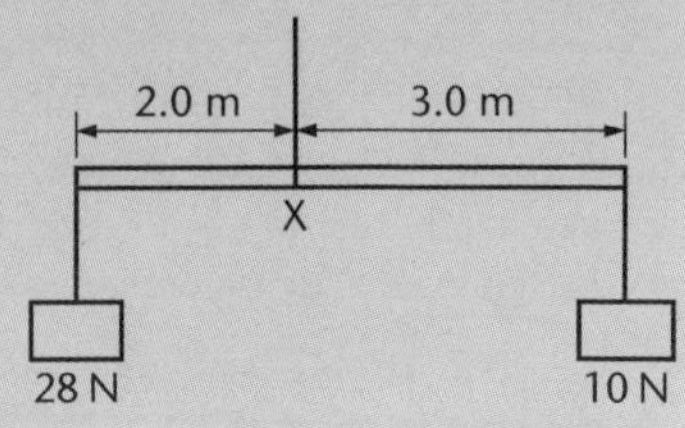

12. A student balances a 1.5 kg broom by placing her finger 1.4 m from the end of the broom handle.

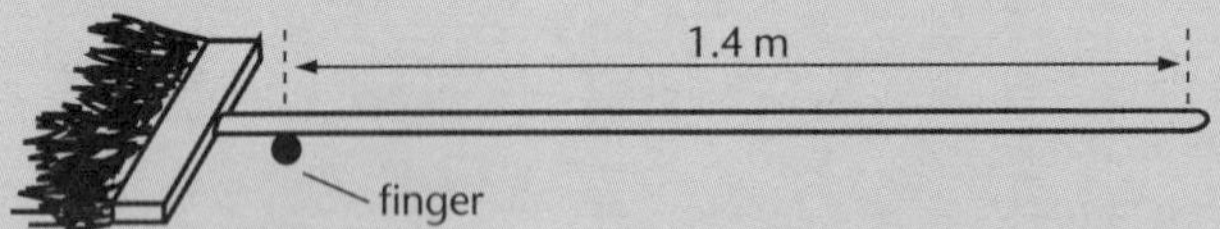

How far from the end of the broom handle would she have to balance the broom if she hung a 400 g mass from the end of the broom handle?

13. Suppose there are three forces acting on a rod XY which is pivoted at X, as shown in the figure. Find the counter-clockwise moment of each force about the pivot or axis of rotation.

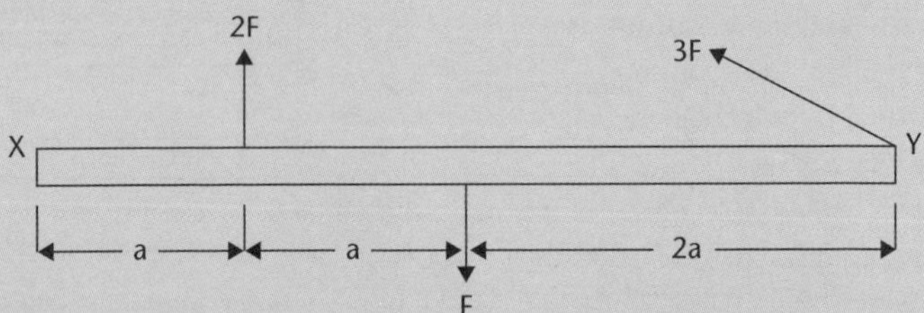

14. Given a rod ABC on which a force F is acting at an angle θ, as shown, if the moment of F about A is 3 Nm clockwise, and about B is 1 Nm clockwise, and if AB = BC = 1 m; find the moment of F about C.

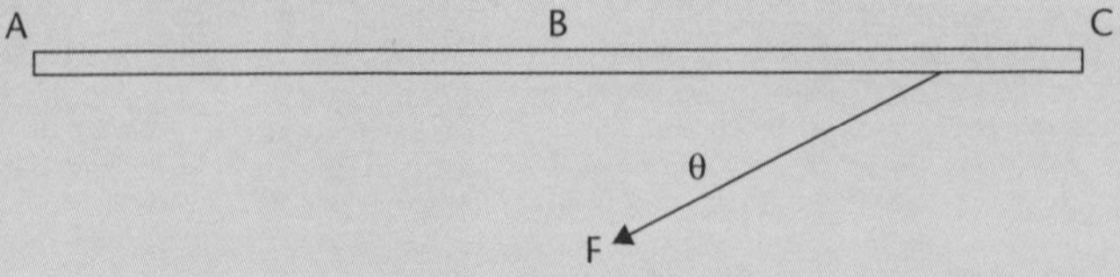

15. In the following figure, XY is a see-saw of 4 metres length. It is pivoted at its midpoint Z. Calculate the anticlockwise moment, about a horizontal axis passing through Z, of a child of weight 230 Newton who sits:

a. at X

b. 0.5 m from X

c. at Y

d. at Z.

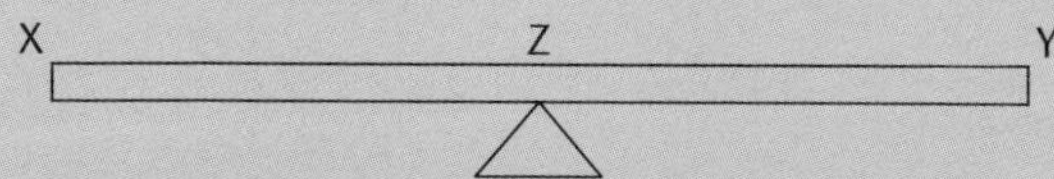

16. Find the moment of the 10 N force about the axis through O and perpendicular to the page in each of the following situations.

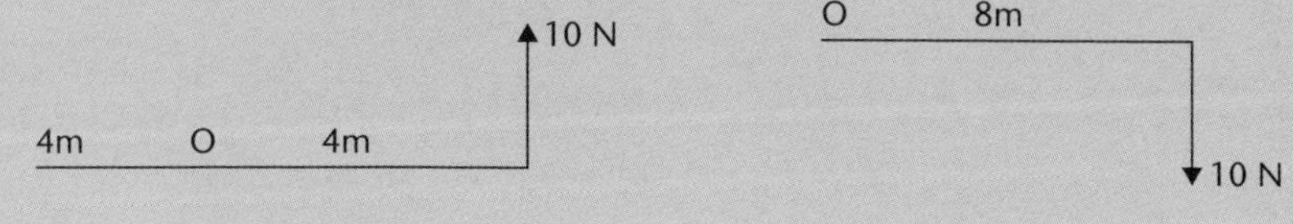

Unit 11.3 Force and Motion (Dynamics)

Topic 5: Momentum and impulse

In this Topic we look at momentum and impulse (Syllabus p.14):

- Momentum.
- Change in momentum and impulse.
- Impulse and force.
- Conservation of momentum.

Introduction

Momentum is a quantity which involves the *velocity* and *mass* of an object. Momentum can be thought of as the amount of 'oomph' a moving object has. The more 'oomph' the object has, the harder it is to stop, and the greater its momentum. Another phrase commonly used to describe momentum is 'quantity of motion'.

Example A

A golf ball is easier to catch than a hockey ball if they are both thrown at the same speed, because the golf ball has *less mass and thus less momentum* than the hockey ball. However, if two hockey balls are thrown at different speeds, the faster one is harder to catch because it has more 'oomph', ie it has *a greater momentum because it is travelling faster.*

Momentum is a useful quantity to consider when **collisions** between different objects occur, or when **explosions** break objects into pieces or push objects apart.

Example B

A controlled 'explosion' can be simulated using two people standing on skateboards who push each other apart– they will end up moving *away from each other* at different speeds. Assuming friction is constant for both people, the person with *greater mass will move with a slower speed.* In fact, if each person's velocity could be measured, the product of mass and velocity for each person immediately after their 'explosion' should be equal and opposite.

Momentum (symbol p) is calculated from the mass, m, and velocity, v, using the formula:

$$p = mv$$

The SI unit for momentum is kg m s^{-1} (the SI unit for mass multiplied by the SI unit for velocity).

Example C

a. A 30 g golf ball travelling at 10 m s^{-1} has momentum:

$p = mv$

$= 0.030 \times 10$ [changing 30 g to 0.030 kg]

$= 0.30$ kg m s^{-1}

b. A shopping trolley of mass 24 kg moving at 0.75 m s^{-1} has momentum:

$p = mv$

$= 24 \times 0.75$

$= 18$ kg m s^{-1}

c. A ship of mass 30 000 tonne moving at 0.2 m s^{-1} has momentum:

$p = mv$

$= 3 \times 10^7 \times 0.2$ [changing 30 000 tonne to 3×10^7 kg]

$= 6 \times 10^6$ kg m s^{-1}

Momentum is a vector quantity. The direction of an object's momentum is always the same as the direction of the object's velocity.

Change in momentum

When a force acts on an object and changes its motion, the object's momentum will also change. The change in momentum ($\Delta \underset{\sim}{p}$) is calculated from the formula:

change in momentum = final momentum – initial momentum

$$\Delta \underset{\sim}{p} = \underset{\sim}{p}_f - \underset{\sim}{p}_i$$

Subscripts are used (i for initial, f for final) with momentum $\underset{\sim}{p}$.

The change in momentum is a vector quantity because it is calculated from a subtraction of two vectors $\underset{\sim}{p}_f$ and $\underset{\sim}{p}_i$.

Example D

A cricket ball of mass 500 g is bowled at a speed of 25 m s^{-1}. A batter hits the ball directly back to the bowler at 35 m s^{-1}. Calculate the ball's change in momentum.

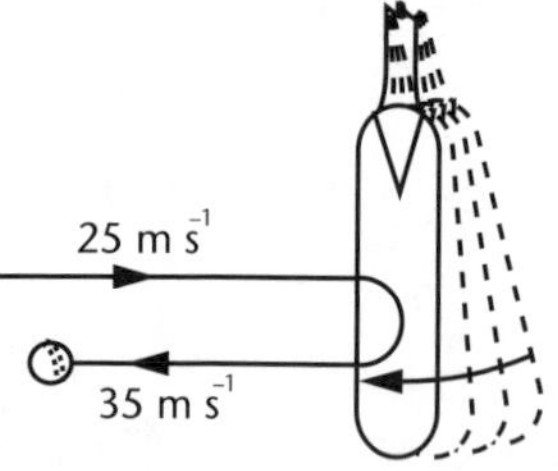

Solution:

Taking the direction away from the cricketer (to the left) to be the positive direction, the cricket ball's initial velocity is –25 m s^{-1} and its final velocity is 35 m s^{-1}.

initial momentum is $p_i = 0.50 \times -25$

$= -12.5$ kg m s^{-1}

final momentum is $p_f = 0.50 \times 35$

$= 17.5$ kg m s^{-1}

change in momentum is $\Delta p = p_f - p_i$

$= 17.5 - -12.5$

$= 30$ kg m s^{-1} away from the cricketer.

The action of the cricket bat reduces the ball's momentum to zero momentarily and then gives the ball momentum in the opposite direction. Thus the change in momentum is larger than when just stopping the ball.

Impulse

Example D shows that the cricket bat exerts a force on the cricket ball and changes its momentum. Clearly force must play a part in changing an object's momentum. When a resultant force acts on an object for a period of time, it accelerates the object and changes its momentum.

Example E

A girl throws a 0.10 kg ball with a steady force of 6.0 N for 0.40 seconds. What momentum does the ball gain, assuming the ball starts from rest?

Solution:

The ball's acceleration is: $a = \frac{F}{m}$ [$F = ma$ rearranged]

$= \frac{6.0}{0.10}$

$= 60 \text{ m s}^{-2}$

The ball accelerates for 0.40 s so its final velocity as it leaves the girl's hand is:

$v_f = v_i + at$

$= 0 + 60 \times 0.40$

$= 24 \text{ m s}^{-1}$

The ball's change in momentum is:

$\Delta p = p_f - p_i$

$= 0.10 \times 24 - 0.10 \times 0$

$= 2.4 \text{ kg m s}^{-1}$

It is interesting to note that, in Example E, the product of force and time, $F \times t$,

gives $F \times t = 6.0 \times 0.40$

$= 2.4$ N s (newton seconds)

The connection between this and the change in momentum can be shown by examining Newton's second law of motion.

$F = ma$ can be expressed as:

$F = m\frac{\Delta v}{\Delta t}$ [since acceleration is defined as $a = \frac{\Delta v}{\Delta t}$]

ie $F = m\frac{(v_f - v_i)}{\Delta t}$ [since $\Delta v = v_f - v_i$]

Rearranging the equation gives:

$F\Delta t = mv_f - mv_i$

and finally, $F\Delta t = \Delta p$ [since $\Delta p = p_f - p_i = mv_f - mv_i$]

The right-hand side of the equation is the change in momentum and the left-hand side is the product of the two factors responsible for causing the change in momentum.
The product, $F\Delta t$, is known as **impulse** and its units are N s (newton seconds).

Impulse equals the change in momentum:

$$F\Delta t = \Delta p$$

Newton's second law of motion, $F = ma$, can be written as $F = \frac{\Delta p}{\Delta t}$,

ie the resultant force acting on an object is equal to the rate of change in the object's momentum. This was the form in which Newton originally wrote his second law of motion.

The units for momentum are kg m s^{-1}, while the units for impulse are N s. (It can be shown (as an exercise) that these two sets of units are identical.)

Example F

How long must a 300 kg satellite, in orbit, fire its thruster rocket in order to increase its speed from 5 000 m s^{-1} to 6 000 m s^{-1}? The force exerted by the thruster when firing is 1 500 N.

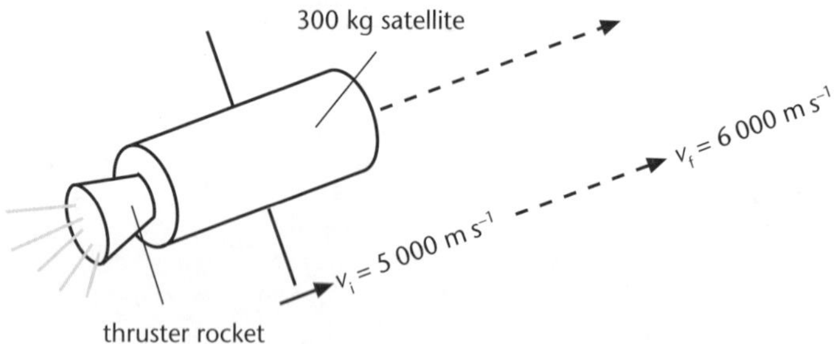

Solution:

The desired change in momentum is:

$$\begin{aligned}\Delta p &= mv_f - mv_i \\ \Delta p &= m(v_f - v_i) \\ &= 300(6\,000 - 5\,000) \\ &= 300\,000 \text{ kg m s}^{-1}\end{aligned}$$

The impulse needed to cause this change in momentum is:

$$\begin{aligned}F\Delta t &= \Delta p \\ 1\,500 \times \Delta t &= 300\,000 \\ \Delta t &= \frac{300\,000}{1\,500} \\ \Delta t &= 200 \text{ s}\end{aligned}$$

The thruster rocket must be fired for 200 s.

Example G

A 0.50 kg softball is pitched at 30 m s^{-1}; a batter hits it at 30 m s^{-1} at an angle of 45° to the pitch direction. If the bat and ball were in contact for 0.20 s, find the average force the batter exerted to make this hit.

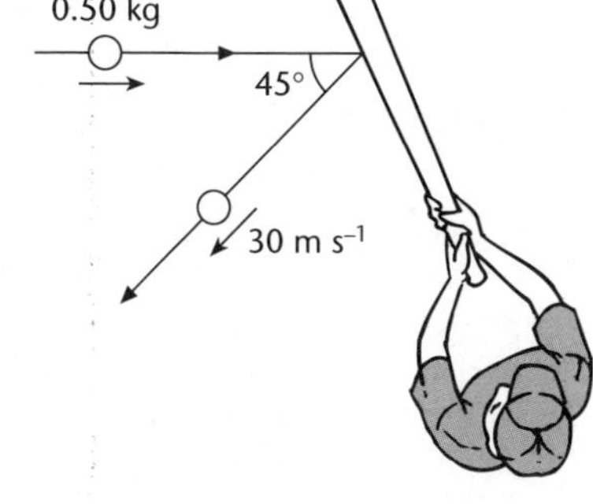

Solution:

$p_i = 0.50 \times 30$ [substituting into $p = mv$]

$= 15$ kg m s^{-1} →

$p_f = 15$ kg m s^{-1} ↙

The desired change in momentum is:

$\Delta p = p_f - p_i$

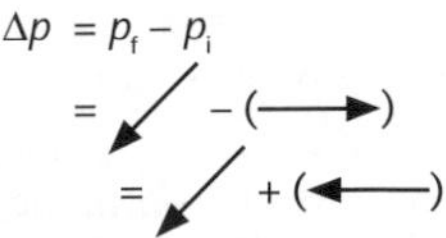

which can be accurately drawn to scale. The resultant vector can be measured off the scale diagram as 28 kg m s^{-1} at 22.5° to the pitch direction.

22.5°
22.5°
Δp
$p_f = 15$ kg m s^{-1}
$p_i = 15$ kg m s^{-1}
Scale
10 kg m s^{-1}

Alternatively, using trigonometry:

$$\frac{\frac{\Delta p}{2}}{15} = \cos 22.5°$$

$\Delta p = 30 \cos 22.5°$ [rearranging]

$= 28$ kg m s^{-1} at 22.5° to pitch direction

$F\Delta t = \Delta p$ [impulse = change in momentum]

$F \times 0.20 = 28$

$F = 140$ N at 22.5° to the pitch direction

The force exerted is in the same direction as the impulse direction. In this case, the force exerted by the softball bat when in contact with the softball will vary. The force calculated above will be the *average* force exerted.

Unit 11.3 Activity 5A: Momentum and impulse

1. Calculate the momentum, in SI units, of the object described in each of the following problems:

a. A 0.50 kg bird flying at 7.0 m s^{-1}.

b. A 20 g snail moving at 1.0 mm s^{-1}.

c. An 85 g tennis ball moving at 30 m s^{-1}.

d. A 1.5 tonne utility van speeding at 120 km h^{-1}.

2. **a**. A lump of plasticene of mass 100 g falls and hits the ground at 4.0 m s^{-1} and stops. Calculate its change in momentum.

b. A 200 g ball hits the ground at 5.0 m s^{-1} and rebounds at an initial speed of 3.0 m s^{-1} upwards. Calculate the ball's change in momentum.

3. A ball of mass 3.0 kg travelling at 2.0 m s^{-1} hits a wall at an angle and rebounds at 2.0 m s^{-1}, as shown in the diagram. Draw a vector diagram to show the *change* in momentum of the ball.

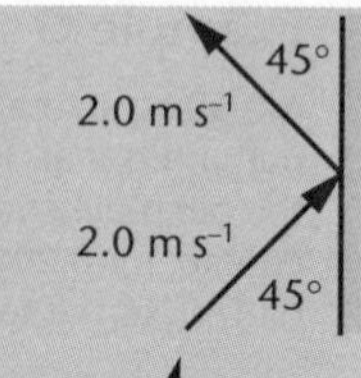

4. A ball of mass 2.0 kg strikes an irregular surface at 20° to the horizontal and rebounds at an angle of 70° to the horizontal, as shown in the diagram. The initial and final speeds are 6.0 m s^{-1} and 8.0 m s^{-1} respectively.

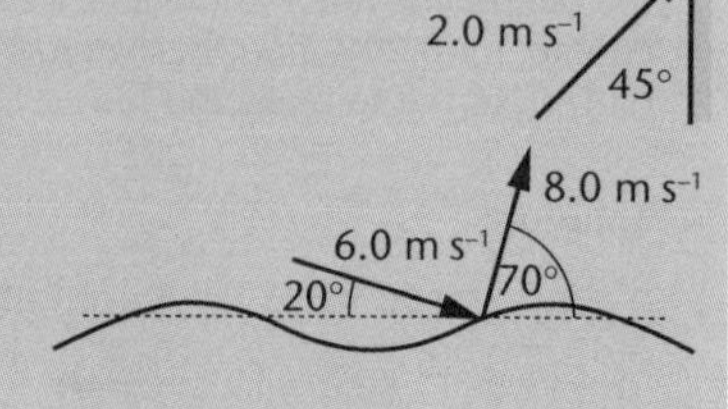

   a. Draw a vector diagram to show the ball's change in momentum.
   b. Determine the magnitude of the ball's change in momentum.
5. Show that the units of impulse, N s, are equivalent to the units of momentum, kg m s^{-1}.
6. A driver tries to slow down a runaway car by pushing against its motion. If the car's mass is 1 800 kg and its initial velocity is 2.0 m s^{-1}, will the driver manage to stop the car if he pushes with a resultant force of 500 N for 5.0 seconds?
7. The graph shows a record of the force applied by a racquet to a squash ball of mass 25 g, initially travelling towards the racquet at 20 m s^{-1}.

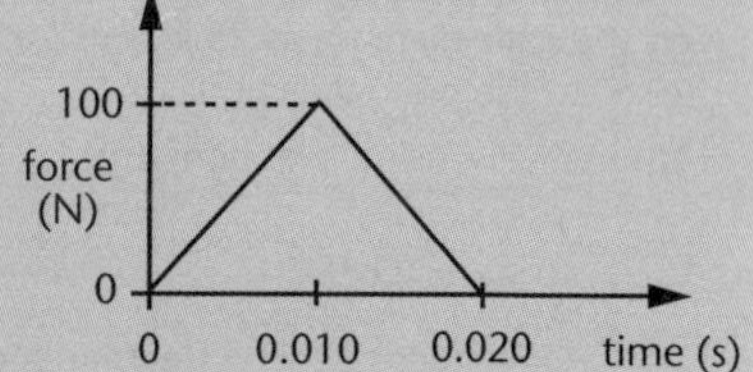

   a. Determine the impulse applied to the ball by calculating the area under the graph.
   b. Calculate the squash ball's final velocity (as it leaves the racquet).
8. Sam sprays water from a hose onto a window-pane. Measurements taken show that the water hits the window at 30 m s^{-1} and then virtually stops (and then runs down the window).

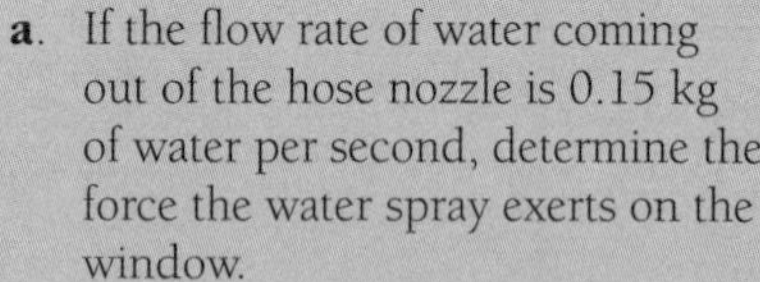

   a. If the flow rate of water coming out of the hose nozzle is 0.15 kg of water per second, determine the force the water spray exerts on the window.
   b. Would this force be likely to break the window?

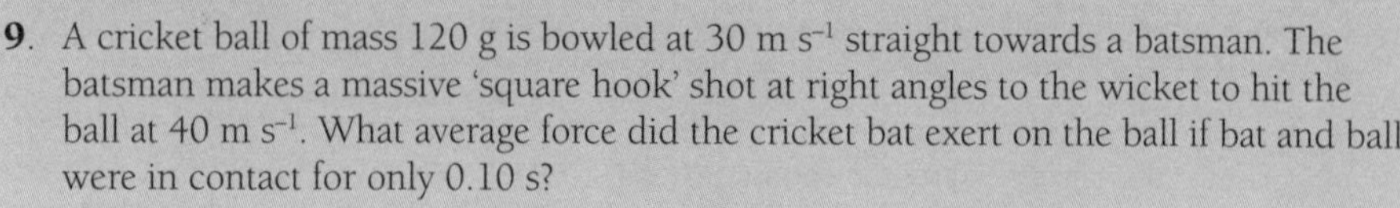

9. A cricket ball of mass 120 g is bowled at 30 m s^{-1} straight towards a batsman. The batsman makes a massive 'square hook' shot at right angles to the wicket to hit the ball at 40 m s^{-1}. What average force did the cricket bat exert on the ball if bat and ball were in contact for only 0.10 s?

Conservation of momentum

Momentum is **conserved** in collisions between objects and in explosions; ie the total momentum of two or more objects *before* a collision or explosion equals their total momentum *after* the collision or explosion.

The above relationship applies only when no resultant *external force* is present, such as gravity and friction. Momentum will be conserved in situations where these forces do not apply or where they have been balanced by reaction forces.

Conservation of momentum is the only way to solve problems where there are collisions or explosions. **Kinetic energy** is not usually conserved in these interactions and cannot be used to solve them.

Conservation of momentum in a straight line

Example H

A shunting locomotive of mass 5.0×10^4 kg travelling at 3.0 m s^{-1} bumps into and locks onto a stationary carriage of mass 3.0×10^4 kg. What combined speed do the locomotive and carriage have *after* the collision?

Solution:

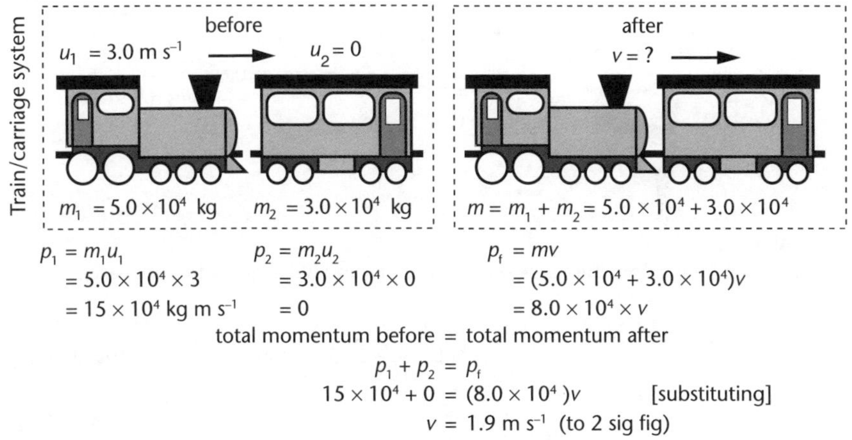

$$p_1 = m_1u_1 = 5.0 \times 10^4 \times 3 = 15 \times 10^4 \text{ kg m s}^{-1}$$

$$p_2 = m_2u_2 = 3.0 \times 10^4 \times 0 = 0$$

$$p_f = mv = (5.0 \times 10^4 + 3.0 \times 10^4)v = 8.0 \times 10^4 \times v$$

total momentum before = total momentum after

$$p_1 + p_2 = p_f$$

$$15 \times 10^4 + 0 = (8.0 \times 10^4)v \quad \text{[substituting]}$$

$$v = 1.9 \text{ m s}^{-1} \text{ (to 2 sig fig)}$$

Problems involving conservation of momentum in a straight line for two objects can be solved using the formula:

$$m_1u_1 + m_2u_2 = m_1v_1 + m_2v_2$$

where m_1 and m_2 are the masses of objects 1 and 2, u_1 and u_2 are the velocities of objects 1 and 2 *before*, and v_1 and v_2 are the velocities of objects 1 and 2 *after* the collision or explosion.

The following example shows how to solve a collision problem using the conservation of momentum in a straight-line formula.

Example I

In Example H:

$$\text{total momentum before} = \text{total momentum after}$$
$$m_1u_1 + m_2u_2 = m_1v_1 + m_2v_2$$
$$(5.0 \times 10^4 \times 3.0) + (3.0 \times 10^4 \times 0) = (5.0 \times 10^4 \times v) + (3.0 \times 10^4 \times v)$$
$$5.0 \times 10^4 \times 3.0 = (5.0 \times 10^4 + 3.0 \times 10^4)v$$
$$15 \times 10^4 = 8.0 \times 10^4 \times v$$
$$v = \frac{15}{8.0}$$
$$= 1.9 \text{ m s}^{-1}$$

The following Example shows how to solve an explosion problem using the conservation of momentum in a straight-line formula.

Example J

A rifle of mass 4.0 kg fires a bullet of mass 20 g at a speed of 400 m s^{-1}. What is the recoil speed of the rifle?

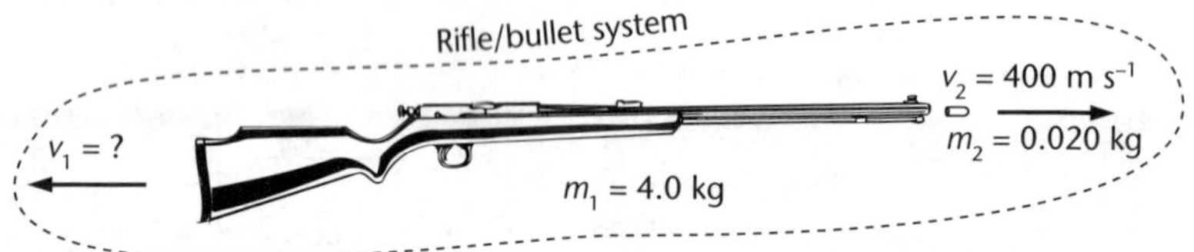

Solution:

Initially, the total momentum is zero (since the rifle and bullet are both stationary).
Thus $m_1u_1 + m_2u_2 = 0$.

Using $m_1u_1 + m_2u_2 = m_1v_1 + m_2v_2$ gives:

$$0 = 4.0 \times v_1 + 0.020 \times 400 \quad \text{[substituting]}$$
$$0 = 4.0 \times v_1 + 8.0$$
$$v_1 = \frac{-8.0}{4.0} \quad \text{[rearranging]}$$
$$= -2.0 \text{ m s}^{-1}$$

The negative sign in –2.0 m s^{-1} shows that the rifle moves *backwards*.

The bullet, being much *lighter* than the rifle, moves much *faster* than the rifle (as experience confirms).

The velocities calculated in Examples I and J are velocities *immediately after* the event. Friction may quickly reduce these velocities.

In Examples H and J there are dashed lines (labelled Train/carriage system and Rifle/bullet system respectively). These enclose the bodies that are interacting and set the areas within which momentum conservation is considered.

Conservation of momentum in two dimensions

Momentum is conserved in *all directions*. If one object moves at an angle to another object, then that angle must be taken into account. A momentum vector diagram can be drawn for problems involving momentum in two dimensions.

Example K

A bowling ball, A, of mass 1.5 kg and travelling to the right at 3.0 m s^{-1} hits an identical ball, B, which is stationary. Ball A moves off at 2.0 m s^{-1} at an angle of 90° to the direction in which B moves.

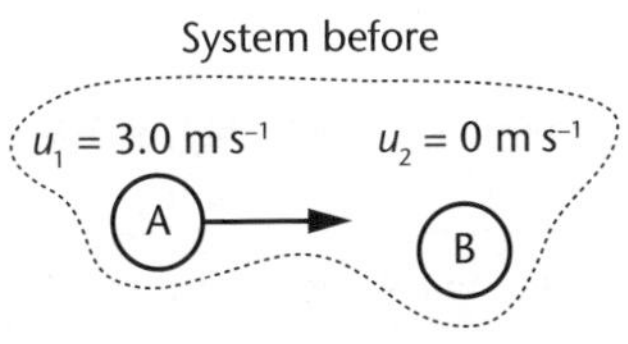

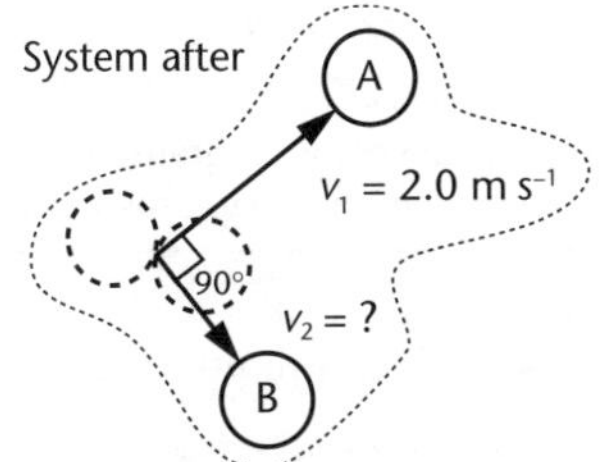

a. Find the speed of ball B after the collision.
b. Find the direction of ball A after the collision.
c. Find the direction of ball B after the collision.

Solution:

a. The total momentum before the collision is $1.5 \times 3.0 = 4.5$ kg m s^{-1} to the right.
This is equal to the total momentum after the collision since momentum is conserved.

The vector diagram shows the momentum vectors of the two balls after the collision added together as vectors. They should be equal to the momentum before the collision.

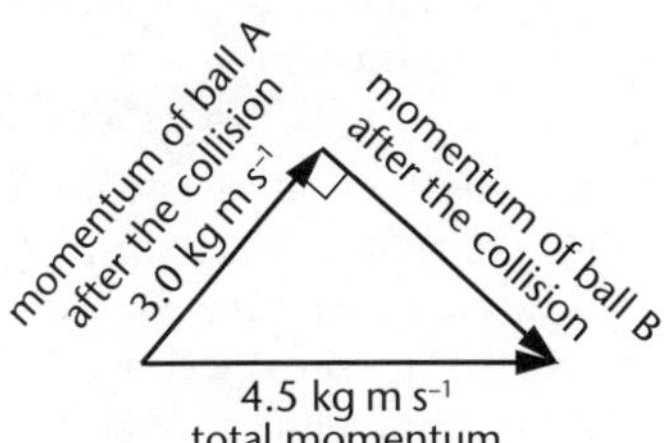

The magnitude of the momentum of ball B is:

$p_B = \sqrt{4.5^2 - 3.0^2}$ [using Pythagoras]

$= 3.35$ kg m s^{-1}

and the speed of ball B is: $v = \frac{p}{m}$ [rearranging $p = mv$]

$= \frac{3.35}{1.5}$

$= 2.2$ m s^{-1} [2 sig figs]

b. Ball A, after the collision, moves at angle θ anticlockwise from the 'rightwards' direction.

$$\sin\theta = \frac{3.35}{4.5}\left[\frac{\text{opposite side}}{\text{hypotenuse}}\right]$$

$$\theta = 48° \text{ (approximately)}$$

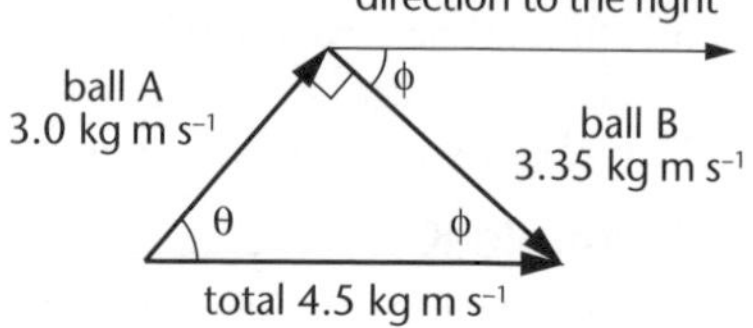

c. Ball B moves at an angle ϕ clockwise from the 'rightwards' direction but $\phi = 90 - \theta$, as ϕ is also an internal angle of the right-angled triangle.

So, $\phi = 90° - 48°$
$= 42°$

In this example, a vector diagram of *velocities* would form a *similar* triangle to the momentum triangle shown. This only happens when the two objects have the *same mass*.

When two colliding (or exploding) objects have *unequal* mass, momentum will be conserved but velocities will not be conserved.

Identify the system and *always draw momentum vector diagrams*.

Unit 11.3 Activity 5B: Collisions and explosions

1. A Morris Minor car (mass 750 kg) is travelling at 30 m s^{-1} and collides head-on with a Mercedes Benz car (mass 1 600 kg) travelling at 20 m s^{-1} in the opposite direction.

  a. Calculate the magnitude of the momentum of each car before crashing.

  b. If the two cars lock together in the crash, in which direction will they be moving immediately after the collision?

  c. **i.** Calculate the initial speed of the combined wreck immediately after the collision.

  ii. Why does this speed quickly reduce to zero?

2. Two balls A and B are about to collide and the momentum of each ball is shown in the diagram. After the collision, both balls move in opposite directions, ball A having a momentum of magnitude 1.5 kg m s^{-1}.

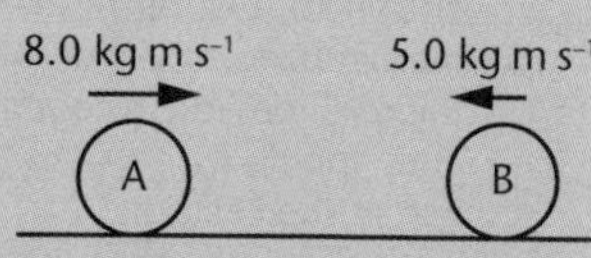

Calculate the magnitude of the momentum of ball B after the collision.

3. Four identical railway trucks, each of mass 2 000 kg, are coupled together and are at rest on a smooth horizontal track. A fifth truck of mass 4 000 kg and moving at 5.0 m s^{-1} collides and joins with the stationary trucks. Calculate the speed of the five trucks after impact.

4. **a.** A trolley rolls along a bench at a constant speed. As it passes a student, she drops a block of wood, at time t_1, vertically down onto the trolley. Draw a graph of the speed of the trolley against time.

  b. A railway cart full of sand is rolling along a horizontal frictionless surface at constant speed. After time t_1, a hole appears in the bottom of the cart and the sand falls out. Draw a graph that represents the speed of the cart against time.

5. Bill and Ted are skating on an ice rink. They stop and then push each other apart. Bill, whose mass is 58 kg, ends up moving to the left at 3.0 m s^{-1}, and Ted (mass 87 kg), ends up moving to the right at 2.0 m s^{-1}.

a. Show that momentum is conserved in this action.

b. In the action mentioned above, Bill pushed Ted for 0.30 s.

  i. Use the concept of impulse to calculate the size of the average force that Bill exerted on Ted.

  ii. If Bill pushed Ted for 0.30 s, how long did Ted push Bill?

  iii. Determine the average force that Ted exerted on Bill.

  iv. Compare the average force of Bill on Ted and Ted on Bill and relate this to Newton's 3rd law of motion.

6. A stationary object explodes into three unequal pieces. One piece of mass 1.0 kg moves north at 40 m s^{-1}, and another piece of mass 2.0 kg moves east at 20 m s^{-1}. In which direction would the third piece move?

7. A 6.0 kg ball travelling east at 2.0 m s^{-1} collides with a ball of mass 4.0 kg travelling at 3.0 m s^{-1} in a direction S 30° W (bearing 210°), as shown in the diagram. After colliding they stick together. Using a vector diagram, or otherwise, calculate the:

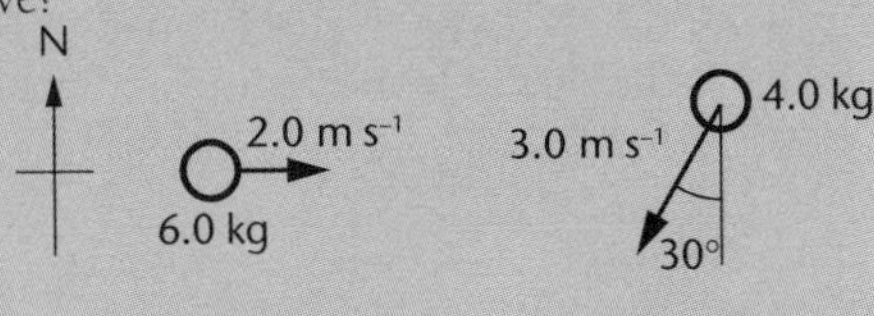

a. Total momentum of the balls before the collision.

b. Magnitude and direction of the momentum after the collision.

c. Speed of the balls after the collision.

8. A bomb of mass 8.0 kg is moving horizontally when it suddenly explodes into two pieces, of mass 6.0 kg and 2.0 kg respectively, which move off at equal angles as shown.

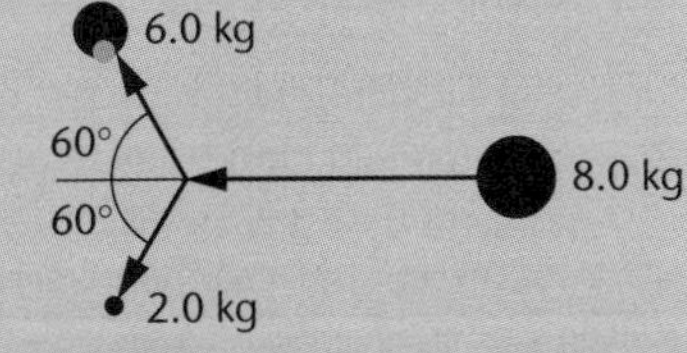

a. For this explosion, which of the two pieces must have the greater speed? Explain your answer.

b. If the speed of the larger piece is v, what is the speed of the smaller piece in terms of v?

c. If the speed of the original bomb was 20 m s^{-1}, calculate the value of v.

9. A firework explodes into three unequal fragments in the sky. One fragment, mass 0.20 kg, flies off due north at 20 m s^{-1}. A second fragment, mass 0.1 kg, flies off due east at 30 m s^{-1}. Find the velocity of the third fragment if its mass is 0.050 kg. (Assume that the firework is at the top of its vertical flight in the sky and so is initially stationary when it explodes.)

10. A frictionless ice hockey puck A (mass 2.0 kg) collides at 2.0 m s^{-1} with an identical puck B moving at 4.0 m s^{-1} as shown in the left-hand side diagram below. After the collision, both pucks move off, as shown on the right-hand side diagram.

   a. Draw a vector that represents the total momentum before the collision.
   b. Draw a vector that represents the total momentum after the collision.
   c. Draw a vector diagram that would be used to find the change in momentum, $\Delta \underset{\sim}{p}$, of puck B.
   d. Determine the change in momentum of puck A.

11. The following figure shows an overhead view of a tennis ball bouncing from a wall without any change in its speed. Consider the change of momentum in the ball's linear momentum.
   a. Is the change of momentum (component on x) positive, negative or zero?
   b. Is the change of momentum (component of y) positive, negative or zero?
   c. What is the direction of the change of momentum?

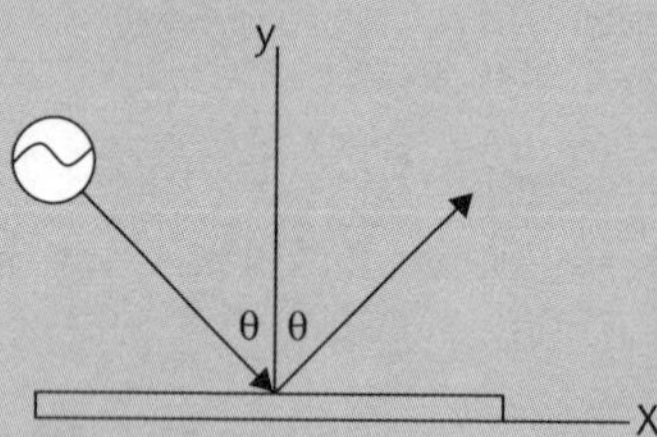

(From *Fundamentals of Physics* by Halliday/Resnick (John Wiley))

12. A paratrooper's chute fails to open and he lands in snow; he is slightly hurt. If he had landed on bare ground, the stopping time would have been 10 times shorter and the collision lethal. Does the presence of the snow increase, decrease, or leave unchanged the values of
   a. the paratrooper's change in momentum?
   b. the impulse stopping the paratrooper?
   c. the force stopping the paratrooper?

(From *Fundamentals of Physics* by Halliday/Resnick (John Wiley))

Unit 11.4 Work, Power and Energy

Topic 1: Work and power

Unit 11.4 covers the fundamentals of mechanical energy, work, power and simple machines. The content of Topic 1 relates to the bullet points on pp. 15–16 of the Syllabus under the headings:

- Work.
- Power.

Work

Work is a term often used in everyday conversation – people say "my calculator doesn't work" or "I worked hard on my homework last night". In physics, work has a special meaning – work is said to be done on an object *when a force causes the object to move*. The work done by a force that acts on an object may be calculated using the formula:

$$W = F \times d$$

work = force × distance moved in the direction of the force

The SI unit for work is the **joule**. This unit was named after the physicist James Joule (1818–1889), famous for his experiments in understanding heat and heat transfer.

1 J is equivalent to 1 N m – the amount of work done when a force of 1 newton moves an object a distance of 1 metre.

Example A

James push-starts a 1 500 kg racing car with a force of 800 N. The car moves a distance of 5 m. Calculate the work done on the car.

Doing work by pushing.

Answer:

$W = F \times d = 800 \times 5 = 4\,000$ J

Assuming that there is no friction, the work done by James in pushing the car is equal to the kinetic energy gained by the car.

Example B

A 1.8 m tall weightlifter lifted a 80 kg mass off the ground to a height 0.5 m above his head. Calculate the work done by the weightlifter on the weight.

Work done by a weightlifter.

Answer:

The weight of the mass = mg

= 80 × 10

= 800 N (therefore the force = needed to lift the mass = is 800 N).

The height the mass was lifted is:

1.8 m (weightlifter's height) + 0.5 m (height above weightlifter's head) = 2.3 m

Work done = $F \times d$
$= 800 \times 2.3$
$= 1\,840$ J

The work done on the mass is equal to the gain in energy of the mass. The mass has gained 1 840 J of gravitational potential energy.

Example C:

Maria does 84 J of work to move a 3.0 kg mass a distance of 4.0 m from rest.

a. Calculate the size of the force she applied.
b. What is the acceleration of the mass?

Answers:

a. $F = \frac{W}{d} = \frac{84}{4.0} = 21$ N

b. $a = \frac{F}{m} = \frac{21}{3.0} = 7.0$ m s^{-2}

Work is done whenever energy is transformed from one form to another.

If you push hard against a wall and the wall does not move, you have not done any work on the wall, even though you may start to feel tired after a while. Inside your body chemical potential energy (from food) is converted to heat and small movements in your muscle fibres – although energy is being transformed, work done on the wall is nevertheless zero, because the wall does not move.

To calculate the work done on an object when the force acting on the object is not constant, a graph of 'Force versus distance moved' is plotted – *the work done is equal to the area under the graph.*

Example D

The graph below shows how the force acting on a trolley changes while the trolley is travelling. Calculate the work done by the force on the trolley.

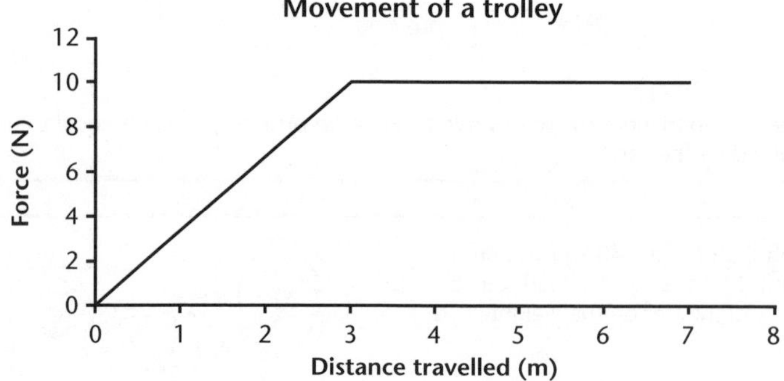

Force versus distance graph for a trolley.

Answer:

The work done is equal to the area under the graph:

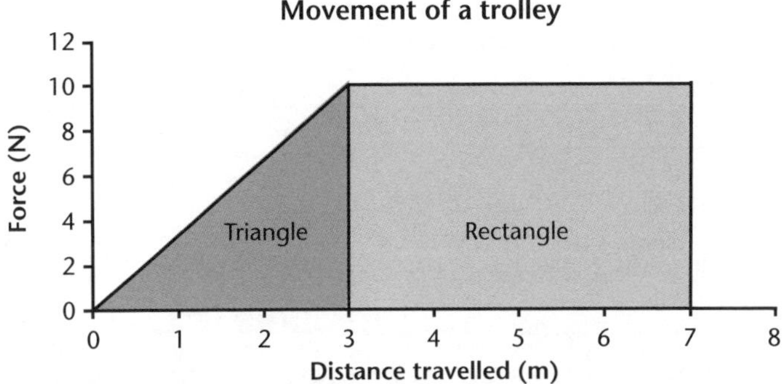

Work done equals area under a graph.

The total area is equal to the area of the triangle plus the area of the rectangle:

$$\text{Area of triangle} = \frac{1}{2} \times \text{base} \times \text{height}$$
$$= 0.5 \times 3 \times 10$$
$$= 15$$

$$\text{Area of rectangle} = \text{length} \times \text{width}$$
$$= (7 - 3) \times 10$$
$$= 4 \times 10$$
$$= 40$$

$$\text{Work done} = 15 + 40$$
$$= 55 \text{ J}$$

Unit 11.4 Activity 1A: Work

1. Complete the following table by calculating **a** to **g**:

Work	Force	Distance
a	200 N	9 m
b	6.5 N	0.05 m
c	10 000 N	1.5 km
100 J	**d**	1 m
350 J	**e**	50 cm
1 500 J	50 N	**f**
280 kJ	700 N	**g**

2. A piano of mass 250 kg is moved 10 m along the floor by two people with a total force of 4 000 N.

a. What is the work done in shifting the piano the 10 m?

The piano is now lifted up some steps, of total height 1.5 m.

b. What force is needed to lift the piano?

c. How much work is done to lift the piano?

d. What is the total amount of work required to shift the piano 10 m along the floor *and* up the stairs?

3. A motorbike of mass 150 kg is lifted 2 m onto the back of a truck.

a. What is the weight of the bike?

b. What is the work done to lift the bike onto the back of the truck?

c. Has the gravitational potential energy of the bike increased or decreased?

d. What is the change in the gravitational potential energy of the bike?

4. A girl climbs to the top of some stairs. She wants to calculate how much work she does in climbing the stairs. Which of the three distances, shown on the diagram, would she use?

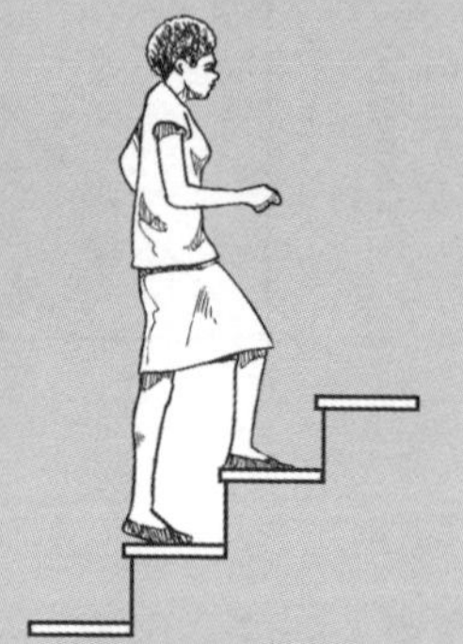

Power

Power is the rate of doing work and may be calculated using the formula:

$$P = \frac{W}{t} \qquad \text{Power} = \frac{\text{work done}}{\text{time taken}}$$

The SI unit for power is the **watt** (W). 1 W = 1 J s^{-1}.

Other units for power are the kilowatt (kW) and the megawatt (MW).

Example E

Eron and Peter each weigh 75 kg. They both climb up steps to a 5 m high diving board. It takes Eron 10 s to reach the top, whereas it takes Peter 15 s.

Calculate the power exerted by:

a. Peter. b. Eron.

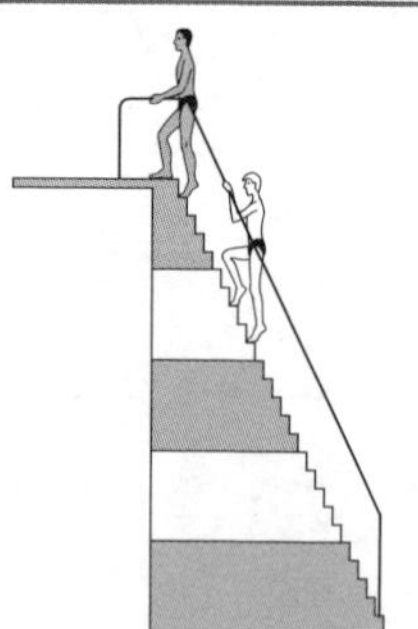

Effect of time on power – Eron is more powerful than Peter.

Answers:

a. Peter's weight = $mg = 75 \times 10 = 750$ N
Work done by Peter = $F \times d = 750 \times 5 = 3\,750$ J
Power exerted by Peter = $\frac{W}{t} = 3\,750 \div 15 = 250$ W

b. Eron's weight = $mg = 75 \times 10 = 750$ N
Work done by Eron = $F \times d = 750 \times 5 = 3750$ J
Power exerted by Eron = $\frac{W}{t} = 3\,750 \div 10 = 375$ W

When they reach the top, Peter and Eron have each done the same amount of work but since it took Eron less time, Eron is said to be more powerful than Peter.

Example F

A microwave has a power rating of 650 W. How much work can it do in 2 minutes?

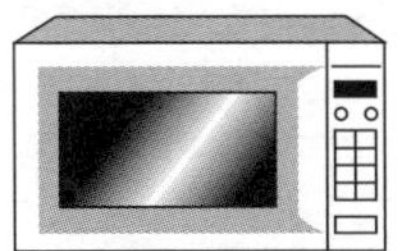

A microwave oven.

Answer:

$W = P \times t$

$= 650 \times 120$ (Always convert to seconds; 2 minutes = 120 s)

$= 78000$ J or 78 kJ

Unit 11.4 Activity 1B: Power

1. Complete the following table by calculating values **a** to **f**:

Power	Work done	Time
a	800 J	40 s
b	0.5 J	2.5 s
1 500 W	c	10 s
2 kW	d	100 s
50 W	100 J	e
0.5 kW	0.5 J	f

2. A crane lifts 800 kg of concrete a height of 25 m in 20 s.

a. What is the weight of the concrete?

b. What is the work done by the crane?

c. What is the power of the crane?

The following information relates to questions 3 to 5 on the next page.

An electric motor is used to raise a lift full of passengers. The total weight of the lift and passengers is 15 000 N. The lift is raised a height of 20 metres in 10 seconds.

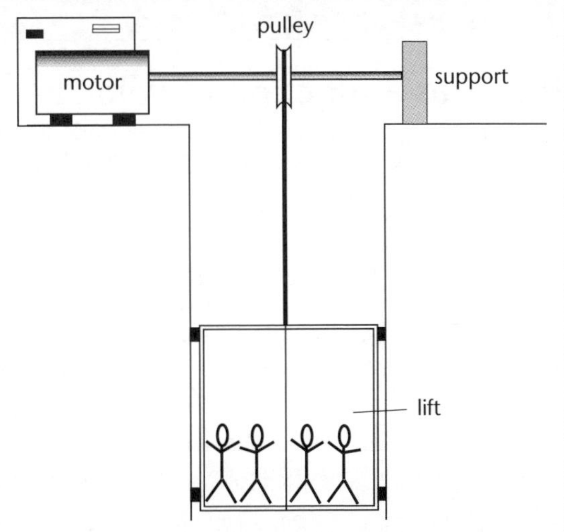

3. Describe the energy transformations that occur.
4. Describe what happens to the energy that is wasted.
5. Calculate the useful power output of the motor.
6. A 1.8 m tall weightlifter lifts a 110 kg barbell from the floor to 600 mm above his head and holds it there for 2 minutes. (Assume $g = 10 \text{ m s}^{-2}$.)
   a. Find the work the weightlifter has done.
   b. If the process of lifting the weight took 4 seconds, find the power the weightlifter exerted.

(Source: NCEA Examination paper)

7. Julius and Limu both race up a set of stairs (as shown).
   In 15 seconds, Julius (weight = 750 N) runs to a level 10 metres above ground level, while Limu (weight = 500 N) runs to a level which is 12 metres above ground level.
   a. Find the work done by Limu against gravity.
   b. Find the work done by Julius against gravity (measured in Joules).
   c. After the race, the class compare Julius's performance with Limu's.
      What can you say about Julius's speed and his power compared with Limu's?

(Source: NCEA Examination paper)

Power as a function of velocity

Power must be considered as the amount of work done during a time, Δt. Therefore, we can define an **average power** due to the applied force on an object during that interval of time as:

$$\overline{P} = W/\Delta t$$

where P is the average power given in watts (W), and $W/\Delta t$ is the rate of doing work given in Joules per second (J s^{-1}). Then 1 Watt = 1 W = 1 J s^{-1} = 1 Joule/second. Sometimes you will be required to determine power using units in the British system, in which the unit for power is the foot-pound per second, and very often horsepower is used as the unit for power. Therefore, you may consider the following relationships:

1 watt = 1 W = 1 J/s = 0.738 ft-lb/s

1 horsepower = 1 hp = 550 ft-lb/s = 746 W

Another useful unit is the Kilowatt-hour. This is a unit for work (= power × time), and it is basically used to specify or identify electrical units, eg your utility bill sent to you by PNG Power (formerly ELCOM) is expressed in kilowatt-hours.

By considering the basic definition for work in terms of the force applied to an object, and the definition for power as the rate at which work is done, then we can express the rate at which a force does work on an object in terms of that force, and the state of motion of that object in terms of its velocity. Thus, if an object moves in a straight line, eg along the x axis, and if a force F directed at some angle θ to that line acts on the object, then we can express power as:

$$P = W/\Delta t = F\text{Cos}\theta\ (\Delta x/\Delta t) = F\upsilon\ \text{Cos}\theta \text{ or}$$

$$P = F \bullet \upsilon \text{ (vector equation: instantaneous power)}$$

This is a general definition for power in terms of a force applied to an object and the velocity of the object, but most of the time we will be dealing only with cases where the applied force is parallel (same direction) to the motion of the object. Therefore, for those situations we will be determining power as the product of the force and the velocity as follows:

$$P = F\upsilon \text{ (instantaneous power)}$$

In this case, P is usually the power output of a machine doing work by applying a force F and moving the point of application of the force with velocity υ.

Example G

You are driving a car of mass 700 kg along a horizontal road at a speed of 40 m/s and you bring the car to rest (stop) by somehow applying a constant retarding force of 4500 Newton. Determine the distance the car moves while coming to rest before stopping.

Solution:

Let us suppose the car travels a distance d before stopping, so by applying the basic equation ($W = Fd$), the work done by the car against the retarding force is:

$$W = 4500d$$

and the kinetic energy ($1/2m\upsilon^2$) lost by the car in coming to stop is determined as:

$$K = ½(700 \text{ Kg})\ (40 \text{ m/s})^2 = 560\ 000 \text{ Joules}$$

By definition of work you conclude that the work done against the retarding force must be equal to the kinetic energy lost by the car, hence:

$$4500d = 560\ 000;\ \text{Thus } d = 124.44 \text{ metres.}$$

Example H

Calculate the required power of a 1500 kg vehicle under the following conditions:

i. The vehicle climbs a hill whose inclination is 12°, at a steady speed of 70 km/h. The force of friction on the vehicle is 700 Newton.

ii. The vehicle accelerates from 90 km/h to 100 km/h in 5 seconds to pass another vehicle while on a level road. The frictional force on the vehicle is 700 Newton.

Solution:

i. You can consider the following figure to illustrate the situation:

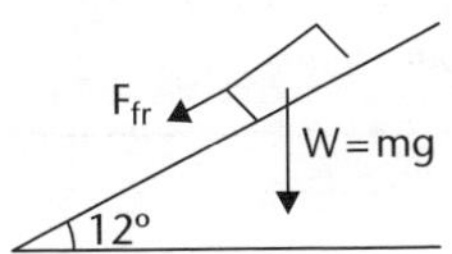

To move at a steady speed uphill, the vehicle must exert a force equal to the sum of the frictional force and the component of gravity **parallel** to the hill, **mgSin12° = 3056 N.**

Since υ = 70 km/h = 22 m/s, Then

$P = F\upsilon = (3056 \text{ N} + 700\text{N})(22 \text{ m/s}) = 8.26 \times 10^{-4} \text{ W}$

$= 8.26 \times 10^{-4} \text{ Watt} = 110.766 \text{ hp}$

ii. The vehicle accelerates from 25.0 m/s to 27.78 m/s (90 km/h to 100 km/h); so, the vehicle must exert a force to overcome the 700 N frictional force plus a force required to give it the acceleration (a = 27.78 m/s – 25.0 m/s)/5.0 s = 0.556 m/s²). Now, since the mass of the vehicle is 1500 kg, the force required for acceleration is $F = ma$ = (1500 kg)(0.556 m/s²) = 834 Newton, therefore, the total force required is 834 N + 700 N = F_T. Since this force is constant, the power required will be greatest when the velocity is greatest ($P = F$ times υ). Therefore, the maximum power needed can be calculated as:

$$P = (1534\ \text{N})(27.78\ \text{m/s}) = 4.26 \times 10^4\ \text{watt} = 57.12\ \text{hp}$$

Unit 11.4 Activity 1C: Power and Speed

1. A heavy vehicle (eg a train) of mass 200 tonnes has a maximum speed of 20 m/s up a hill inclined at 1.15° to the horizontal when the engine is working at 800 kwatts. Determine the resistance to motion of the truck.
2. A big vehicle (eg a train) has a maximum speed of 80 km/h on a level ground, and the resistance of motion is estimated to be 50 000 Newton. Determine the power of the engine of this vehicle.
3. A car of mass 750 kg has a maximum power of 30 KW and moves against a constant resistance to motion of 800 N. Calculate the maximum speed of the car in the following situations:
   a. on level ground
   b. up an incline (hill) of 5.74° to the horizontal.
   c. down the same hill
4. A car of mass 1500 kg has a maximum speed of 150 km/h on a level road, when working at its maximum power against a resistance to motion of 60 Newton. Determine the acceleration of the car when it is travelling at 60 km/h on a level road, with the engine working at maximum power, assuming that the resistance to motion remains constant.
5. An engine of mass 100 tonne pulls a train of mass 400 tonne. If the resistance to motion of the engine is estimated at 1000 N and the resistance to motion of the train is 20 000 N; determine the tension in the coupling between the engine and the train at the instant when the speed of the train is 80 km/h and the engine is exerting a power of 4000 kw (consider the following diagram).

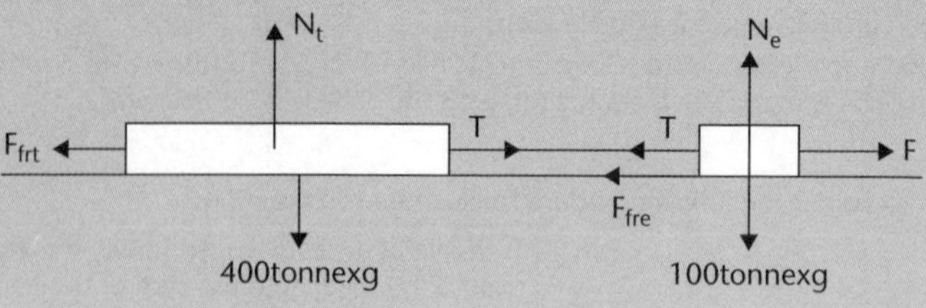

Unit 11.4 Work, Power and Energy

Topic 2: Energy

Unit 11.4 covers the fundamentals of mechanical energy, work, power and simple machines. The content of Topic 2 relates to the bullet points on pp. 16 of the Syllabus under the heading 'Energy', in particular:

- Types of energy.
- Mechanical energy (kinetic and gravitational potential).
- Conservation of energy.

Energy is the capacity to do **work**. There are many different forms of energy.

Type of energy	Definition	Situations where energy may be found
Radiant energy	Energy of the **electromagnetic spectrum**, eg heat, light, etc	Television or radiant heater
Nuclear energy	Energy stored in the **nucleus** of an **atom**	**Nuclear reactor**
Kinetic energy	Energy of a moving object	Transport or sport
Gravitational potential energy	Energy stored in an object because of its position in the Earth's gravitational field	Standing on a diving board
Elastic potential energy	Energy stored in a compressed or stretched spring or stretched rubber band	A slingshot, car suspension
Chemical potential energy	Energy stored in food, fuel or chemicals	Petrol or a **battery**
Sound energy	Energy emitted by a vibrating source	Loudspeaker or tuning fork

Types of energy.

Energy cannot be created or destroyed, but is always conserved. Energy can, however, be transformed (changed) from one form to another. A **transducer** is an instrument that transforms energy from one form to another; eg, an electric heater changes electrical energy to **heat** energy. Energy transformations may be described using subscript notation:

E_{heat} means energy associated with heat

$E_{electrical}$ means electrical energy

Example A

The energy transformations in climbing and diving from a high diving board are:

$E_{p\ (chemical)} \longrightarrow E_{p\ (gravitational)} \longrightarrow E_k$

Chemical potential energy – from food burnt in muscle action

Gravitational potential energy – from increase in height as climbed ladder

Kinetic energy – dive from the board

Example B

State the energy transformations involved in a **hydroelectric power** station.

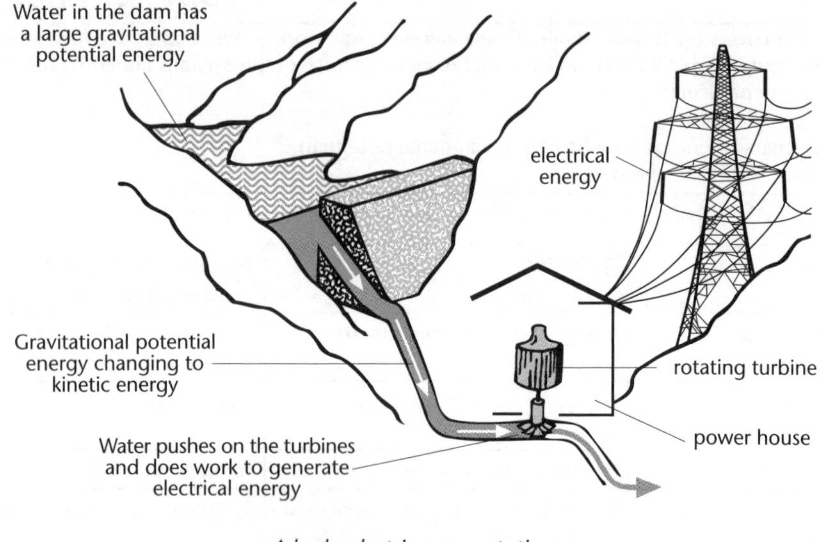

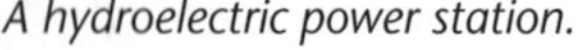
A hydroelectric power station.

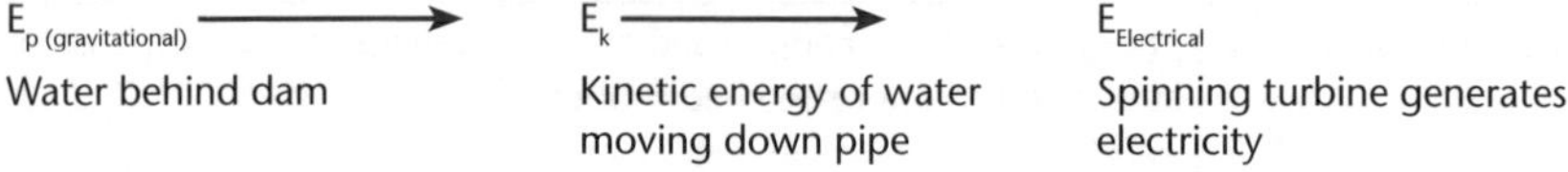

Mechanical energy

Kinetic energy and **gravitational potential energy** are examples of **mechanical energy**.

Kinetic energy

Kinetic energy is the energy of motion. The kinetic energy E_k of an object of mass m that travels at a speed v can be calculated using the formula:

$$E_k = \frac{1}{2} mv^2$$

Kinetic energy is measured in Joules (J).

Example C

Calculate the kinetic energy of a 1 500 kg car moving at 2 m s^{-1}.

Answer:

$E_k = \frac{1}{2} mv^2 = \frac{1}{2} \times 1\,500 \times 4 = 3\,000$ J = 3 kJ

Unit 11.4 Activity 2A: Kinetic energy

1. Solomon is riding his skateboard. The combined mass of Solomon and his skateboard is 60 kg. Solomon moves down a ramp at a constant speed of 3 m s^{-1}. What is the kinetic energy of Solomon and his skateboard?
2. Calculate the kinetic energy of each of the following.
   A. A 10 kg dog racing at 10 m s^{-1}.
   B. A 40 kg child running at 6 m s^{-1}.
   C. An 80 kg athlete jogging at 3 m s^{-1}.
   D. A 120 kg person walking at 1 m s^{-1}.
3. Calculate the kinetic energy of a 3 kg bullet moving at 400 m s^{-1}.
4. Two cars, P and Q, have equal mass. Car P is travelling at 50 km h^{-1} and car Q is travelling at 100 km h^{-1}. How does the kinetic energy at Q compare with that of P?
5. Complete the table by filling in the kinetic energy values **a** to **d**. The first line has been done for you.

Mass	Speed	Kinetic Energy
2 kg	4 m s^{-1}	16 J
4 kg	4 m s^{-1}	a
6 kg	4 m s^{-1}	b
2 kg	8 m s^{-1}	c
2 kg	12 m s^{-1}	d

Gravitational potential energy

Gravitational potential energy is the energy stored in an object because of the object's position in the Earth's gravitational field. The gravitational potential energy (E_p) of a mass m at a height h above a chosen reference level can be calculated using the formula:

$$E_p = mgh \qquad g = 10 \text{ m s}^{-1}$$

Example D

Tahi (mass 75 kg) stands on a 2 m high diving board. What is his gravitational potential energy?

Answer:

$E_p = mgh = 75 \times 10 \times 2 = 1\ 500$ J

Unit 11.4 Activity 2B: Gravitational potential energy

1. Solomon pushes his skateboard with a force of 35 N. Solomon and his skateboard have a combined mass of 60 kg. There is a friction of 5 N. What is Solomon's acceleration?

2. Complete the table by filling in the values **a** to **e**.

Gravitational potential energy	Mass	Height
a	10 kg	2 m
b	0.5 kg	20 m
500 J	c	10 m
1 500 kJ	500 kg	d
20 kJ	e	500 cm

Conservation of energy

The *law of conservation of energy* states that 'energy cannot be created or destroyed, but can be transformed from one form to another'.

Example E

Michael climbs the steps to the high diving board at a swimming pool. His muscles convert chemical potential energy (from the food he has eaten) to kinetic energy and gravitational potential energy. Standing on the high diving board, he has gravitational potential energy. When he dives off, the gravitational potential energy changes to kinetic energy. *In all of these energy transformations, the total amount of energy is conserved.*

When an object of mass, m, falls from a height, h, above the floor, the speed of the object just before it hits the ground can be calculated using the law of conservation of energy:

Gravitational potential energy at the top = Kinetic energy at the bottom

$$mgh = \frac{1}{2}mv^2$$

$$v = \sqrt{2gh}$$

Example F

Michael (mass 75 kg) dives off a diving board that is 2 m above the surface of a swimming pool. Calculate Michael's speed on entering the water.

Answer:

$v = \sqrt{2gh} = \sqrt{2 \times 10 \times 2} = 6.3 \text{ m s}^{-1}$

Unit 11.4 Activity 2C: Conservation of energy

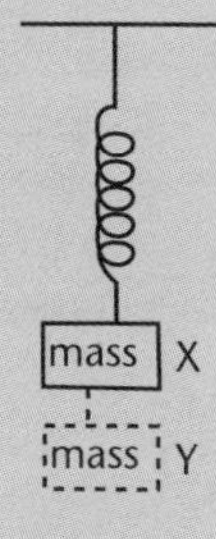

1. The diagram shows a mass moving up and down on the end of a spring. **X** and **Y** are the highest and lowest positions of the mass. Which one of the following is true when the mass is at **X**?

	Kinetic energy of the mass	Potential energy of the mass
A.	maximum	minimum
B.	maximum	zero
C.	zero	maximum
D.	same as potential energy	same as kinetic energy

2. In a hydroelectricity generating station, water is stored in a reservoir behind a dam. The water then falls through pipes to turbines which drive the generators.
   a. What energy changes occur when water flows from the reservoir to the turbines?
   i. energy to **ii**. energy.
   b. What energy changes occur when the turbine spins the generator?
   i. energy to **ii**. energy.
   c. In one hydroelectricity generating station, 2 tonnes (2 000 kg) of water passes through the turbines every second after falling through a vertical height of 100 metres. (The gravitational field strength is 10 N kg^{-1}. Assume no energy is wasted.) What is the weight of the 2 000 kg of water?
3. A steel ball-bearing of mass 0.1 kg falls from a height of 2 m onto the ground.
   a. What is the loss of gravitational potential energy when the ball reaches the ground?
   b. What will be its kinetic energy as it reaches the ground?
   c. Find the speed of the ball-bearing when it reaches the ground.
4. Mona sits on a chairlift as shown in the diagram.
   The combined mass of Mona and her snowboard is 88 kg.

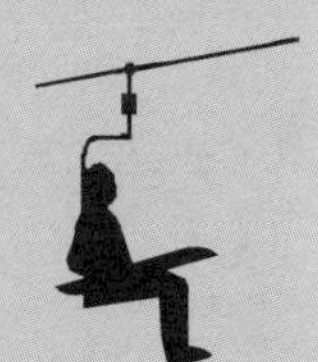

   a. Calculate the force provided by the seat to support Mona and her snowboard. State the direction of this force.

   The chairlift carries Mona from the base to the ski slope, which is 18 m above the base level as shown in the diagram below. Her mass is 85 kg.

   b. Calculate her gain in gravitational potential energy at this height.

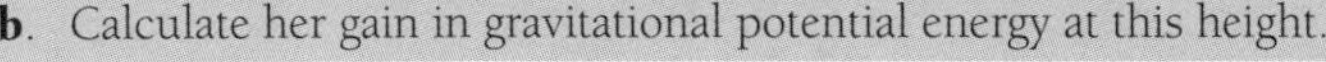

Mona now snowboards down the slope and she levels off onto the flat section. The flat section is 12 m from the base level. Assume that the energy loss due to friction down the slope is zero.

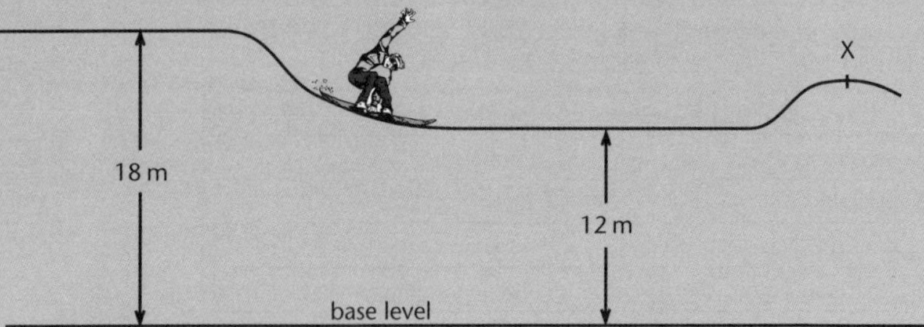

c. Calculate her speed at the bottom of the slope as she begins to level off onto the flat section.

d. On the flat section of the slope, Mona slows herself down by turning her feet so the snowboard is almost at right angles to the direction of her motion. Then she goes up a small hill and stops at the point X in the diagram above. Explain what happened to her kinetic energy.

(Source: NCEA Examination paper)

5. Suppose you throw an object of mass 3 kg so that it just clears the top of a 2 m–high fence when its speed is 4 m/s. Calculate its total mechanical energy as it passes over the fence.

6. Identify the following statements as true (if the statement is always true) or false.
   i. The energy stored in an elastic string is proportional to the extension.
   ii. As long as no external forces act on a system the kinetic energy must be constant.
   iii. Some external forces which act on a moving body do not do any work.
   iv. A spring obeys Hooke's law when it is stretched but not when it is compressed.

7. Consider a ramp as illustrated in the following figure:

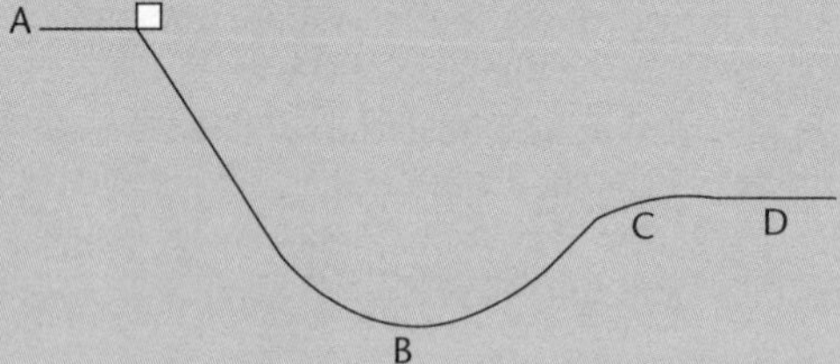

An object slides from A to C. A to C can be considered frictionless. The object then goes through a horizontal portion CD, where a frictional force acts on that object. Determine whether the kinetic energy of the object is increasing, decreasing or remains constant in:

   i. region CD
   ii. region BC
   iii. region AB

Unit 11.4 Work, Power and Energy
Topic 3: Energy and mechanics

The content covered in the Topic extends the concepts introduced in Topics 1 and 2 about work, power, energy and mechanics, in particular:

- Elastic and inelastic collisions.
- Work done by a changing force.
- Conservation of mechanical energy.
- Power.

Introduction

The concepts of work and power, gravitational potential energy, kinetic energy, and the conservation of mechanical energy are revised below before using them in the NCEA Level 2 Physics aspects of energy that follow.

Revision

Work

Work is the process that *transfers energy from one form to another.* The amount of work done depends on the forces involved and the distance through which those forces act. The relationship between the force, F (in newtons), the distance moved, d (in metres), and the work done, W (in joules, symbol J), is:

work = force × distance moved *in the direction of the force*

$$W = Fd$$

One joule of work is done when a force of one newton moves an object through a distance of one metre.

Work is a scalar quantity – although vectors are used in its calculation, the resulting work done has no specific direction.

Potential energy

If an object of mass m is lifted against gravity a distance Δh from one height to a higher one, then work is done.

The force needed to do the lifting, F, is equal in size to the weight of the mass:

F = weight

$= mg$

The work done is:

$W = Fd$ [formula]

$= (mg) \times \Delta h$ [substituting Δh for d]

$= mg\Delta h$

Lifting a mass

The lifting process has done $mg\Delta h$ joules of work. This energy has been transferred *from* **chemical energy** in a person's body *to* **gravitational potential energy**, E_p. The object's mass has not changed and the object has not heated up. The only difference is that the object is now in a *different position* in the earth's gravitational field.

The word *potential* means that the energy transferred is *stored* and available for doing work at a later time. As a result of the work done, the object now has $mg\Delta h$ joules of gravitational potential energy with respect to the ground level.

If ΔE_p is the change in potential energy in joules, m is the object's mass in kg, g is the acceleration due to gravity in m s^{-2}, and Δh is the height lifted in metres, then:

$$\Delta E_p = mg\Delta h$$

There is usually some reference level of zero potential energy. Often ground or floor level is taken to be at zero potential energy.

Kinetic energy

Objects moving at speed have energy *due to their motion*. This energy is called **kinetic energy**, E_k. When moving objects slow down and stop, their kinetic energy is transferred into other forms of energy.

The kinetic energy, E_k, of an object with mass m, travelling at speed v, can be calculated using the formula:

$$E_k = \frac{1}{2} mv^2$$

Energy, like work, is a scalar quantity and the direction of the motion does not matter.

Power

Power, P, is the *rate at which work is done,* ie power measures how quickly energy is transferred.

Power can be calculated from the work done and the time taken, using the formula:

$$P = \frac{W}{t}$$

where P is the power, W is the work done in joules, and t is the time taken in seconds.

The unit for power is the **watt**, W. One watt is one joule per second, ie 1 W = 1 J s^{-1}. Other units for power are the **kilowatt** (symbol kW) and **milliwatt** (symbol mW).

Conservation of mechanical energy

In all situations where work is done, energy is transferred from one form to another. Sometimes the energy is transferred into several forms of energy, such as heat, sound and **light**. This means that energy is *not lost or destroyed* but rather it is **conserved** (ie the total energy before the work is done equals the total of all the forms of energy produced afterwards).

Hooke's Law

Work is done in compressing or stretching a spring. When this is done, energy is stored in the spring and can be released later. Clocks and catapults use this form of stored or **elastic potential energy**.

The force needed to extend (stretch) an ordinary spring (eg the spring in a ball-point pen), increases as the spring extends. This is also true if the spring is compressed instead of stretched.

A graph of the extending force against extension shows a linear relationship. If F is the extending force in N, and x is the extension in m, then:

$$F = -kx$$

where k, the slope of the graph, is called the **spring constant**. The units for k are N m^{-1}.

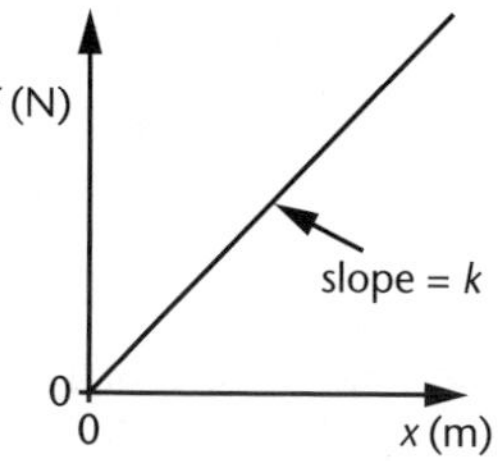

A force/extension graph

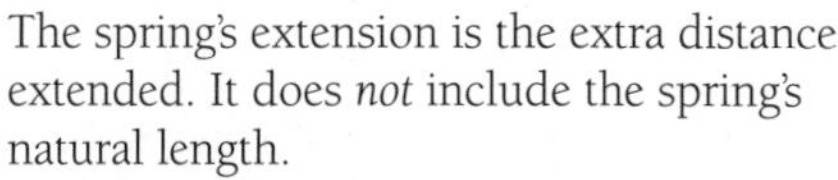

The spring's extension is the extra distance extended. It does *not* include the spring's natural length.

The elastic potential energy, E_p, stored when the spring is extended (or compressed) a distance x, is equal to the area under the graph from zero to x.

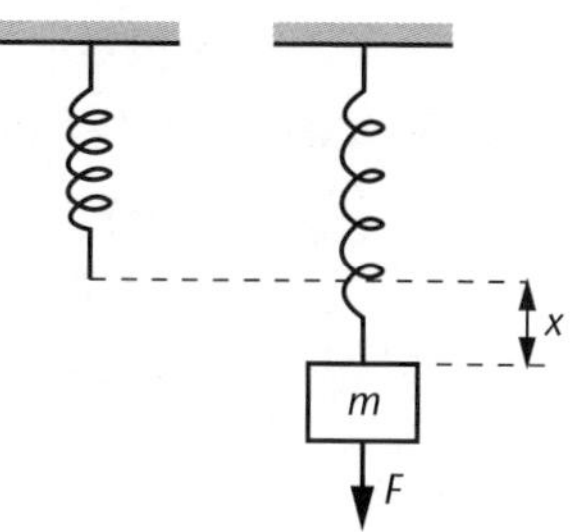

Explaining extension

The area under the graph has units of joules. This comes from multiplying the unit of force (newtons) by the unit of distance extended (metres).

The elastic potential energy is shown as the shaded area in the graph below:

$E_p = \frac{1}{2}Fx$ [area = $\frac{1}{2}$ × height × base]

But $F = kx$, so substituting into $E_p = \frac{1}{2}Fx$ gives:

$$E_p = \frac{1}{2}(kx)x$$
$$= \frac{1}{2}kx^2$$

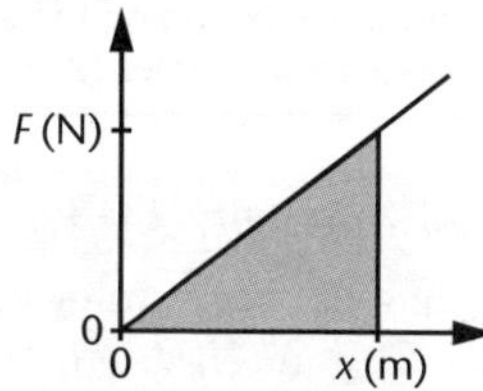

Finding elastic potential

$$E_p = \frac{1}{2}kx^2$$

The x^2 in the formula $E_p = \frac{1}{2}kx^2$ shows that, proportionally, much more (a squared function) elastic potential energy is stored in a spring as it is stretched (or compressed) further.

Example A

A mass of 0.5 kg hung from the end of a spring extends the spring by 25 cm.

a. Calculate the spring constant.

b. How much elastic potential energy is stored in the spring?

Another mass of 0.5 kg is added to the first mass.

c. What is the new extension?

d. How much energy is stored in the spring now?

Solution:

a. The extending force is the weight force on the 0.5 kg. The size of the force is:

$$F = mg$$
$$= 0.5 \times 10 \qquad [g = 10 \text{ m s}^{-2}]$$
$$= 5 \text{ N}.$$

Rearranging $F = kx$, gives the spring constant as:

$$k = \frac{F}{x}$$
$$= \frac{5}{0.25}$$
$$= 20 \text{ N m}^{-1}$$

b. At an extension of 25 cm, the potential energy stored is:

$$E_p = \frac{1}{2}kx^2$$
$$= \frac{1}{2} \times 20 \times (0.25)^2 \qquad [25 \text{ cm} = 0.25 \text{ m}]$$
$$= 0.625 \text{ J}$$

c. Doubling the mass doubles the extending force to 10 N, which doubles the extension, ie $x = 50$ cm.

d. At an extension of 50 cm, the potential energy stored is:

$$E_p = \frac{1}{2}kx^2$$
$$= \frac{1}{2} \times 20 \times (0.50)^2 \quad [50 \text{ cm} = 0.50 \text{ m}]$$
$$= 2.5 \text{ J}$$

Doubling the extending force *doubles* the extension, but the elastic potential energy stored increases by *four* times (2^2).

Unit 11.4 Activity 3A: Energy and springs

1. A slingshot has a spring constant of 50 N m^{-1}.

a. What force is needed to pull the elastic band back 0.15 m?

b. What elastic potential energy is stored in the slingshot at that extension?

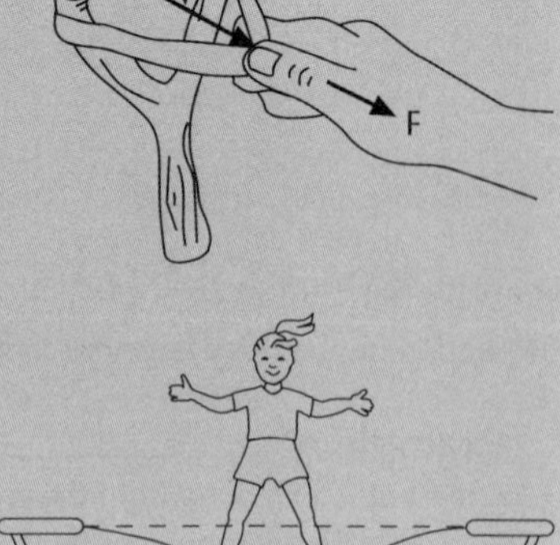

2. A 30 kg child stands on a trampoline and causes the trampoline to sag by 10 cm.

a. What is the child's weight?

b. What is the trampoline's spring constant?

c. How much elastic potential energy is stored in the trampoline?

d. The child now bounces on the trampoline and at maximum extension she presses the surface down 30 cm below its normal level. If the trampoline's spring constant is assumed to remain constant, how much *more* elastic potential energy is stored in the trampoline at this maximum extension?

3. The diagram shows how the force F needed to compress a spring varies as the spring is compressed.

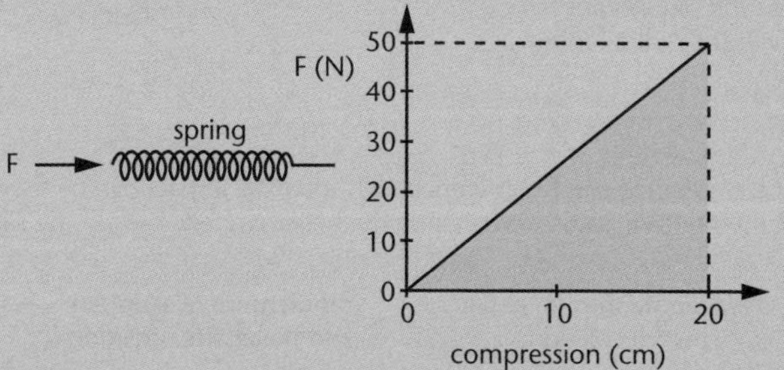

a. How much work is done to compress the spring 20 cm?

b. What is the value of the spring's force constant, k?

c. Draw a sketch graph of how the energy stored in the spring varies with **compression**.

Combined mechanics

Most situations involve several aspects of physics working together.

Example B

Studying the motion of a bicycle ride may involve measuring time and distance and calculating the resultant displacement and acceleration. The bicycle doesn't move without forces being exerted on the pedals by the cyclist, resulting in a torque.

Riding up a steep hill, the cyclist and bicycle will gain gravitational potential energy and probably lose kinetic energy.

Circular motion occurs in the bicycle wheels and when the cyclist is turning a corner.

A gravitational weight force keeps the bicycle on the road and a vector component of the weight force acts to accelerate the cyclist when cycling downhill.

The useful effects of friction, through the brakes, will stop the bicycle at the end of the journey.

Hopefully, the only mechanics topic not involved in a bicycle ride is projectile motion!

This section contains questions on situations which involve several interrelated aspects of physics.

Example C

A bullet of mass 30 g is fired with a speed of 400 m s^{-1} into a sandbag. The sandbag has a mass of 10 kg and is suspended by two ropes so that it can swing.

What is the maximum vertical height, h, that the sandbag rises as it recoils with the bullet lodged inside?

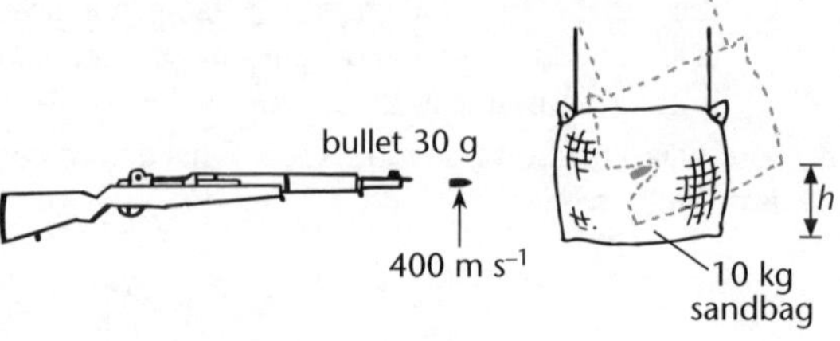

Solution:

The speed of the *sandbag and bullet* immediately after the impact can be found by applying the idea that momentum is conserved when the bullet embeds itself in the sandbag.

momentum of bullet before collision = momentum of sandbag and bullet after collision

$$0.030 \times 400 = (10 + 0.030) \times v$$

[v is the speed of sandbag and bullet immediately after the collision, 30 g = 0.030 kg]

$$12 \approx 10 \times v$$

[since 30 g is negligible compared with 10 kg]

$$v = \frac{12}{10}$$

$$= 1.2 \text{ m s}^{-1}$$

After the collision, the sandbag swings upward by a height h, and the kinetic energy of the sandbag immediately after the collision is transferred to gravitational potential energy. At the maximum height of swing, all the kinetic energy has been transferred to gravitational potential energy and:

loss in kinetic energy = gain in gravitational potential energy

$$\frac{1}{2}mv^{-2} = mgh$$

$$h = \frac{v^2}{2g}$$

[rearranging and cancelling m]

$$= \frac{(1.2)^2}{2 \times 10}$$

[$g = 10$ m s^{-2}]

$$= 0.072 \text{ m}$$

ie the sandbag rises a vertical height of 7.2 cm.

A suspended sandbag used as in Example C is called a ballistic pendulum. This technique was used to determine the muzzle velocity of bullets before the development of high-speed photography, or digital timers.

An approximation (symbol ≈) was used in Example C. Approximations can be used when they do not change the answer significantly. If the actual values for v were used in Example C, then the final answer would have been 7.16 cm, which rounds off to 7.2 cm. The value taken for gravity (10 m s^{-2}) is another example of an approximation.

The collision between a bullet and sandbag is called an **inelastic** collision because the two objects stick together and kinetic energy is lost as they collide. At the other extreme, an **elastic** collision involves no loss of kinetic energy. Most collisions are inelastic to some extent.

The sport of bungi jumping involves conservation of energy between gravitational potential energy, kinetic energy and elastic potential energy.

Example D

A bungi jumper of mass 75 kg jumps off a bridge over a river. The spring constant of the rubber bands that make up the bungi rope is adjusted so that the jumper's head just touches the river at maximum stretch (30 m) before springing back up into the air. If the natural length of the rope is 10 m, calculate the rope's spring constant. Assume no energy is lost due to friction.

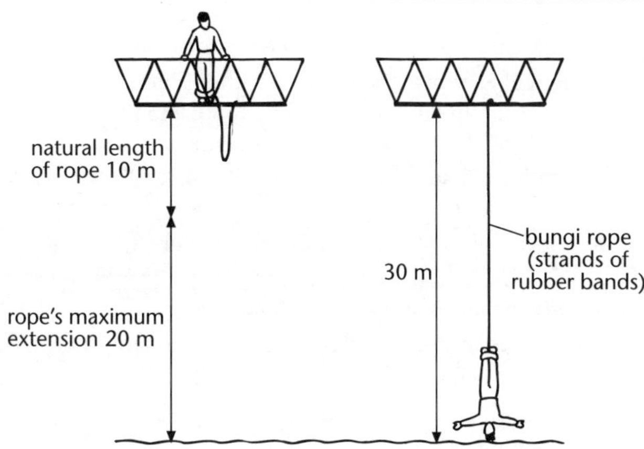

Solution:

Using conservation of energy, all of the jumper's original gravitational potential energy above the river level is converted into elastic potential energy when the jumper reaches the river level:

$$\begin{aligned}\Delta E_p &= mg\Delta h \\ &= 75 \times 10 \times 30 \qquad [\Delta h = 30\text{ m}] \\ &= 22\,500\text{ J}\end{aligned}$$

At the river level, the rope's extension from its natural length is:

$$\begin{aligned}x &= \text{stretched length} - \text{natural length} \\ &= 30 - 10 \\ &= 20\text{ m}\end{aligned}$$

$$\begin{aligned}\text{So } \Delta E_p &= \tfrac{1}{2}kx^2 \\ 22\,500 &= \tfrac{1}{2} \times k \times 20^2 \qquad [\text{substituting}] \\ k &= \frac{2 \times 22\,500}{20^2} \qquad [\text{rearranging}] \\ \text{ie, } k &= 112.5\text{ N m}^{-1} \\ &= 110\text{ N m}^{-1}\text{ (2 significant figures)}\end{aligned}$$

The value for *k* would only be approximate since some friction would be present in the rubber strands of the rope, and air resistance would have an effect. The effect of the rope's mass and the extra length of the person hanging at the end would also have to be taken into account.

Unit 11.4 Activity 3B: Combined mechanics

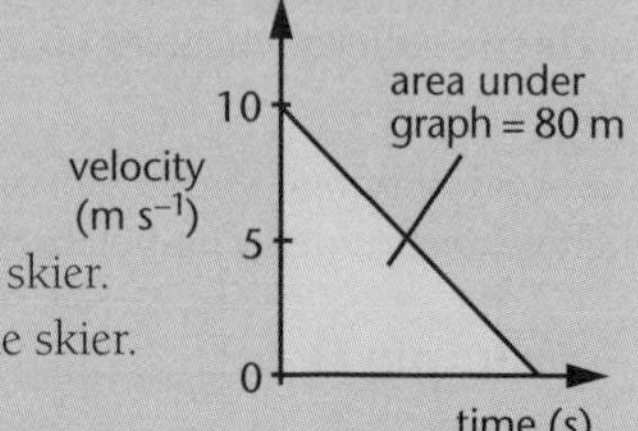

1. The graph shows the motion of a 60 kg skier who is travelling at 10 m s^{-1} before being brought to rest by a constant force. The distance required to stop the skier is 80 m.
   a. Calculate the amount of kinetic energy lost by the skier.
   b. Determine the magnitude of the force acting on the skier.
   c. Calculate the deceleration of the skier.
   d. Determine the time for which the force acts.

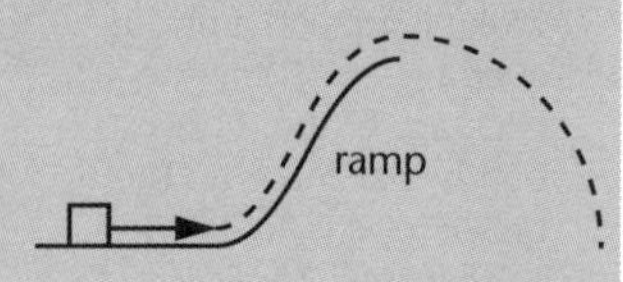

2. An ice cube, of mass 50 g, slides without friction and with an initial velocity of 5 m s^{-1} up a ramp. The ramp is 1.2 m high and is horizontal on top.
   a. With what speed does the ice cube slide off the ramp into the air?
   b. How far from the ramp does the ice cube hit the ground?
3. A girl of mass 40 kg drops vertically down onto a trolley of mass 20 kg which is already moving at a speed of 1.5 m s^{-1}.
   a. What is the new combined speed of girl and trolley?
   b. How much horizontal momentum does the girl gain?
   c. How much horizontal momentum does the trolley lose?
   d. How much kinetic energy does the trolley lose?
   e. Is the loss in the trolley's kinetic energy equalled by the gain in kinetic energy of the girl? Explain your answer.
4. A netball is thrown and follows a parabolic path (air resistance is neglected). Comment on the correctness of the following statements about the ball in flight.
   a. The higher the ball goes, the greater its gravitational potential energy.
   b. The horizontal component of the ball's velocity is constant.
   c. The total energy of the ball is constant.
   d. The acceleration of the ball is constant.
   e. The momentum of the ball is constant.
5. A trolley of mass 1.0 kg is set up against a spring of spring constant 50 N m^{-1}. The spring has been compressed 0.20 m. When the spring is released, the trolley runs up the ramp shown. Friction can be neglected.

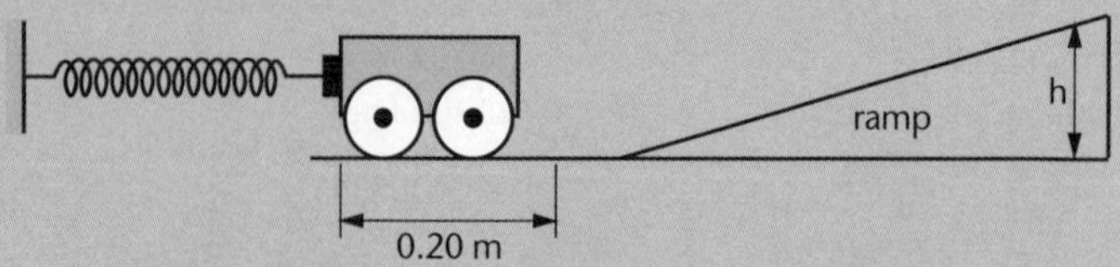

   If there is no energy loss due to friction, calculate:
   a. The maximum speed of the trolley.
   b. The vertical height, h, the trolley will rise up the ramp.

6. The graph represents a plot of velocity against time for a trolley of initial mass 2.0 kg as it travelled in a straight line across a level surface. At some time during the first 10 s, an additional mass was dropped vertically onto the trolley.

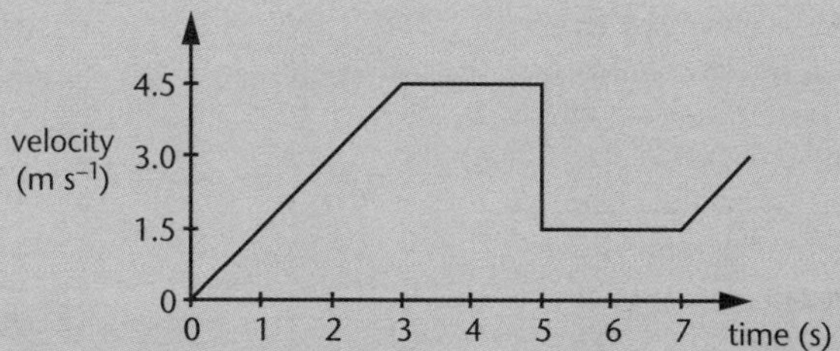

   a. At what time was the mass dropped?
   b. Calculate the size of the additional mass.
   c. What was the size of the resultant force acting on the trolley at $t = 2$ s?
   d. Calculate the kinetic energy of the trolley at $t = 4$ s.
   e. How far did the trolley travel during the first 5 s?

7. A 100 kg satellite is in stable circular orbit about the earth, with a radius of orbit of $r_1 = 10\,000$ km (10^7 m). It is found that the satellite needs to travel at 6.32 km s^{-1} (6 320 m s^{-1}) to maintain that orbit. The satellite then moves to a new orbit, of a radius twice that of the original orbit $r_2 = 20\,000$ km (2.0×10^7 m), and it is found that the satellite's new speed is 4.47 km s^{-1} (4470 m s^{-1}).
   What is the ratio of:

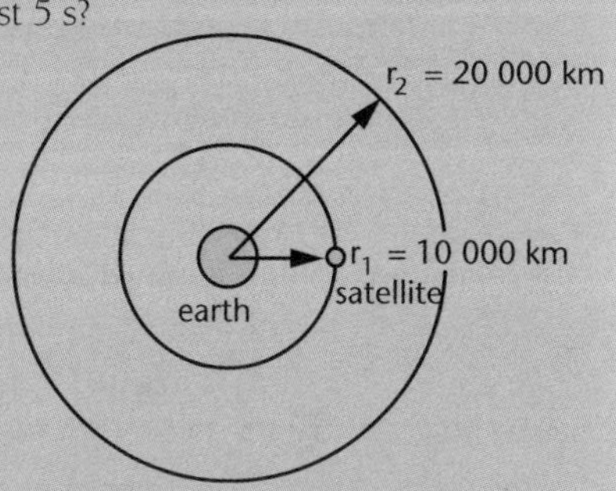

   a. Magnitude of momentum orbit 1 : magnitude of momentum orbit 2?
   b. Kinetic energy orbit 1 : kinetic energy orbit 2?
   c. Period orbit 1 : period orbit 2?
   d. Centripetal acceleration orbit 1 : centripetal acceleration orbit 2?

8. In Example C in this chapter:
   a. Determine how much kinetic energy is 'lost' when the bullet embeds itself in the sandbag.
   b. Where does the 'lost' energy actually go?

9. In Example D in this chapter:

a. Determine at what point below the bridge the diver has maximum speed.

b. Calculate the diver's maximum speed.

10. A suspended steel sphere, **A** (mass 50 g), is pulled to one side and released so that it hits an identical stationary sphere, **B**, at 0.15 m s^{-1}. The collision is elastic, with no loss in kinetic energy, and sphere **A** stops on collision with sphere **B**.

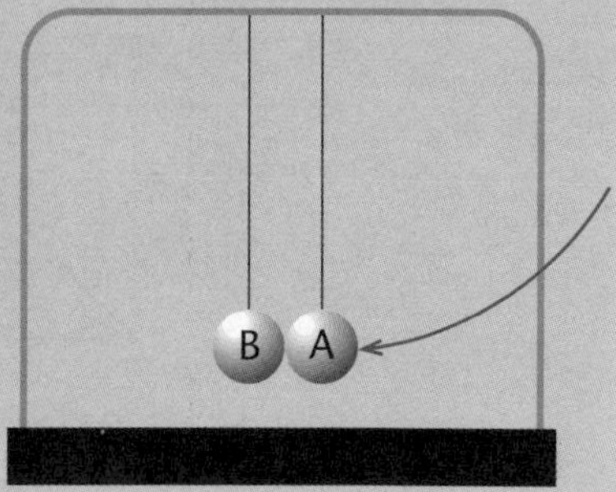

Spheres A and B shown at the instant of collision

a. With what initial speed does sphere **B** move away from sphere **A**?

b. How high does sphere **B** rise vertically before stopping momentarily?

c. Describe the subsequent motion of spheres **B** and **A**.

Suppose that on colliding, sphere **A** stuck to sphere **B** and they both moved off together (this can be made to happen using a small amount of plasticine stuck to sphere **B**'s surface – the plasticine's mass can be ignored).

d. With what initial speed do spheres **B** and **A** move after the collision?

e. How much energy is lost in the collision?

f. Where has this energy gone to?

g. How high do the combined spheres rise vertically before stopping momentarily?

h. Describe the subsequent motion of spheres **B** and **A**.

11. An object of mass m is released from rest and falls under gravity against air resistance. The object reaches a speed v after falling through a height h. Find the work done by the object against the air resistance.

12. A force, acting vertically upwards on an object of mass 10 kg, moves the object vertically from rest to a height 5 m above its starting point and gives it a speed of 6 m/s. Find the work done by the force (you can take g = 10 m/s^2).

13. Suppose there are three different situations where an object is sliding on a 'not frictionless' (ie rough) floor, and the floor has the orientation as illustrated in the following figure:

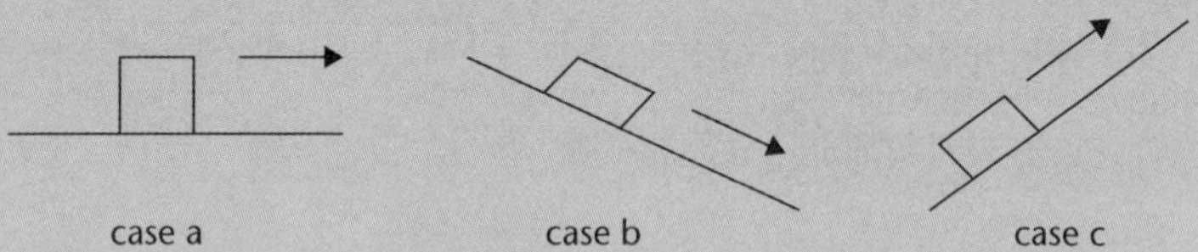

Kinetic frictional force: 3 situations

Suppose the object begins with the same speed in all three cases, and slides until the kinetic frictional force between the object and the floor makes the object stop. In which case is the minimum amount of mechanical energy **dissipated**? In which case is the maximum amount of mechanical energy dissipated?

Unit 11.4 Work, Power and Energy

Topic 4: Simple machines

'Simple machines' is a heading on p. 16 of the Syllabus and Topic 4 aims to cover the concepts and content summarised in the bullet points on pp. 16–17 of the Syllabus:

- Equilibrium.
- Sums of forces.
- Torque.
- Levers and other simple machines.

Forces are often used to rotate or turn objects. The turning effect of a force is called a **torque** or **moment**. The turning of a door handle, using a spanner to tighten a nut and the force applied to the pedal of a bicycle are examples of torques. The torque is always stated relative to a point called the **pivot**. To describe a torque, the size of the force and the perpendicular distance between the line of action of the force and the pivot must be known. Torque is calculated using the formula:

$$\tau = F \times d_{\perp}$$

F = Force d = perpendicular distance between line of action at force and pivot

The SI unit for torque is the Newton-metre, N m. The direction of the torque is usually stated as being clockwise or anticlockwise.

Equilibrium

An object is said to be in **equilibrium** when:

- the sum of the forces (ΣF) acting on the object is zero, and
- the sum of the torques acting on the object is zero.

Sum of the forces acting on an object are zero

When the sum of the forces acting on an object is zero, the object is at rest (**static equilibrium**) or the object is travelling at a constant speed (**dynamic equilibrium**):

$$\Sigma F = 0$$

(Σ means 'sum of')

Example A

A skydiver jumps out of a plane. His parachute is unopened. The forces acting on the skydiver are the weight force and a small upward force due to air resistance. The net downward force acting on the skydiver causes him to accelerate.

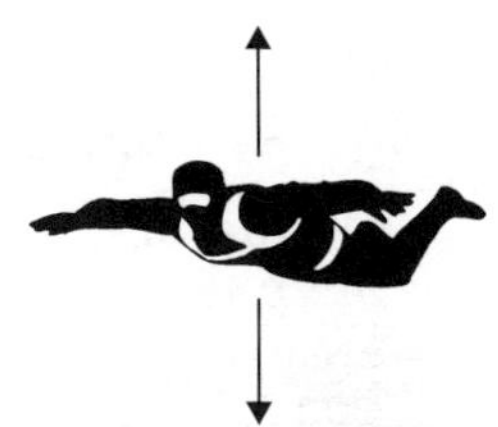

Forces acting on an accelerating skydiver.

After the skydiver opens his parachute, the upward air resistance force increases and the skydiver slows down. Eventually the upward force due to air resistance and the downward weight force are equal and opposite. The skydiver is now in dynamic equilibrium and has reached a constant **terminal velocity**.

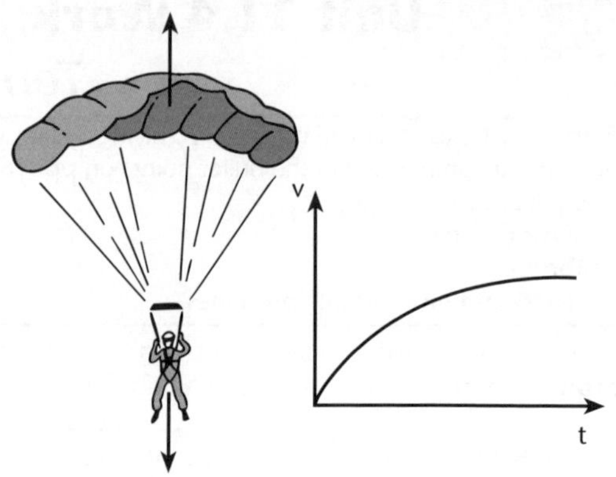

A skydiver who has reached terminal velocity.

Sum of the torques acting on an object is zero ($\Sigma\tau = 0$)

This implies that the sum of the clockwise torques acting on an object are equal to the sum of the anticlockwise torques acting on the object:

$$\Sigma\tau_{clockwise} = \Sigma\tau_{anticlockwise}$$

Example B

A metre rule of negligible mass is pivoted at the 30 cm mark. A mass weighing 20 N is suspended from the 70 cm mark. What weight must be suspended from the 20 cm mark to balance the metre rule?

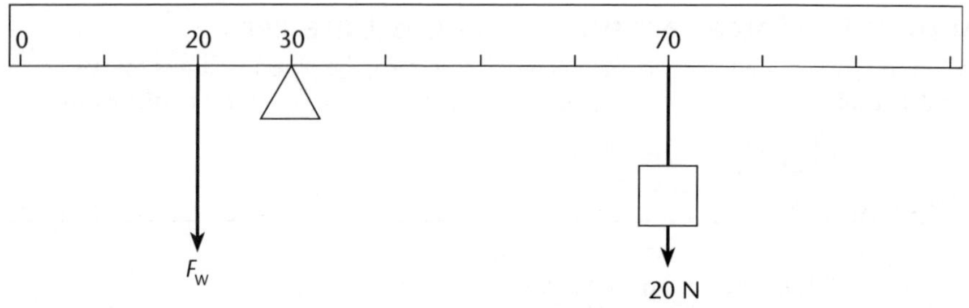

Balancing masses on a pivoted metre ruler.

$\Sigma\tau_{clockwise} = 20 \times (0.7 - 0.3)$
$\Sigma\tau_{clockwise} = 20 \times 0.4$
$\Sigma\tau_{clockwise} = 8$ N m
$\Sigma\tau_{anticlockwise} = 0.1\ F_W$
At equilibrium:
$0.1\ F_W = 8$ N m

$$F_W = \frac{8}{0.1} = 80 \text{ N}$$

Example C

A cyclist exerts a force of 450 N on a pedal. The torque exerted about the axle when the pedal is in the positions shown is calculated underneath each diagram.

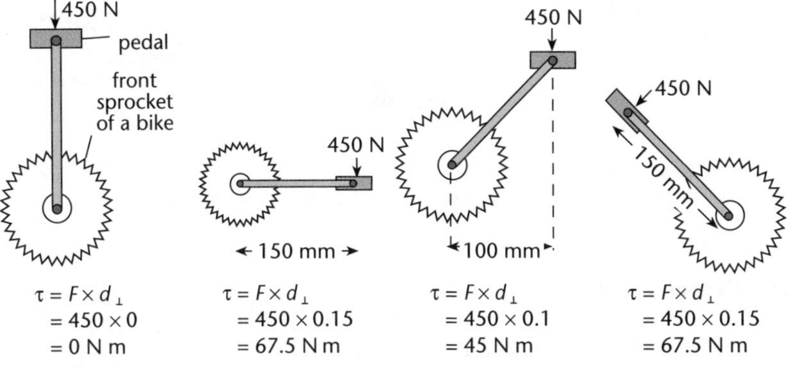

$\tau = F \times d_{\perp}$
$= 450 \times 0$
$= 0$ N m

$\tau = F \times d_{\perp}$
$= 450 \times 0.15$
$= 67.5$ N m

$\tau = F \times d_{\perp}$
$= 450 \times 0.1$
$= 45$ N m

$\tau = F \times d_{\perp}$
$= 450 \times 0.15$
$= 67.5$ N m

Levers

A lever is a simple machine that can be used to reduce the effort required to overcome a load force or to make a job easier to achieve.

There are three types of levers, grouped according to where the turning point (*fulcrum*) is.

First order levers

In a first order lever, the turning point is between the effort and the load. A small effort force can be used to overcome a large load force. *Archimedes was aware of this when he said "give me a lever long enough and I will move the world".*

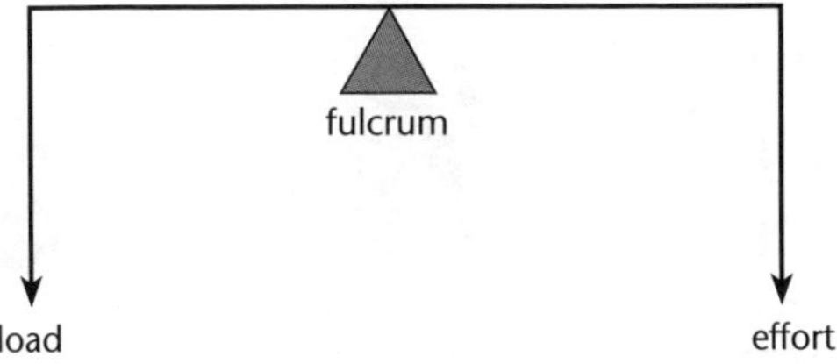

A first order lever.

There are a number of situations where first order levers can be useful:

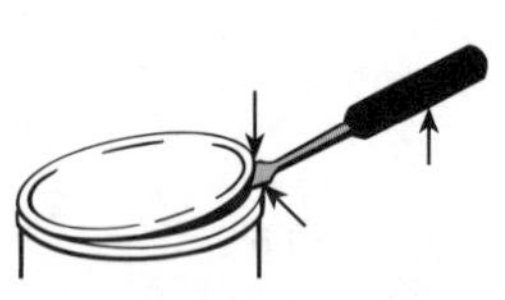

The screwdriver as an example of a first order lever.

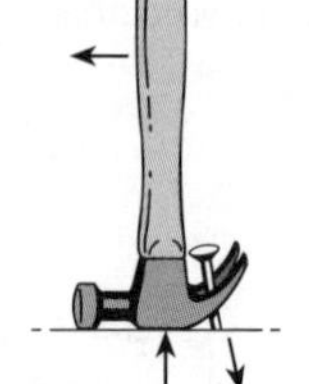

The claw hammer as an example of a first order lever.

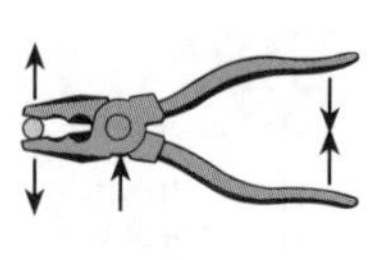

A pair of pliers as an example of a first order lever.

Second order levers

In a second order lever, the load acts between the turning point and the effort. A small effort force can be used to overcome a large load force.

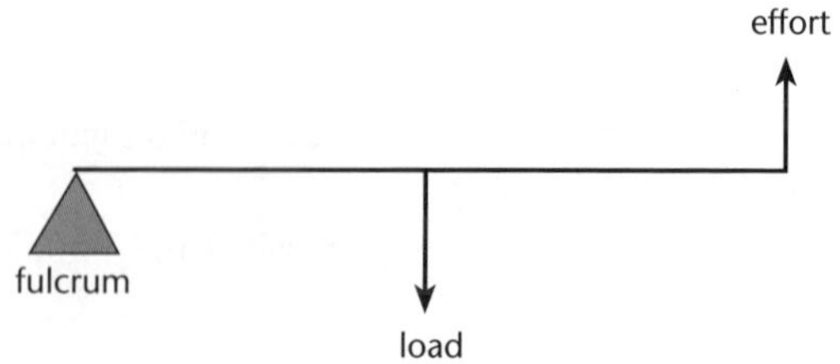

A second order lever.

There are a number of situations where second order levers can be useful:

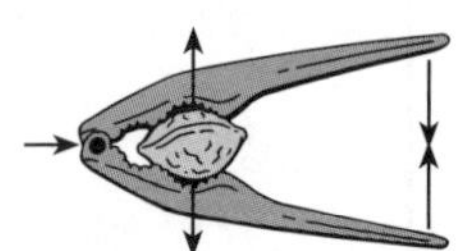

The nutcracker as an example of a second order lever.

The bottle opener as an example of a second order lever.

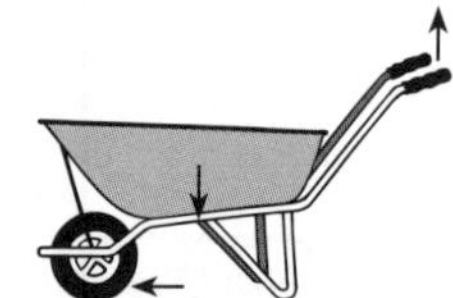

The wheelbarrow as an example of a second order lever.

Third order levers

In a third order lever, the effort acts between the turning point and the load. While the effort is always greater than the load, this type of lever is nevertheless convenient to use.

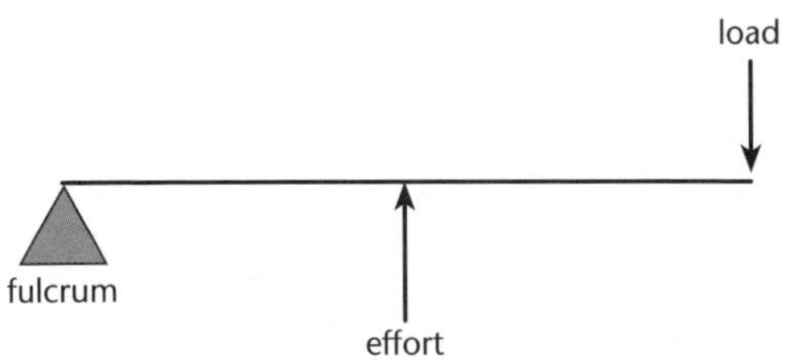

A third order lever.

There are a number of situations where third order levers can be useful:

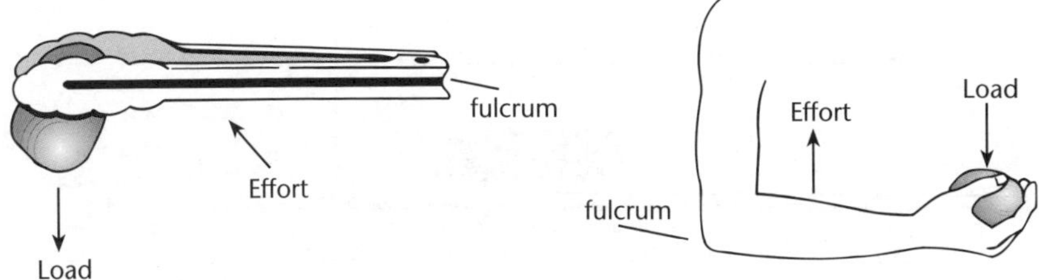

Food tongs as an example of a third order lever.

The human forearm as an example of a third order lever.

Unit 11.4 Activity 4A: Torque and equilibrium

Use the following information and the diagram to answer Questions 1 and 2.

Rose (weight 600 N) stands on a plank of weight 200 N which is supported by two trestles.

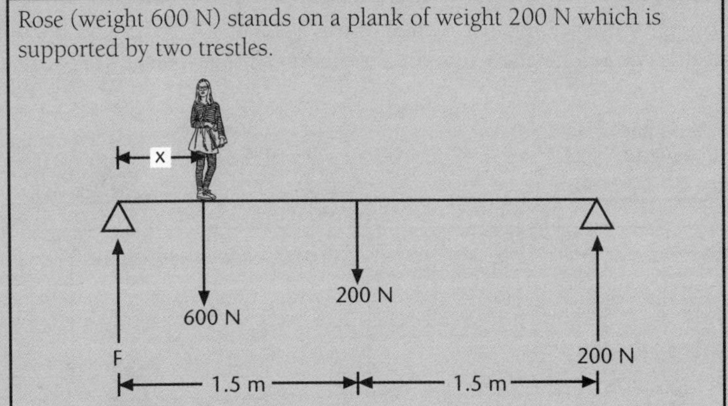

1. Calculate the force with which the left hand trestle supports the plank.
2. What is the distance marked **x** on the diagram?

3. Three balanced beams have been used to compare four different masses, A, B, C and D:

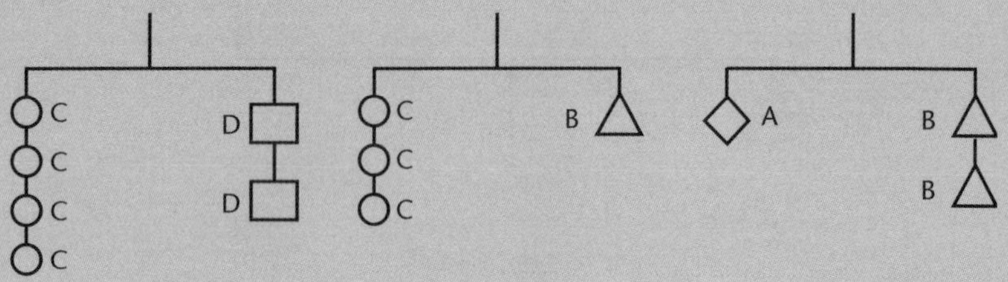

The mass of A is 24 g. Calculate the mass of D.

4. The diagram shows a locomotive on a bridge.

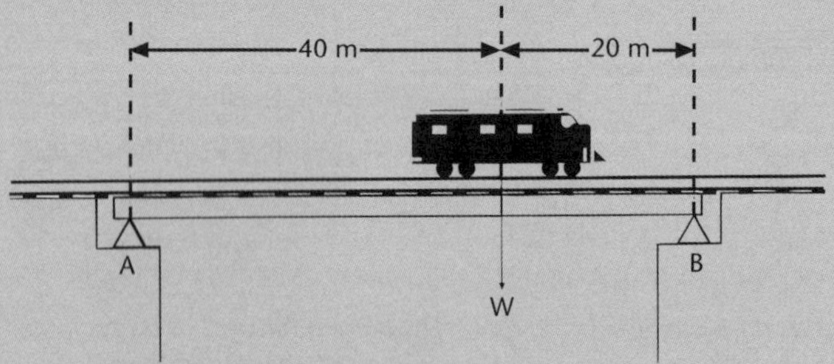

   a. The mass of the locomotive is 45 000 kg. Calculate its weight.
   b. By taking moments about **A**, find the increase in the upward force provided by support **B** when the locomotive is on the bridge in the position shown.
   c. Find the corresponding increase in the force provided by support **A**.

5. The diagram shows a mouse trap which works as follows – as the mouse walks towards the cheese, the plank tips and the mouse falls conveniently near to Boris the cat.

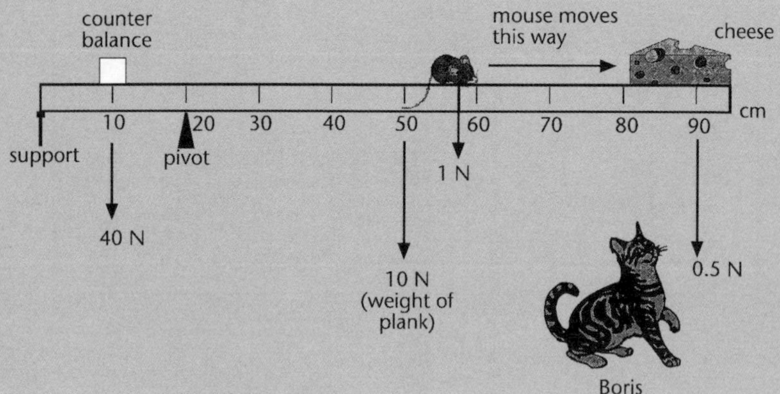

   a. Calculate the turning moment of the counter balance weight about the pivot.
   b. The weight of the plank may be taken to act through the 50 cm mark. Calculate the total turning moment of the cheese and the plank about the pivot.

  c. What is the minimum distance from the support beyond which the mouse will overbalance the trap and make Boris a happy cat?

6. Find the size of the force needed to balance the levers shown.

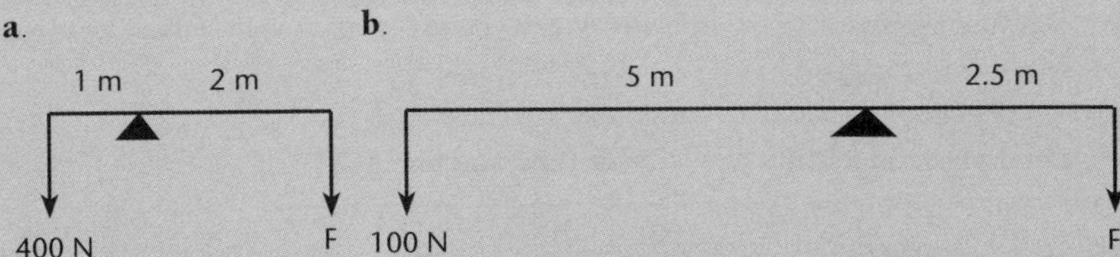

7. The diagrams show some simple machines. In each case, find the size of the effort force needed to lift the load.

a.

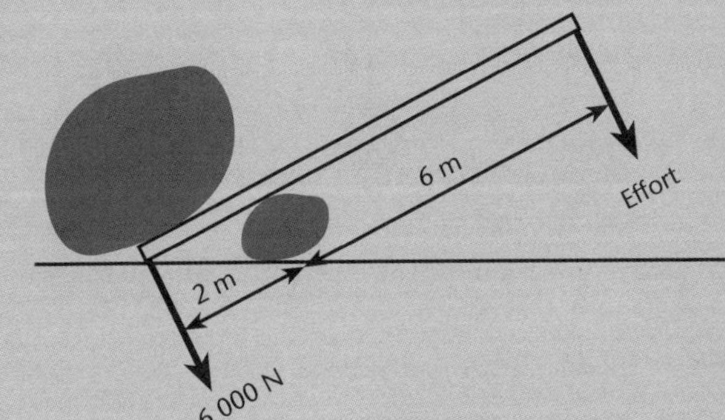

b.

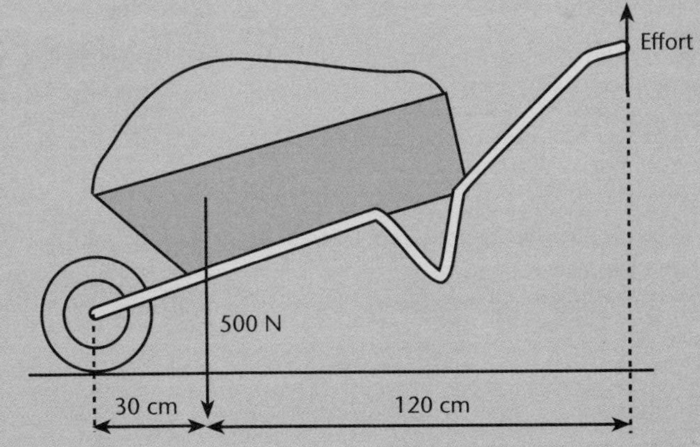

8. A box weighing 500 N is pushed up a ramp which is 4 m long. This lifts the box to a height of 1 m. The force necessary to push the box up the ramp is 160 N.

  a. How much work is done to shift the box up the ramp?

  b. If the box were to be lifted directly up to 1 m, how much work would be done?

  c. What is the:

    i. Advantage of using the ramp?

    ii. Disadvantage of using the ramp?

9. **a**. Why is the wheel brace used to tighten the wheel nuts of a truck larger than the wheel brace used to tighten the wheelnuts for a car?
   b. Explain why it is easier to operate a car jack that has a long handle.
   c. Draw a diagram of a bottle opener which you are familiar with. Label the pivot, effort force and load force. Explain how it works.

10. Write a list of ten simple machines under two headings:
    Inclined plane machines **Lever type machines**
    Choose one of each type and explain with the help of a diagram how it works.
    (**Note:** Often gadgets around the house and in the workshop are not simple machines but combinations of machines, eg the lawnmower.)

11. A front-wheel drive car is travelling at constant velocity. The various forces acting on the car are shown in the diagram.

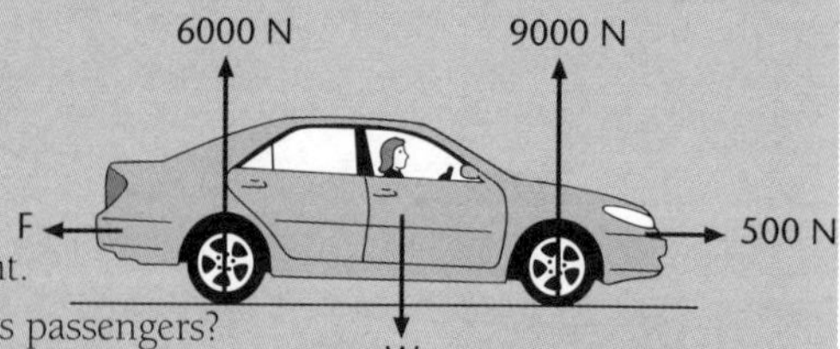

   a. Name the 500 N force on the right.
   b. What is the mass of the car and its passengers?
   c. What is the value of the force F on the left?
   The force shown on the right is now doubled. The other forces remain constant.
   d. Calculate the new resultant force on the car.
   e. Calculate the acceleration of the car.

(Source: NCEA Examination paper)

12. A skydiver jumps out of an aeroplane and begins to fall vertically down to earth.
   a. Complete the diagram alongside to show the forces acting on the skydiver.
   Label the forces.

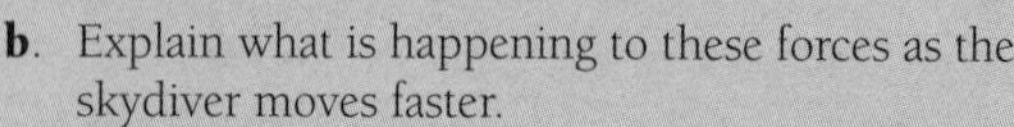

   b. Explain what is happening to these forces as the skydiver moves faster.

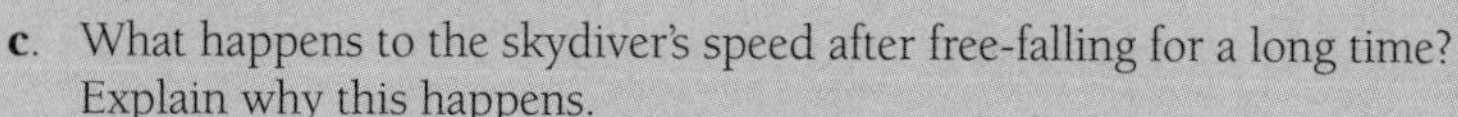

   c. What happens to the skydiver's speed after free-falling for a long time? Explain why this happens.
   d. Sketch a speed-time graph for the skydiver's descent prior to opening his parachute.

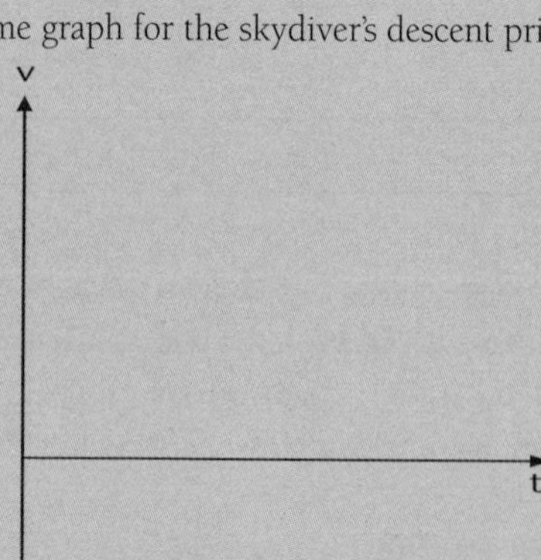

(Source: NCEA Examination paper)

Other machines

The ramp

A ramp is considered a machine. We can pull a load up a ramp using less force that we would need to lift the load vertically. Therefore, a ramp can be considered a force multiplier.

The pulley

A pulley or a system of pulleys is another commonly used machine. It can also be considered a force multiplier. For example, a set of pulleys called a block and tackle is usually used in garages for lifting an engine out of a car. Pulleys are also used in a crane for lifting cargo on to a ship, as shown in Figure 1.

Figure 2 shows an arrangement where a block and tackle is being used to raise a load. By using this arrangement, a heavy load of 90 N can be lifted by a much smaller effort. However, to compensate for this, the effort has to move a great distance to raise the load by a small amount. In this example, as shown, the effort has to move three times as far as the load.

Figure 1

Figure 2

Gears

Now that we know what a lever is, we can describe a gear as follows: linked gearwheels do for torques or turning moments what levers can do for forces; they multiply or reduce their strength.

The gearwheels in the following figure are similar to those used on a bicycle. In the one shown in the following figure, the output shaft turns at twice the speed of the input shaft. Thus, it only delivers half the torque (if we neglect the friction).

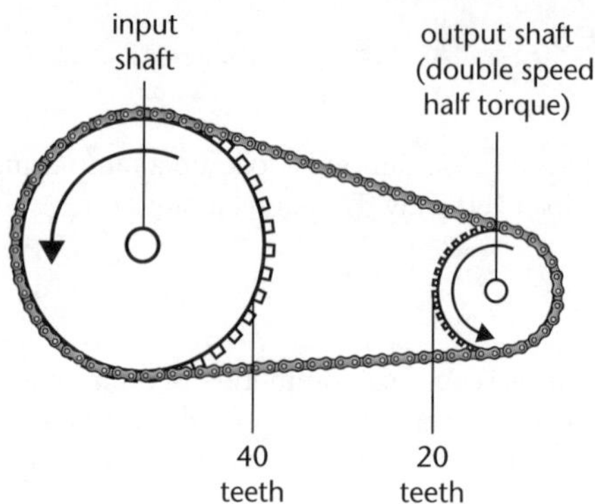

Example D

On many bicycles, when you change gear, the chain is usually moved in such a way that it fits over a different combination or set of gearwheels. When you change down a gear on your bicycle, you need to pedal faster to keep moving at the same speed. However, the torque on the rear wheel decreases so it is easier to cycle uphill.

Example E

On Figure 2 above, a 90 N load is lifted 2 cm by applying an effort, measured by a Newton balance. If the effort moves a distance of 6 cm, determine its value.

Solution:

Useful output energy = load × upward displacement of load = 90 × 0.02 = 1.8 Joules
Input Energy = effort × displacement of Newton balance = F × 0.06 (F = effort)
Assumption: all input energy is transferred to become useful output energy
Input Energy = Useful Output Energy
F × 0.06 = 1.8 then: F = 1.80/0.06 = 30 N

Efficiency

In practice, the effort required for the machine in Example E above is really found to be larger than 30 N. This means that the input energy actually needed is greater than the useful output energy. Not all the input energy is transferred by the pulleys to become useful output energy.

This is true for all machines. The useful energy we get from a machine is never as great as the energy we put in. Some of the input energy is always changed into other energy forms such as heat.

However, it is important to keep in mind that no energy is actually destroyed, so that the total amount of energy stays the same. Therefore, the total energy is always conserved.

We used the term efficiency as a way of expressing how well a machine converts input energy into useful output energy. This is usually expressed as the percentage of the input energy that finally appears as useful output energy.

$$\text{Efficiency (\%)} = \frac{\text{useful out energy}}{\text{input energy}} \times 100$$

Unit 11.5 Electricity Principles

Topic 1: Electrostatics

Unit 11.5 Topic 1 aims to cover the bullet points listed under 'Electrostatics' on p. 19 of the Syllabus:

- Structure of the atom.
- Electrostatics.
- The electroscope.
- The van der Graaf generator.
- Lightning.
- Conductors and insulators.
- Coulomb's Law.

Structure of the atom and electrostatics

In an uncharged or **neutral atom**, the number of positively charged **protons** in the **nucleus** is equal to the number of negatively charged electrons surrounding the nucleus. If a neutral atom gains electrons it becomes negatively charged and if it loses electrons it becomes positively charged.

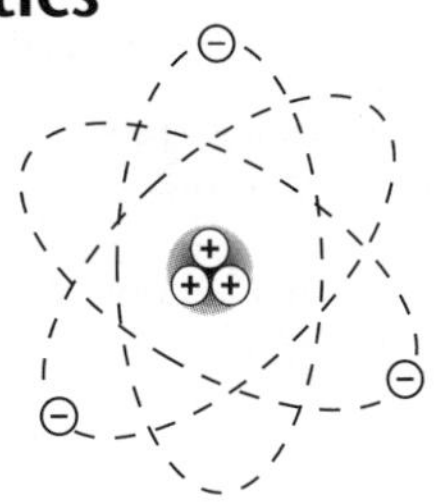

a neutral atom:
3 protons ⊕ 3 electrons ⊖

Structure of a neutral lithium atom.

Objects become charged when electrons are transferred from one object to another. Protons cannot be transferred because they are firmly held inside the nucleus of the atom.

A positively charged object has a smaller number of electrons than protons.
A negatively charged object has a larger number of electrons than protons.

Example A

When a glass rod is rubbed with a silk cloth, electrons are transferred from the glass to the silk. The glass rod becomes positively charged and the silk becomes negatively charged.

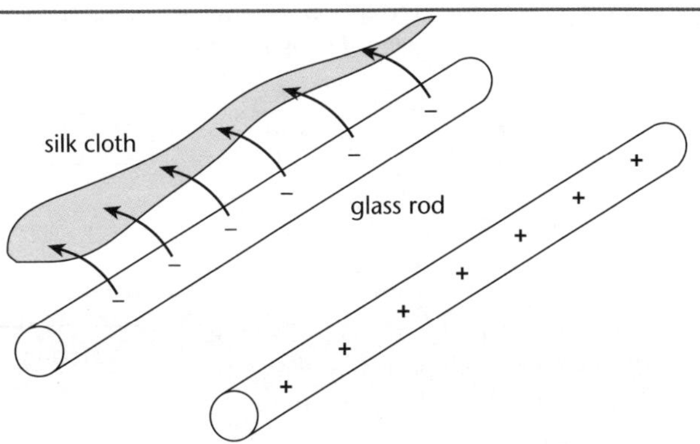

Charging a glass rod positively by rubbing with silk.

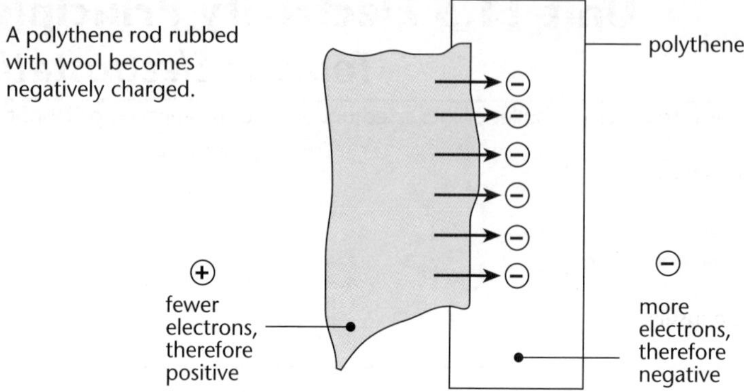

Charging a polythene rod negatively by rubbing it with wool.

Cars or planes may become charged as they move through the atmosphere because friction with the air causes electrons to be transferred.

When two charged objects are placed near each other they exert a force on each other. The direction of this force is given by the rule:

Like charges repel, unlike charges attract.

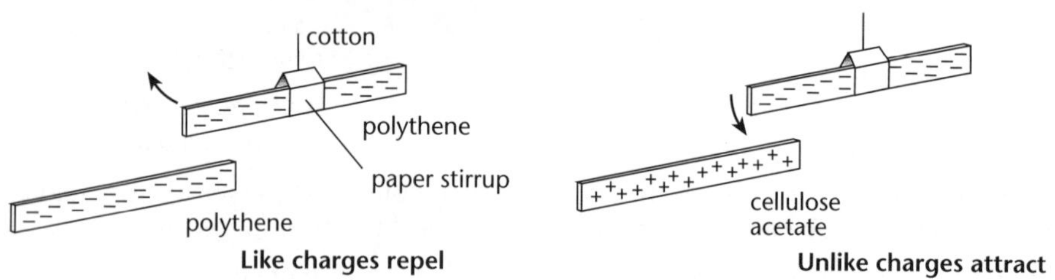

Like charges repel, unlike charges attract.

The ancient Greeks observed that when amber was rubbed with fur, it could pick up bits of straw. *The word electron is derived from the Greek word for amber (elektron).*

A charged object, such as a plastic pen rubbed on a woollen jersey, can pick up light, neutral bits of paper. The pen is positively charged and electrons are attracted to the top of the paper. Since opposites attract, the paper moves, attracted to the pen. Once the paper touches the pen, electrons are transferred to the pen. The pen and paper are now both positively charged. Since like charges repel, the paper drops.

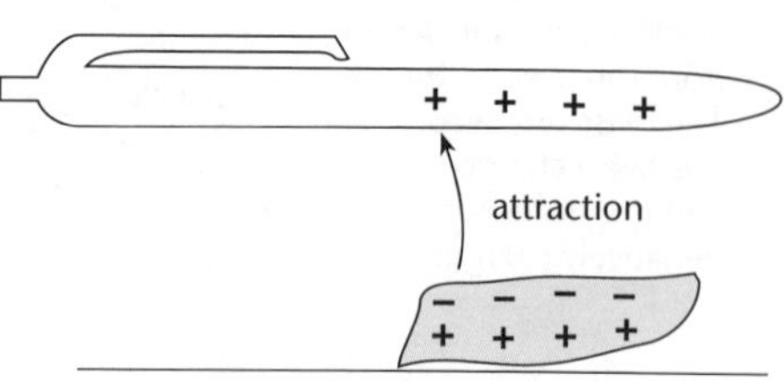

Picking up bits of paper using a charged pen.

Charges spread out over a conducting surface but are more concentrated near sharp corners.

Example B

Stephen Gray (1696–1736) suspended his boy servant from the ceiling on insulating ropes. When he brought a large positively charged object near the boy's feet, the boy's nose picked up bits of paper from the table top below.

Stephen Gray's experiment.

Unit 11.5 Activity 1A: Attraction and repulsion

1. A negatively charged thundercloud passes over a steel-framed building.
   a. What happens to the electrons in the building?
   b. Are the induced charges on the roof of the building negative or positive?
   c. Explain what happens to the charges when the roof is struck by lightning.
   d. What safety precaution can prevent a lightning strike from damaging the building?
   e. A 1 cm spark in air represents a potential difference of 20 000 V. The potential difference between the cloud and the building is 1×10^9 V. What is the length of the spark produced when a lightning strike hits the roof? Express your answer in kilometres.
   f. Tina sees a flash of lightning and hears the sound 3.5 s later. The speed of sound in air is 330 m s^{-1}. How far away is Tina from the source of the lightning?

2. Sally is a manager of a dress shop. It is important that Sally looks presentable for her job. She always brushes her hair thoroughly when she arrives at work but finds on dry days that her hair sticks out after she has brushed it.

   a. Use electrostatics to explain why Sally's hair sticks out after she has removed the brush.
   b. Sally's hair moves towards her brush if she holds it close to her hair after brushing. Explain why.
   c. Why does she not have this problem on damp days?

(Source: NCEA Examination paper)

3. John charges a balloon by rubbing it on his top.

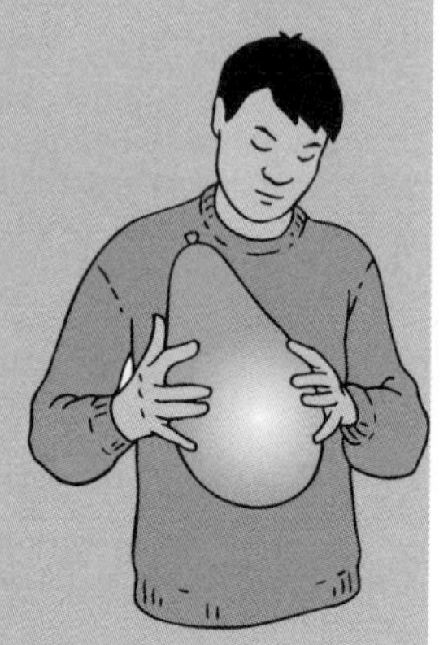

When he does this the balloon becomes negatively charged.

a. What is the resultant charge on John's top: positive or negative?

b. Explain why the balloon becomes negatively charged.

John holds the negatively charged balloon near a thin stream of water from a tap in the school laboratory as shown below.

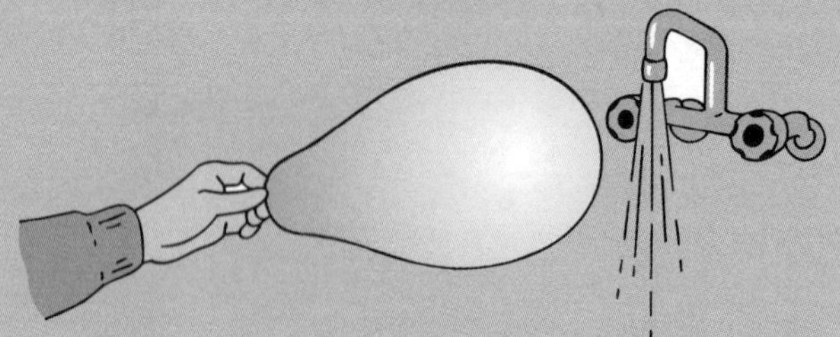

c. Describe what happens to the stream of water and explain your answer.

John then holds the balloon near a conducting rod (on an insulating base) next to an electroscope.

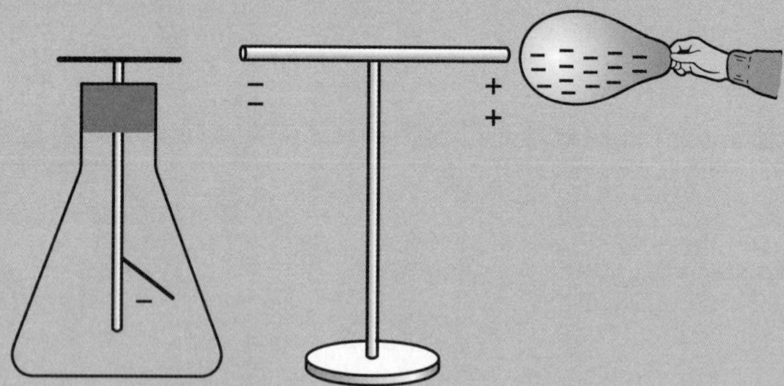

d. On the diagram above, draw the:

i. position of the leaf of the electroscope

ii. net charges at points A, B and C.

Next John holds the negatively charged balloon near a piece of paper on the bench. The paper is neutral.

e. On the diagram below, draw the position of the positive and negative charges on the piece of paper shown in the diagram.

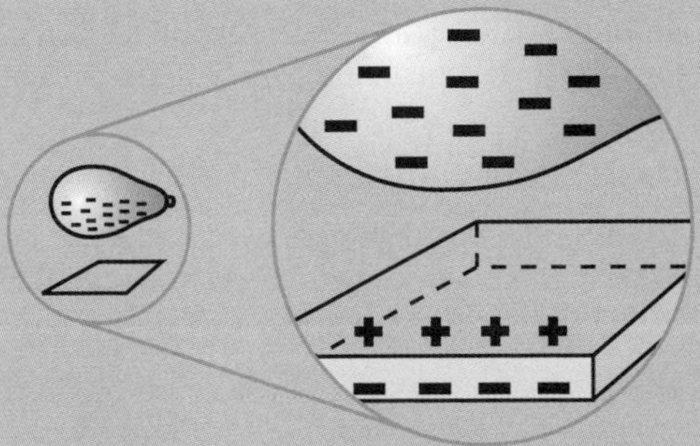

f. Describe what then happens to the piece of paper and explain your answer.

(Source: NCEA Examination paper)

The electroscope

An **electroscope** can be used to detect the type and quantity of charge on an object. It consists of an insulated metal rod and a thin leaf of metal foil placed in a glass jar or glass-fronted container.

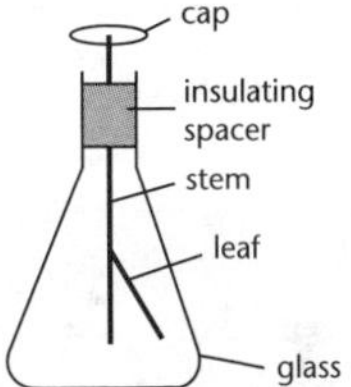

The electroscope.

If a negatively charged rod touches the cap of a neutral electroscope, electrons will travel from the rod to the leaf and stem of the electroscope. Since the leaf and stem are now both negatively charged, they repel each other and the leaf rises. The height to which the leaf rises is an indication of the quantity of charge transferred from the rod.

Likewise, if a positively charged rod touches the cap of a neutral electroscope, it will attract electrons from the electroscope, leaving the stem and leaf positively charged.

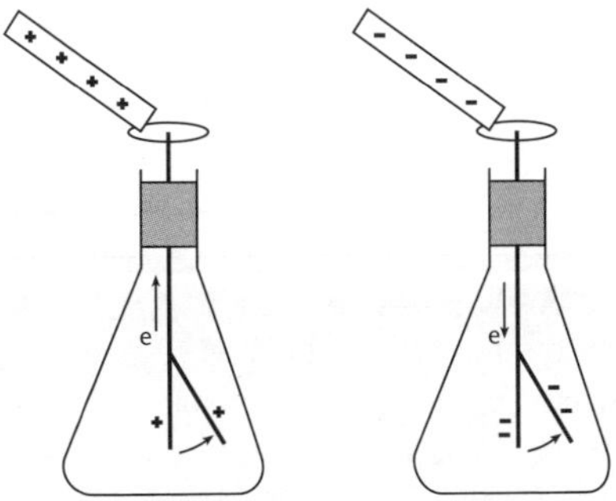

Charging an electroscope.

If a rod of unknown charge is brought close to, but not touching, the cap of a positively charged electroscope and the leaf rises further, the rod must have been positively charged; if the leaf falls, the rod must have been negatively charged.

Unit 11.5 Activity 1B: The electroscope

1. For each of the following situations, describe what will happen if you join the caps of two identical electroscopes that contain the same amount of charge with a conductor:
   a. Both electroscopes are positively charged.
   b. Both electroscopes are negatively charged.
   c. The electroscopes are oppositely charged.
2. A rod of unknown charge is brought close to the cap of a positively charged electroscope. The leaf of the electroscope rises. What does this indicate about the charge on the rod?
3. A rod touches the cap of a positively charged electroscope. The leaf of the electroscope remains in the same position. What does this indicate about the charge on the rod and the type of material the rod is made of?
4. An electroscope is charged by touching the cap with a positively charged rod. Next, the positively charged rod is brought close to the end of two touching metal balls on insulating stands.

   With the rod still in place, the balls are separated. The rod is then removed.

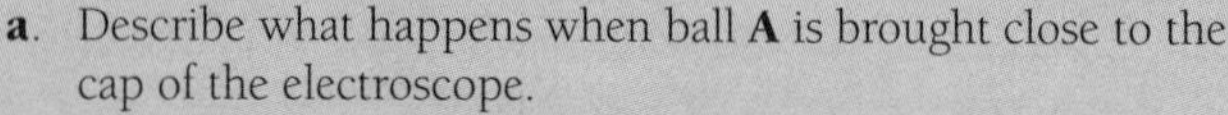

   a. Describe what happens when ball **A** is brought close to the cap of the electroscope.
   b. What does the result in **a** tell you about the charge on ball **A**?
   c. Describe what happens when ball **B** is brought close to the cap of the electroscope.
   d. What does the result in **c** tell you about the charge on ball **B**?

The following information and diagram apply to Questions 5 and 6.

A glass rod is given a positive charge by rubbing with a silk cloth. The rod is then placed close to, but not touching, the cap of an electroscope, as shown.

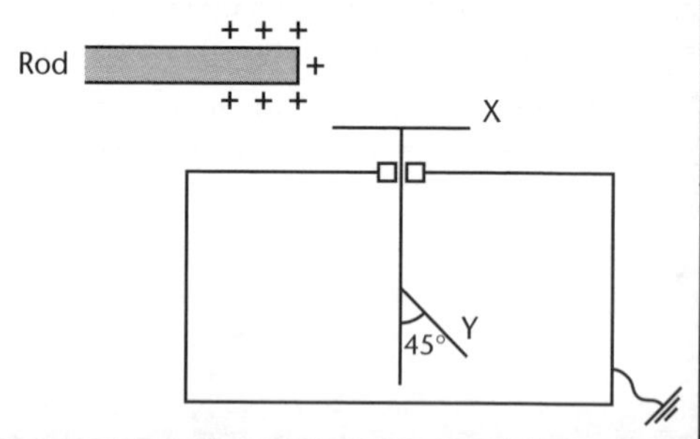

5. Why does the rod become charged?
6. What are the he charges on the cap and the leaf?
7. John frequently gets an electric shock when he gets out of his car.
   a. How does the car become charged?
   b. Why does John get a shock?
   c. What can be done to prevent this problem from recurring?

8. Tina charges a glass rod by rubbing it with a silk cloth. The cloth becomes negatively charged. She then holds the charged rod close to two touching iron rods elevated on insulating polystyrene beakers.

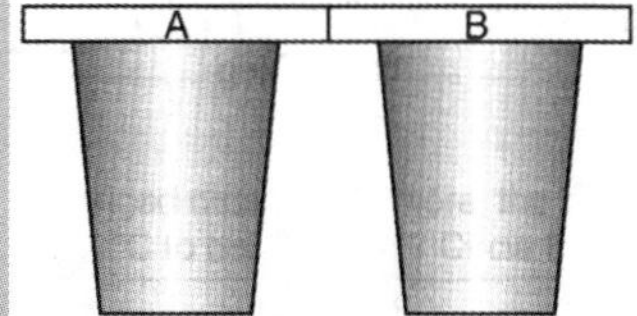

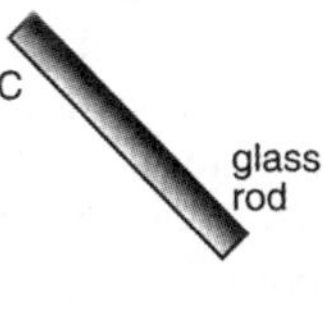

 a. What are the resulting charges on the rods A, B, and C?

 The rods are separated and the positively charged iron rod is touched on the cap of a neutral electroscope.

 b. Copy the diagram alongside and draw the position and charge on the leaf of the electroscope.

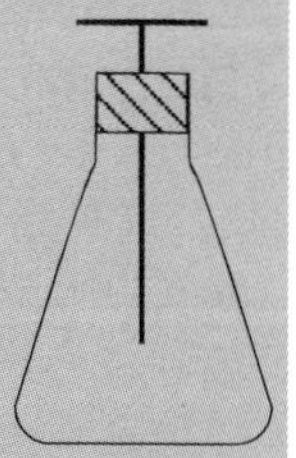

 The other charged iron rod is now brought close to, but not touching, the cap of the electroscope.

 c. What happens to the leaf of the electroscope?

9. A glass rod is rubbed with a silk cloth and becomes positively charged. The charged rod is brought near the cap of an uncharged electroscope, but does not touch it. The leaf deflects as shown in the diagram. What are the charges on the leaf and cap?

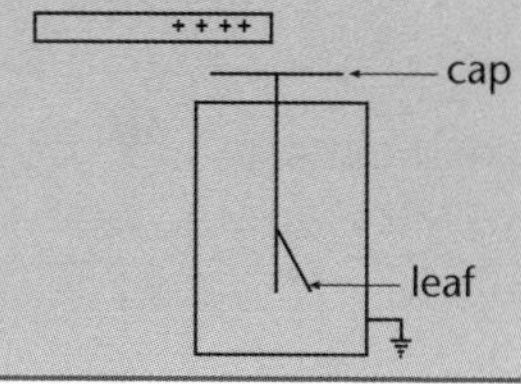

Conductors and insulators

A **conductor** is a material that allows electrons to move easily through it. The best conductor is silver. Connecting wires in circuits are made of copper rather than silver because silver is too expensive. Metals are good conductors because the outer electrons in metal atoms are loosely held and so free to move from atom to atom.

An **insulator** is a substance that does not allow electricity to travel through itself.

Good conductors	Poor conductors	Good insulators
The metals	Water	Dry air
	Cotton	Plastic
	The human body	Rubber
	Wood	Porcelain

The circuit shown can be used to test if a substance is a conductor or an insulator. If the bulb glows the substance is a conductor.

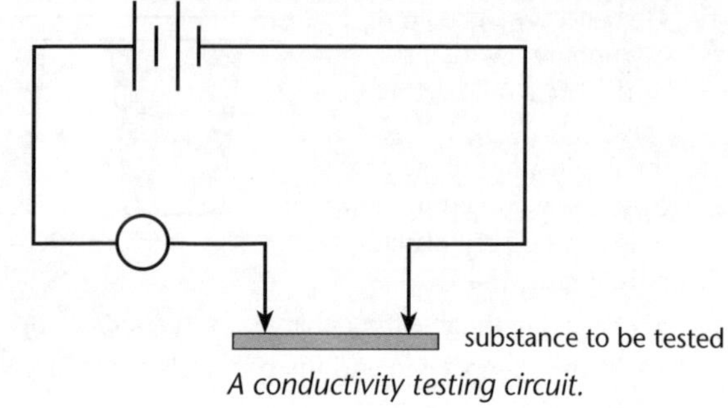

A conductivity testing circuit.

The van der Graaf generator

The **van der Graaf generator** is a machine used to charge objects.

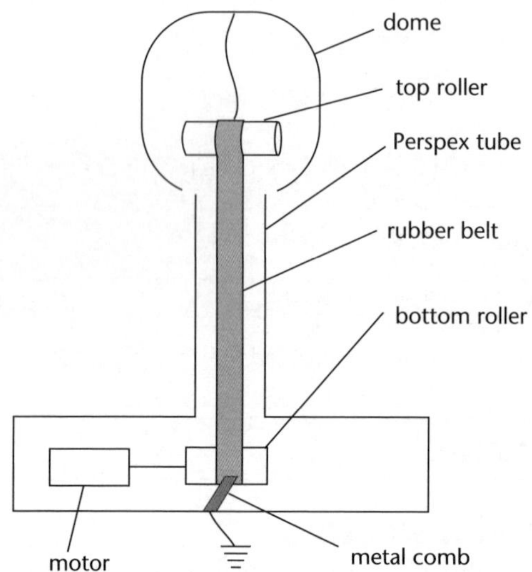

The van der Graaf generator.

The electric motor rotates the bottom roller. This causes the rubber belt to rub past the metal comb that is connected to Earth. The friction between the rubber belt and the comb brushes electrons from the belt onto the comb. This leaves the belt positively charged. At the top of the generator a metal comb connected to the dome brushes against the belt. Electrons pass from the dome of the van der Graaf generator onto the positively-charged belt and are carried away. The dome becomes positively charged.

Example C

If somebody touches the dome of the van der Graaf generator while standing on an insulator (such as a polystyrene slab), electrons from the person will be transferred from the person to the dome. The person's hair will become positively charged, the individual strands repel each other and depending on the type of hair, it may stand up or the person may feel it move.

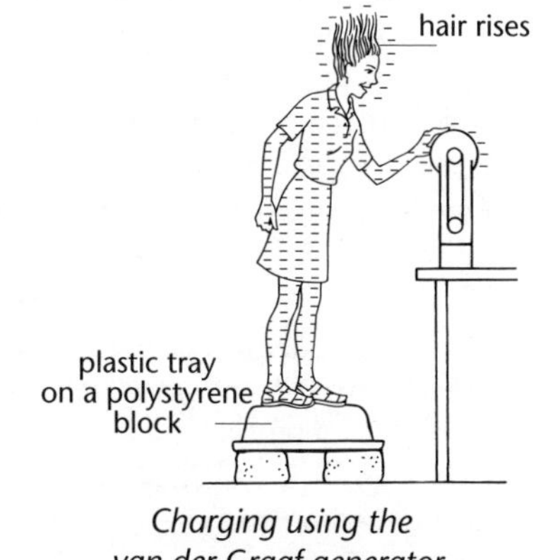

Charging using the van der Graaf generator.

Lightning

When clouds and air currents move past each other, the clouds may become negatively charged through friction. The electrons on the cloud repel each other and a **potential difference** or **voltage** is set up between the cloud and the earth. When electrons discharge to earth through air, lightning occurs. The length of the lightning bolt is an indication of how much electrical energy is discharged – a 1 cm spark in air represents a potential difference of 20 000 V.

Lightning can do a lot of damage. A tall building or tree is most likely to be hit by lightning, so tall buildings are often protected by a lightning rod that acts as a path for charges to safely flow to earth.

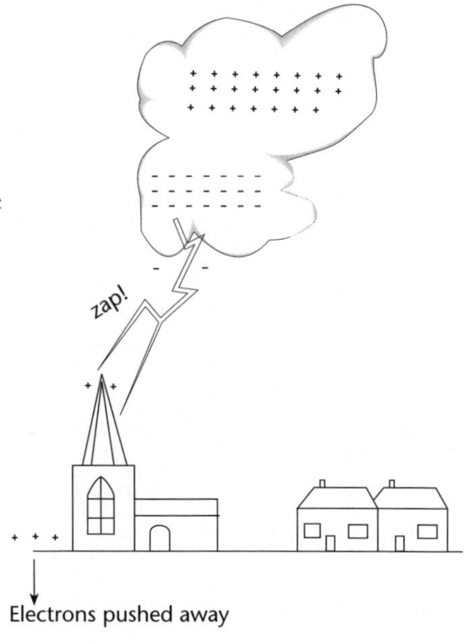

Protection against lightning strikes.

Unit 11.5 Activity 1C: Thunder and lightning

1. Two balloons are found to attract each other. What can be said about the sign of the charge on each of the balloons?
2. Ball A attracts ball B, ball A repels ball C and ball C attracts ball D. Ball D is negatively charged. What are the charges on balls A, B and C?
3. Rose has charged a glass rod positively by rubbing it with silk.
   a. What happened to the silk?
   b. Describe how you could test your answer to question (a).
4. Draw a diagram to explain how a DC power supply can be used to charge an electroscope. Clearly show the signs on the terminals, the connections and the charge on the electroscope.
5. Use diagrams to explain why a positively-charged plastic rod can be used to pick up neutral bits of paper.
6. Takis rubs a pen on his top and holds it near to a thin stream of water from the tap in the physics lab.
   a. Describe what Takis observes.
   b. In terms of charges, explain what happens at each stage of Takis's demonstration.
7. Why should people not seek shelter under a tree in a paddock during a thunderstorm?
8. The diagram below shows a small positively charged ball suspended between two charged rods. In which direction will the ball move?

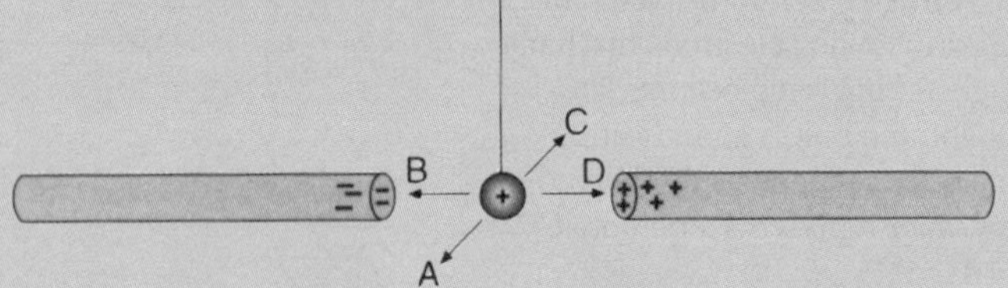

Use the diagram on the right to answer Questions 9 and 10.

9. What charges (if any) are induced at P and Q on the sphere when a positively charged rod is brought near?

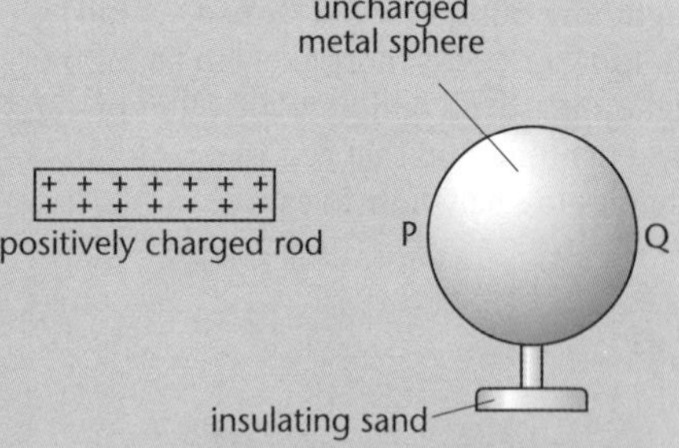

10. The sphere is then earthed by touching it with a finger at P while the positively charged rod remains present. The finger is then removed, while the rod remains. What charges are now present at P and Q?

	Charges at P	Charges at Q
A.	no charge	no charge
B.	no charge	positive
C.	negative	no charge
D.	negative	negative

11. Ruth and Mark are investigating static electricity. Ruth rubs a piece of plastic rod with a cloth. The plastic rod becomes negatively charged.

a. Copy the diagram and on your diagram show what happens to *both* the plastic rod and the cloth.

The charged plastic rod is then brought close to an initially uncharged conducting ball suspended from a cotton thread as shown.

b. Copy and complete the diagram to show the charges on the left and right side of the ball.

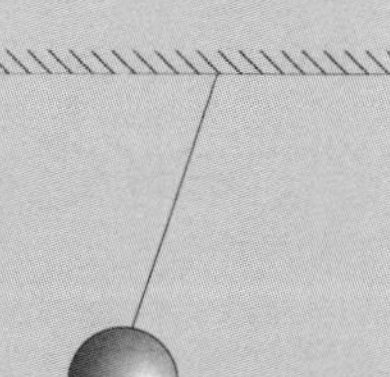

Mark briefly touches the ball with his finger while the plastic rod is still next to it.

c. Describe what happens.

d. What is the name of the process by which the ball is charged?

Ruth removes the plastic rod.

e. Describe the final position and charge on the ball.

12. A plane has just landed and is parked at the airport.

a. Explain why the plane may have become electrically charged.

b. The plane needs to be refuelled. The fuel tank is in the plane's wing.

i. What is the main safety precaution the tanker driver needs to take before connecting the fuel hose to the plane?

ii. Explain what may happen it the safety precaution in **i** is not taken.

Coulomb's Law

An electrically charged particle exerts a force on other electric charge particles. The magnitude of the force exerted depends on several factors.

The French physicist Charles Coulomb (1736–1806) investigated electric forces in the 1780s and found out that when two charged particles were placed at finite distances, the magnitude of the force exerted on to each other was proportional to the distance between them and the quantity of charge possessed by each particle. Coulomb's findings contributed to the formulation of Coulomb's Law.

Coulomb's Law is stated as:

The magnitude of the force (F) between two charged particles, which are small compared with their separation (r), is inversely proportional to r^2, and directly proportional to the product of their charges (Q_1 and Q_2).

Mathematically, Coulomb's Law can be expressed as:

$$F \propto \frac{Q_1 Q_2}{r^2}$$

That is: F is directly proportional to Q_1 and Q_2 but inversely proportional to r^2.

A constant of proportionality is required to turn this relationship into an equation.

$$F = k\frac{Q_1 Q_2}{r^2}$$

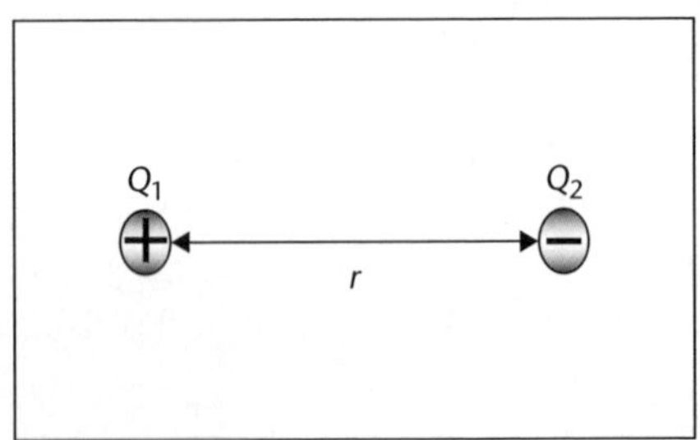

Two charged particles: Q_1 and Q_2 separated by distance r.

where F is the force, Q_1 is the charge on particle 1 and Q_2 is the charge on particle 2, r is the distance between particle 1 and particle 2, and k is a constant of proportionality. The value of k depends on the **medium** in which the charges are situated.

For most calculations at standard conditions the average value of k can be taken as $k = 9.0 \times 10^9\ Nm^2C^{-2}$ in SI unit.

For two particles of the same sign, a repulsive force of magnitude given by the equation $F = k\frac{Q_1 Q_2}{r^2}$ will be exerted on to each other. This means that the two particles are going to experience a repulsive force equal in magnitude but opposite in direction. Since force is a vector quantity, the direction of the force will be along the line joining the two particles.

If, on the other hand, the two particles are of opposite signs, then they are going to exert an attractive force on to each other which is equal in magnitude and directed toward one another.

The example below gives the force between two charged particles. If more than two charged particles are present, then the net force on any one of them, due to the rest of them, is the vector sum of the forces exerted by them (see Example E below).

Example D

Calculate the electrostatic force between two electrons separated by a distance of 10^{-10} m.
Solution:

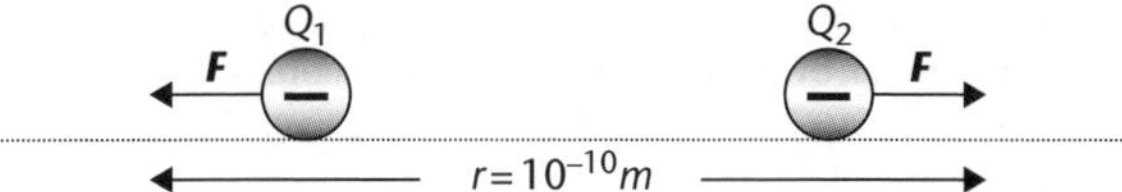

The figure shows the two electrons Q_1, and Q_2 separated by distance *r*. The arrows show the direction of the force on each electron.

The magnitude of the force is given by Coulomb's law as:

$$F = \frac{kQ_1Q_2}{r^2} = \frac{(9.0 \times 10^9 Nm^2C^{-2})(1.6 \times 10^{-19}C)(1.6 \times 10^{-19}C)}{(1 \times 10^{-10}m)^2} = 2.3 \times 10^{-6}N$$

The direction of the force is away from each other as shown in the diagram.

Note: Since force is a vector quantity, you must make sure that you state the magnitude and the direction of the force. In a test situation, you will lose marks if you state only the magnitude of the force without the direction. A vector quantity is completely described by its *magnitude* and *direction*.

Example E

In the following diagram, calculate the resultant force on charge Q_3 due to charges Q_1 and Q_2. The charges are $Q_3 = +65\mu C$, $Q_2 = +50\mu C$, $Q_1 = -86\mu C$. The distances and angles between the charges are as shown in the figure.

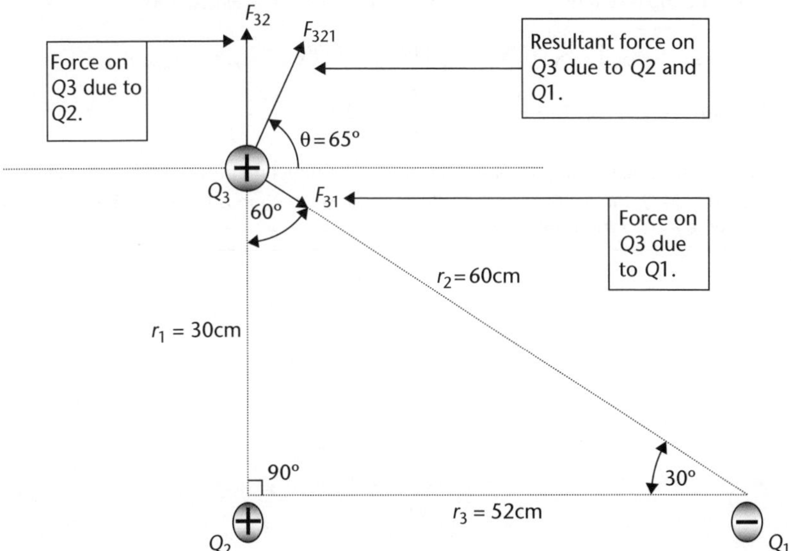

Solution:

Let the force on Q_3 due to Q_2 alone be F_{32}. Let the force on Q_3 due to Q_1 alone be F_{31}. Let the resultant force on Q_3 due to Q_1 and Q_2 be F_{321}. Let the distance between Q_3 and Q_2 be r_1 and that between Q_3 and Q_1 be r_2.

The force on Q_3 due to Q_2 is a repulsive force and that on Q_3 due to Q_1 is an attractive force as shown in the Figure by the arrow directions at Q_3. The resultant force on Q_3 is the **vector sum** of the forces F_{32} and F_{31}. That is: $\mathbf{F}_{321} = \mathbf{F}_{32} + \mathbf{F}_{31}$.

Now calculate F_{32} and F_{31}:

$$F_{31} = \frac{kQ_3Q_1}{(r_2)^2} = \frac{(9.0\times10^9 Nm^2C^{-2})(65\times10^{-6}C)(86\times10^{-6}C)}{(0.60\,m)^2} = 140N$$

$$F_{32} = \frac{kQ_3Q_2}{(r_1)^2} = \frac{(9.0\times10^9 Nm^2C^{-2})(65\times10^{-6}C)(50\times10^{-6}C)}{(0.30\,m)^2} = 330N$$

Note that the signs had been ignored, since we know the directions of the forces F_{32} and F_{31}.

Next, resolve F_{31} into its components along the *x* and *y* directions.

$F_{31x} = F_{31}\cos 30° = 120N$. This is the *x* component of F_{31}.

$F_{31y} = -F_{31}\sin 30° = -70N$. This is the y component of F_{31}.

Next, resolve F_{32} into its components.

$F_{32x} = F_{32}\cos 90° = 0N$. This is the *x* component of F_{32}.

$F_{32y} = F_{32}\sin 90° = 330N$. This is the *y* component of F_{32}.

Now calculate the components of the resultant force F_{321}.

The *x*-component of the resultant (F_{321x}) is the vector sum of the individual *x*-components.

$F_{321x} = F_{32x} + F_{31x} = 0N + 120N = 120N$. This is the *x* component of the resultant force.

The *y*-component of the resultant (F_{321y}) is the vector sum of the individual *y*-components.

$F_{321y} = F_{32y} + F_{31y} = 330N + -70N = 260N$. This is the *y* component of the resultant force.

Now calculate the resultant force F_{321} from its components calculated above.

$$F_{321} = \sqrt{(F_{321x})^2 + (F_{321y})^2} = \sqrt{(120N)^2 + (260N)^2} = 290N.$$

The direction is given by: $\tan\theta = \frac{F_{321y}}{F_{321x}} = \frac{260N}{120N} = 2.2$

Therefore θ = 65°. See the figure on p. 209 to see the direction of the resultant force.

The net force on Q_3 due to Q_2 and Q_1 is $290N$ and the direction is 65° from the positive *x*-axis anti-clockwise as shown in the figure.

Unit 11.5 Electricity Principles

Topic 2: Current Electricity

Unit 11.5 aims to cover the content on 'Electricity Principles' as set out in the Syllabus and this Topic 2 aims to cover the bullet points listed under 'Current electricity on p. 19 of the Syllabus:

- Voltage.
- Current.
- Circuit diagrams.
- Series circuits.
- Parallel circuits.

A complete **electric circuit** consists of a power source, an unbroken pathway for current and electrical **components** that use or control energy. It may also include meters, such as ammeters, **voltmeters** or multimeters.

The power source supplies the electrons with energy to move around the circuit. Examples of power sources are the **dry cell** or **power pack**. The positive terminal of the supply is normally coloured red and the negative terminal coloured black. Two or more cells connected in **series** are called a **battery**.

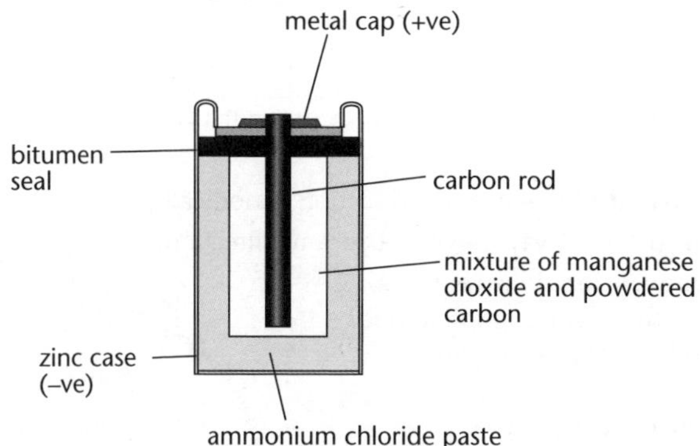

Diagram is of a Leclanchè cell – the cell produces a voltage of 1.5 V.

A primary cell.

Voltage

Voltage, V, or potential difference (pd) is measured in **volts** (V), using a voltmeter. A voltmeter can be used to measure the energy that is supplied by the power source or the energy used up in a component. Voltmeters are always connected in parallel across a component or power source. The positive terminal of the voltmeter must be connected to a wire that leads back to the positive terminal of the supply.

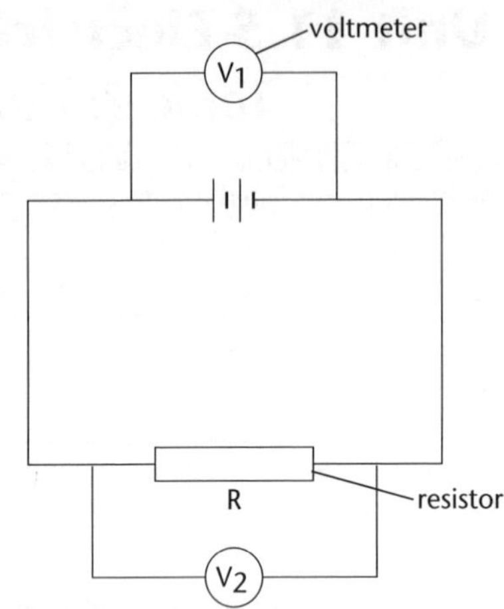

Energy in an electric circuit.

Current

Electric current, *I*, consists of a flow of electrons through a circuit. There are two types of current:

- **Alternating current** (AC) – changes direction periodically.
- **Direct current** (DC) – always flows in the same direction.

The direction of **conventional current** in a circuit is away from the positive terminal (red) of the power supply towards the negative terminal (black) of the supply. Electrons flow in the opposite direction – away from the negative terminal towards the positive terminal.

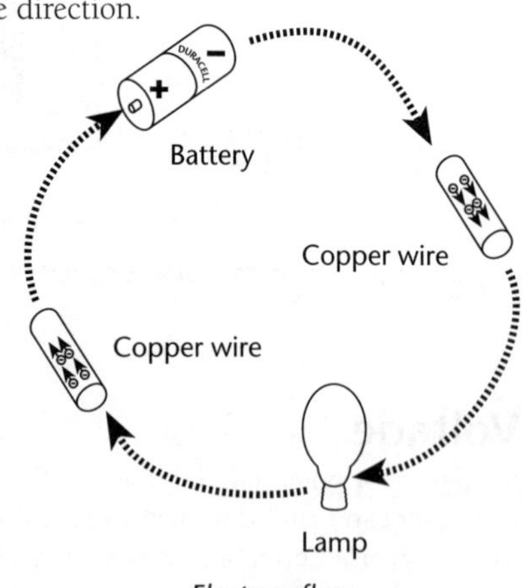

Electron flow.

Current flow is measured in **amperes** (amps). A current of one ampere means 6×10^{18} electrons pass a point in a circuit every second.

The current at a point in the circuit can be measured using an **ammeter**. Ammeters are always connected in series. The positive terminal of the ammeter must be connected to a wire that leads back to the positive terminal of the supply.

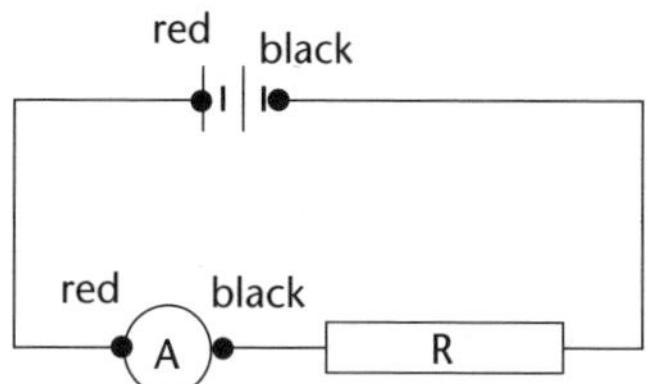

Connecting an ammeter.

Circuit diagrams

Circuit diagrams use symbols to represent electrical components. Diagrams must be neatly drawn using a ruler and without any gaps in the connecting wires. The table below shows the electrical symbols and describes the function of each component.

Component	Symbol	Function
Cell		Supplies energy to the electrons in the circuit
Battery		Two or more cells in series
Wire		Conducts electricity around a circuit
Wires crossing (not joined)		
Wires crossing (joined)		
Resistor	or	Changes electrical energy to heat energy
Rheostat	or or	A variable resistor used to control the current in a circuit
Lamp		Converts electrical energy to light and heat energy
Fuse		A wire of low melting point that breaks a circuit if the current exceeds a certain value
Ammeter	A	Meter that measures current in a circuit
Voltmeter	V	Meter that measures the potential difference between two points in a circuit
Diode		Allows current to pass through it in one direction only

Electrical components, symbols and functions.

There are two ways of joining electrical components in a circuit – *series* or *parallel*.

Series circuits

In a **series circuit** there is only one pathway for current to flow.

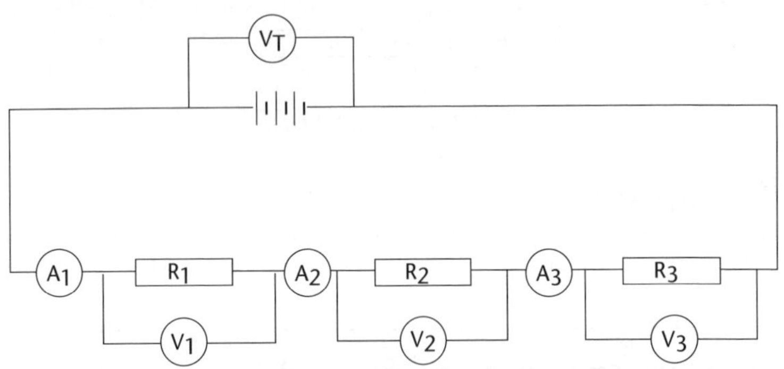

A series circuit.

The current passing through each component is the same:

$I_1 = I_2 = I_3$.

The sum of the voltages across each component adds up to the supply voltage:

$V_T = V_1 + V_2 + V_3$.

Example A

Christmas tree lights are often connected in a series circuit. If one lamp blows, the circuit is broken and none of the lamps will glow. Since all the lamps are connected in series, they share the energy of the supply. Adding additional lamps to a series circuit will reduce the brightness of all the individual lamps.

Parallel circuits

In a **parallel circuit** there are two or more pathways for the current.

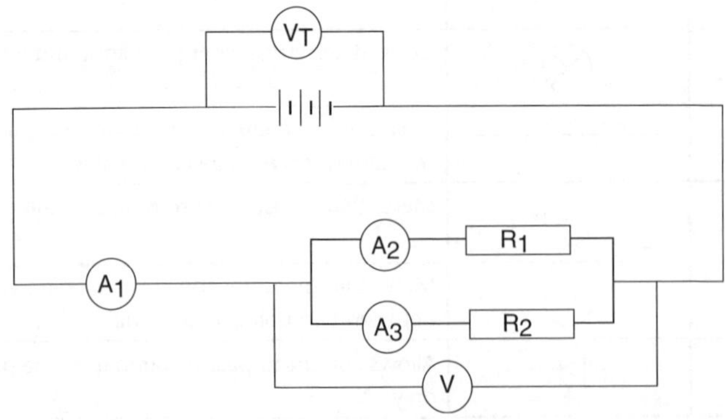

A parallel circuit.

Current splits up when it reaches a junction in the circuit:

$I_T = I_1 + I_2$.

The voltage across each branch is the same:

$V_T = V_1 = V_2$.

Example B

A house electrical lighting circuit is a parallel circuit. All the lamps are connected in parallel. If one lamp blows, the other lamps will still glow. Since all the lamps are connected in parallel they each receive the same amount of energy (230 V). Adding additional lamps to a parallel circuit will *not* affect the brightness of the individual lamps.

Unit 11.5 Activity 2A: Circuits

1. Complete the following conversions:

a. 230 ma = A

b. 23 kV = V

c. 8.5 A = mA

2. State the readings on the following meters:

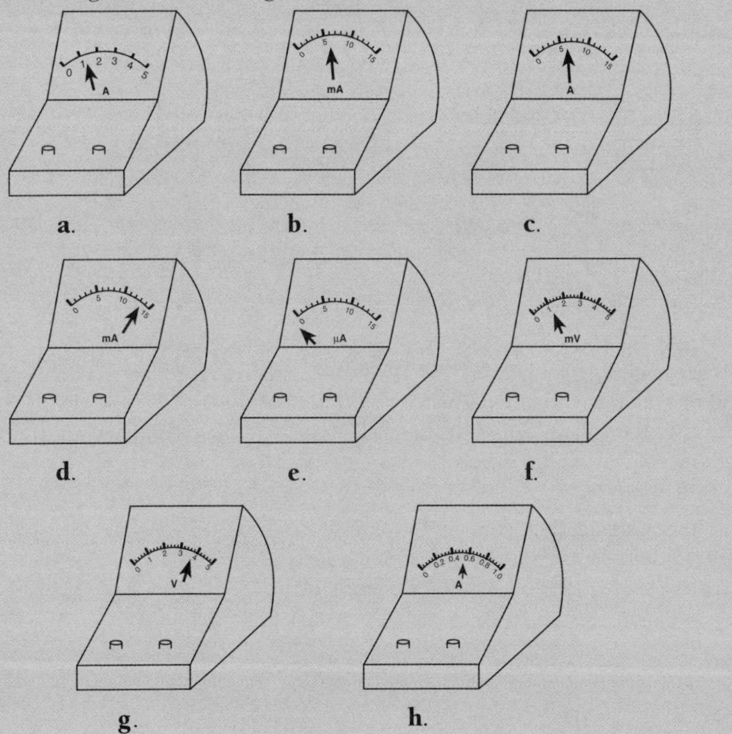

3. Draw a circuit diagram for each of the following circuits.

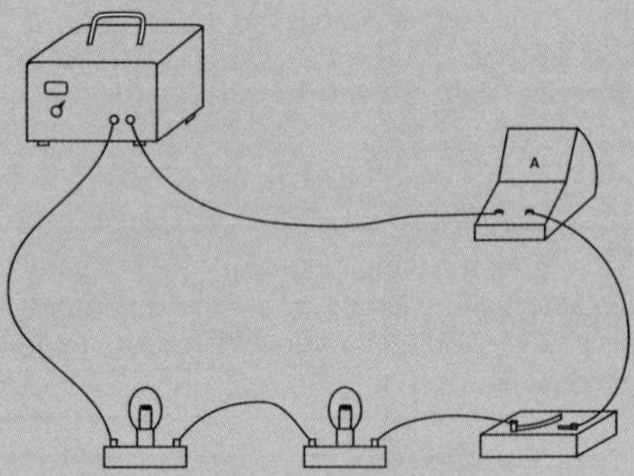

a. 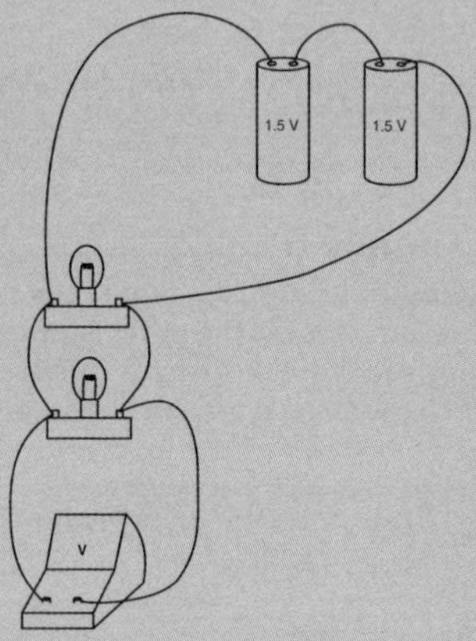 **b**.

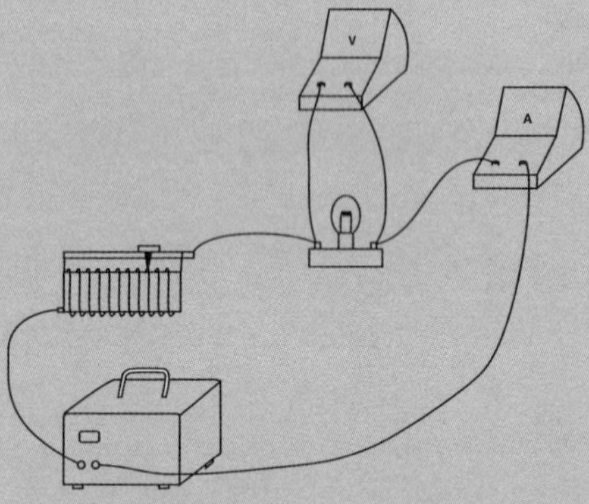

c. 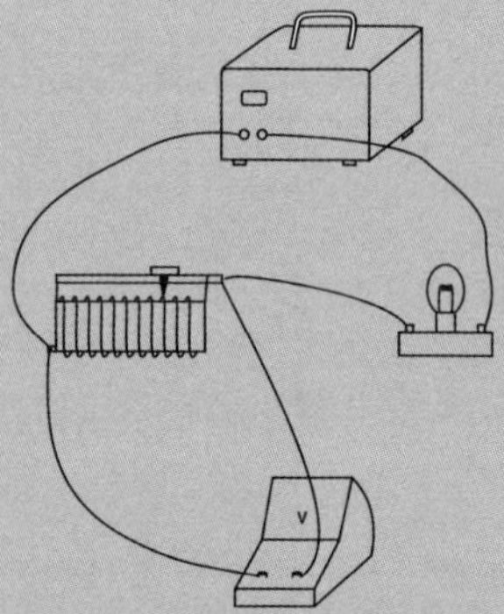 **d**.

4. Draw a circuit diagram for a 12 V DC supply circuit that contains 3 lamps in parallel and one lamp in series. One switch controls all of the lamps in the circuit. A voltmeter measures the voltage across the parallel lamp combination and an ammeter measures the current leaving the supply.

5. Copy the diagram then connect the lamps, switches and battery shown so that either lamp may be switched on and off independently of the other.

6. Copy the diagram then connect the 2-volt cells and other circuit components shown so that the 6-volt lamp lights up and the current through the lamp can be measured and varied.

7. State whether each of the meters in the following circuit is a voltmeter or an ammeter.

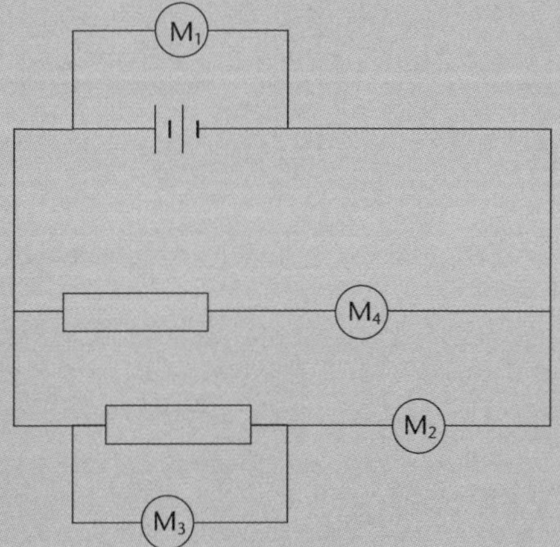

Meter	Type of meter
1	
2	
3	
4	

8. The following diagram shows a junction in a circuit that connects 4 ammeters:

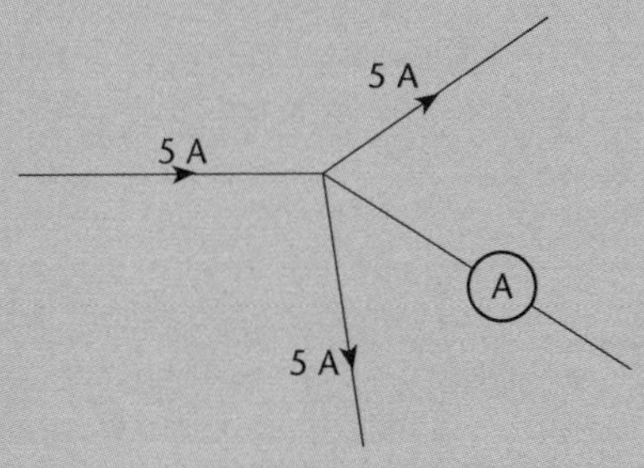

What is the reading on ammeter A?

9. The lamps in the following circuit are identical. The current at E is 0.5 A. Calculate the current at points A, B and E.

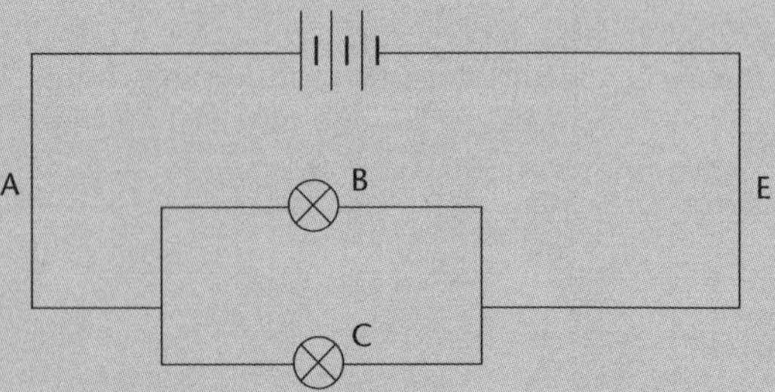

10. A battery and three similar lamps are connected as shown.

Describe the relative sizes of the currents I_1, I_2 and I_3.

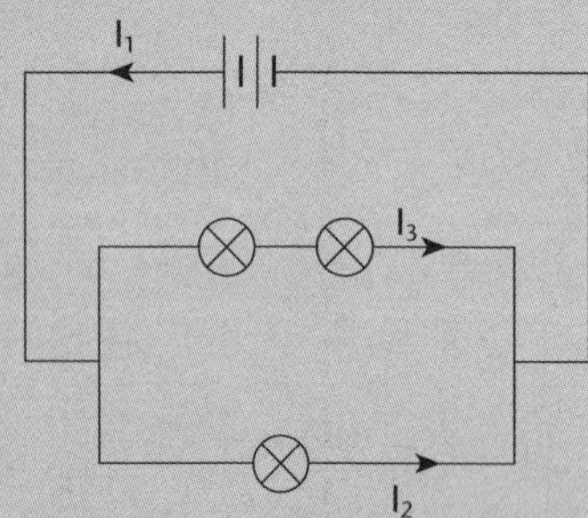

11. Explain what happens to the identical lamps A and B in the circuit alongside, when:

   a. Switch 1 *only* is closed.

   b. Switch 2 *only* is closed.

   c. Both Switch 1 and Switch 2 are closed.

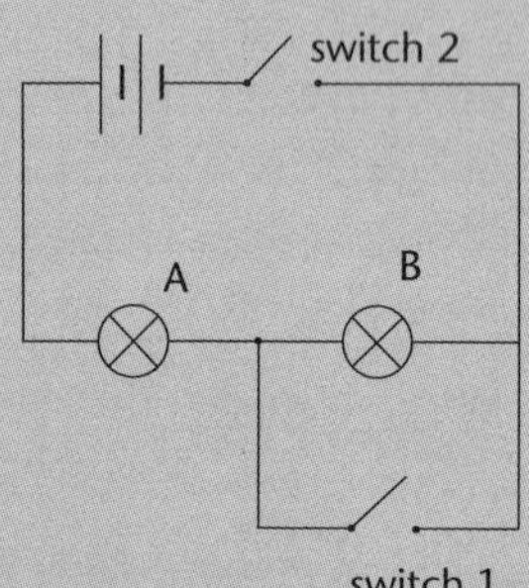

12. Draw a circuit diagram to show how three lamps can be lit using a power pack so that two of the lamps are controlled by the same switch while the third can be switched on and off independently of the other two.

13. In the diagram alongside are two identical lamps L_1 and L_2. Ammeter A_2 reads 1.5 A.

   a. What are the readings on the other ammeters A1, A3 and A4?

   b. Redraw the diagram to include a switch that can control both lamps L1 and L2 at the same time.

   c. Redraw the diagram to include two switches to control both lamps separately.

   d. Redraw the diagram to include a variable resistor to control lamp L1 only.

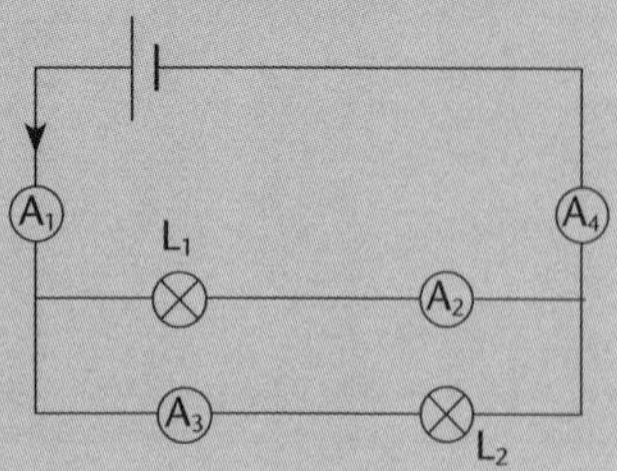

Use the following information to answer Questions 14 and 15.

In the circuit shown, the three identical lights are shining when the circuit is first connected. After some time the fuse blows.

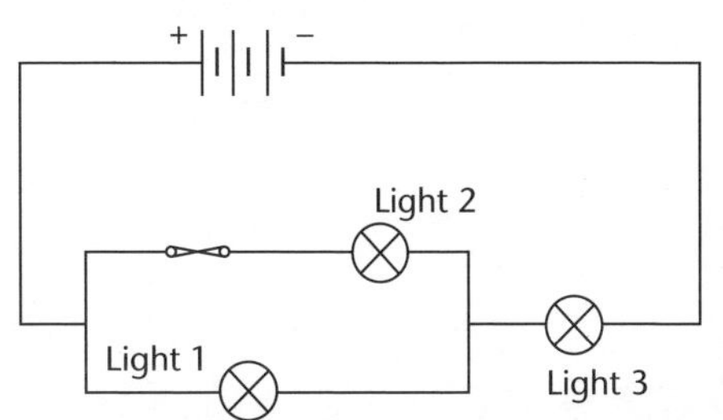

14. Before the fuse blows describe the relative brightness of the lights.

15. How does the brightness of the lights change when the fuse has blown?

16. 1.5 volt dry cells can be connected together in various ways to provide a power supply for toys and torches.

The diagrams show three possible connections.

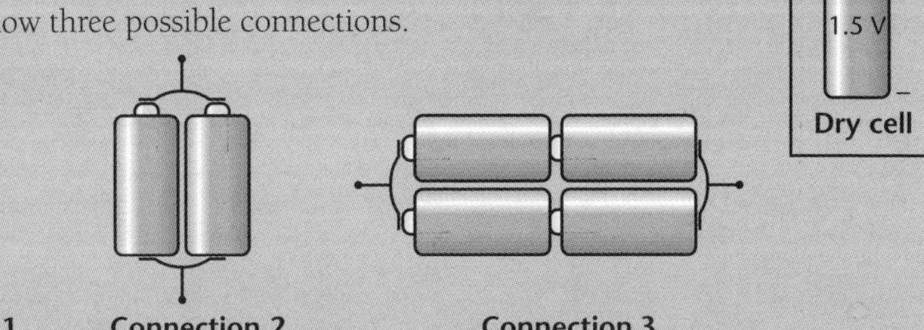

State the voltage of each of the three different connections.

17. The lamps and batteries in the circuits below are identical. Which circuit contains the two brightest lamps?

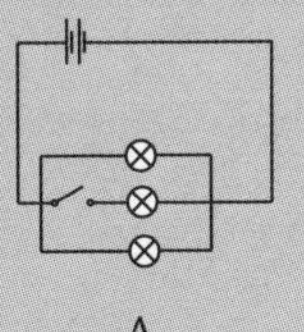

A.

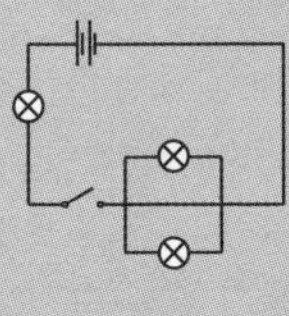

B.

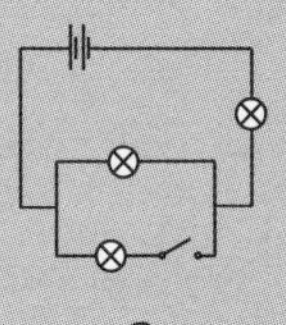

C.

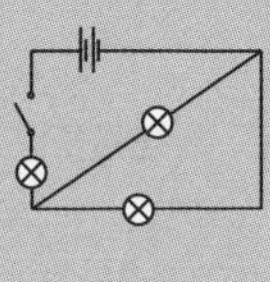

D.

18. In the circuit shown on the right, both the lamps are identical.

a. State how the voltmeter readings are related.

b. State how the ammeter readings are related.

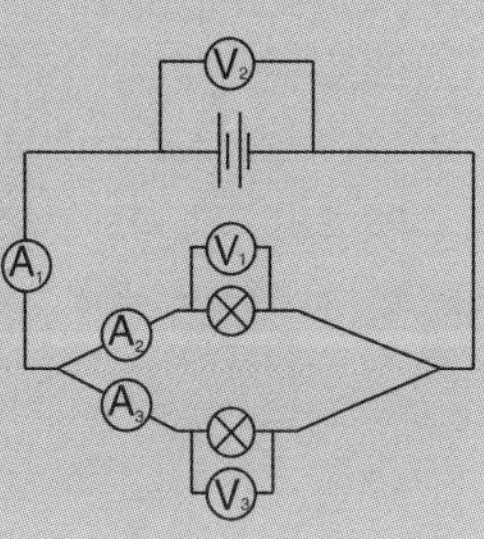

19. The drawing on the right shows an electric circuit. Sketch a circuit diagram for this circuit.

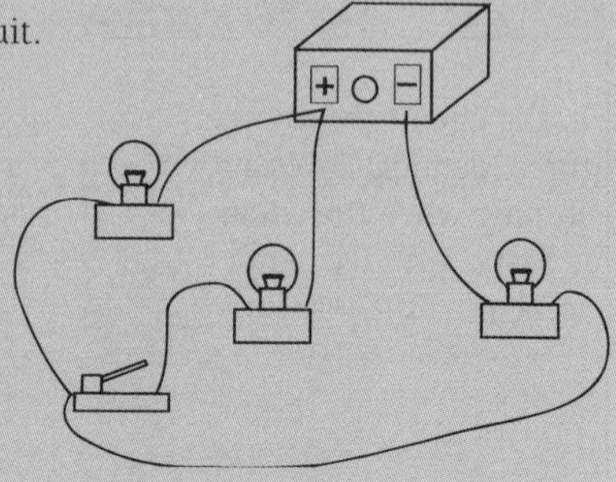

20. Draw a circuit diagram for each of the following circuits:

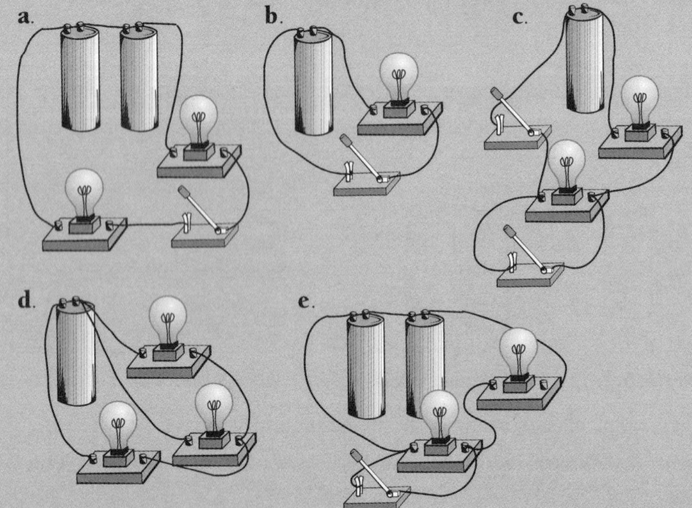

21. In the circuit diagram shown alongside, which switch(es) must be closed to:

a. Allow only the lamp to glow?

b. Allow only the bell to ring?

c. Allow both the lamp to glow and the bell to ring?

22. a. Draw the symbol for a diode.

b. Explain what a light-emitting diode is.

c. Give two uses of diodes.

23. The following circuits contain LEDs and a resistor. Decide on the direction of the current in each circuit and hence which LEDs will light up.

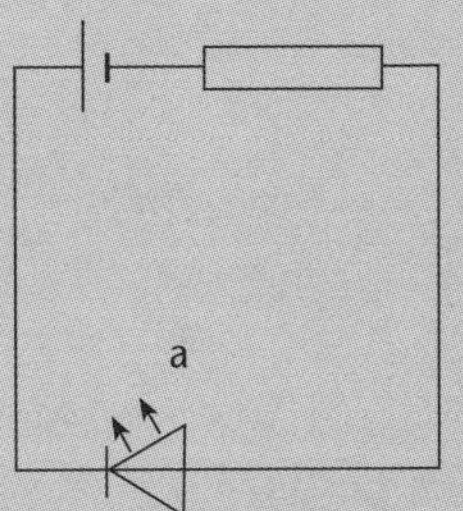

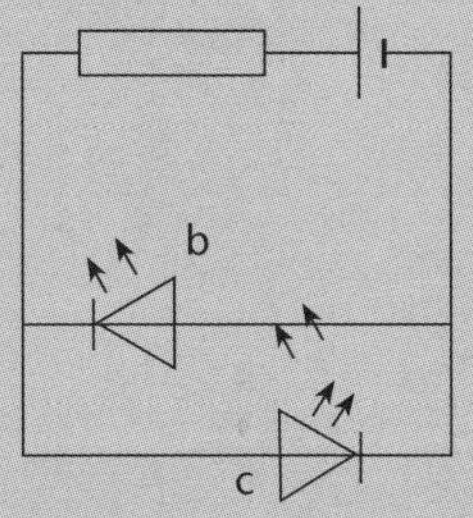

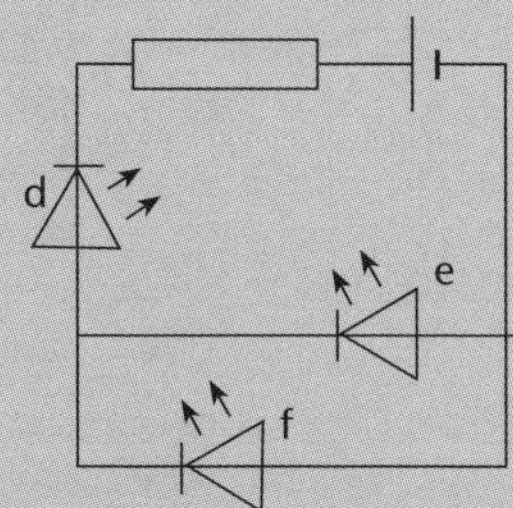

24. A diagram of the face of a milliammeter is shown. What is the reading on the meter?

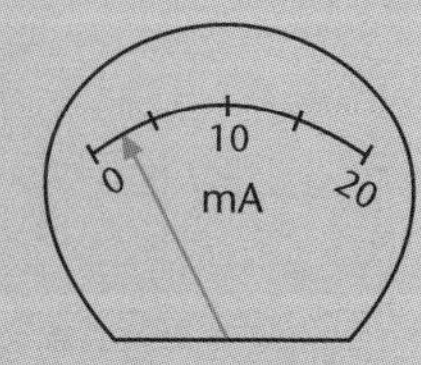

25. In the electrical circuit shown, M_1 and M_2 are meters. What does the meter labelled M_2 measure?

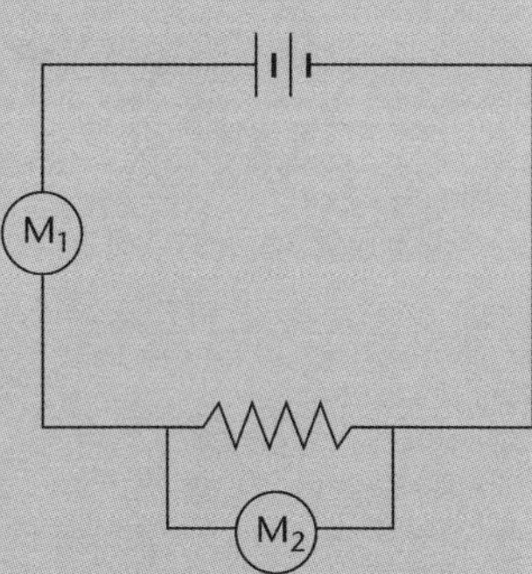

26. The diagram shows a circuit containing two identical lamps, a cell and two ammeters.

a. Draw a circuit diagram for this circuit.

b. The reading on ammeter A1 is 1.0 A. What is the reading on ammeter A2?

c. One lamp blows.

i. Why does the other lamp continue to glow?

ii. What change will now occur on ammeter A1?

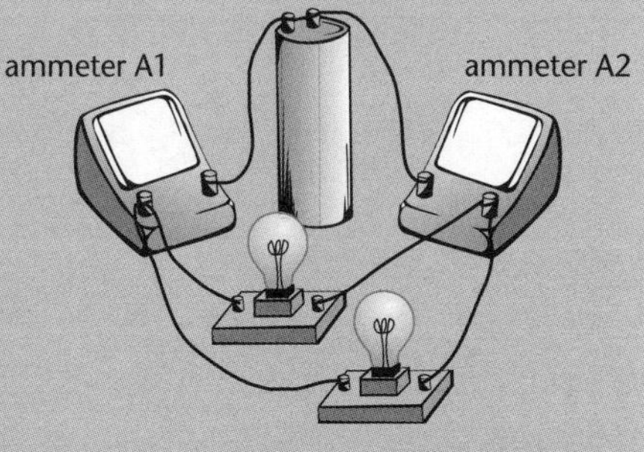

27. Draw circuit diagrams for each of the following circuits.

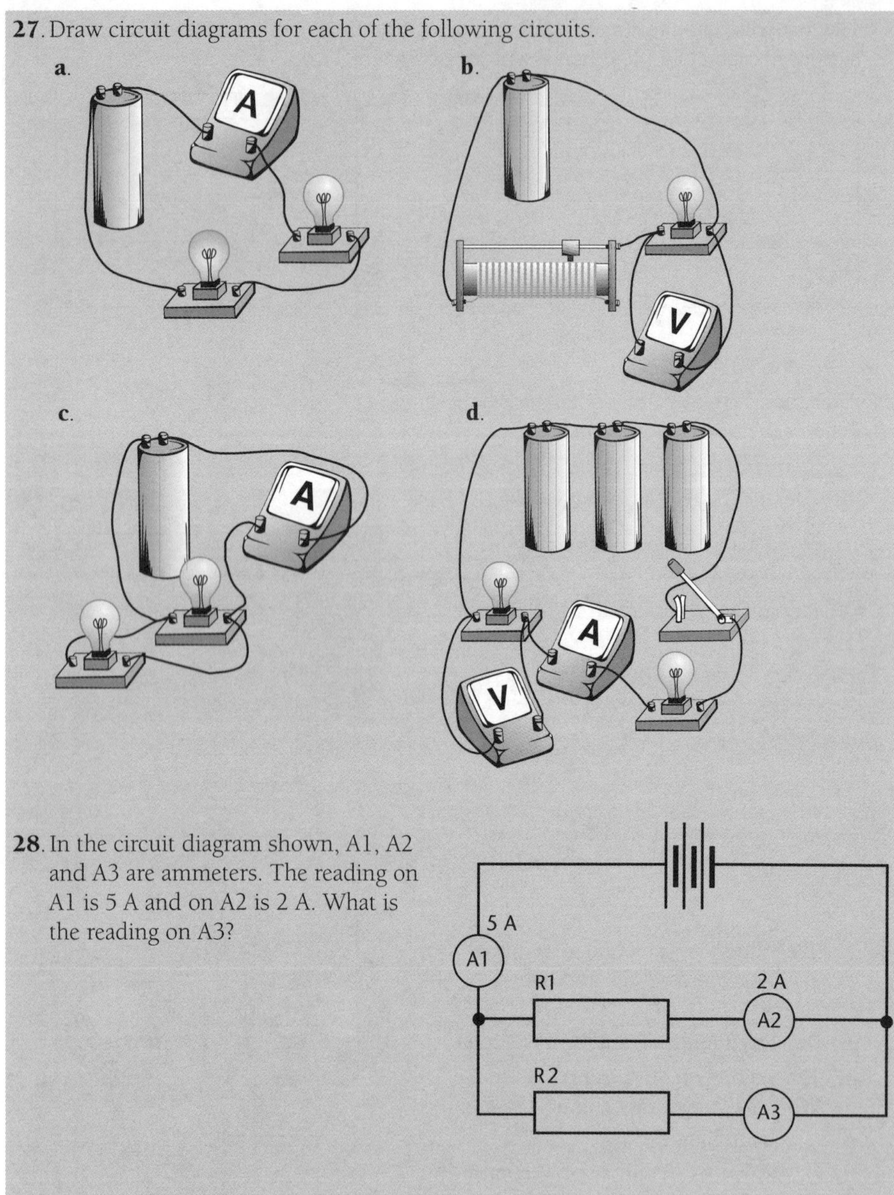

28. In the circuit diagram shown, A1, A2 and A3 are ammeters. The reading on A1 is 5 A and on A2 is 2 A. What is the reading on A3?

29. Jenifa and Rachel connect the circuit shown alongside.
The switch is open and the voltmeter reading is 12 V.

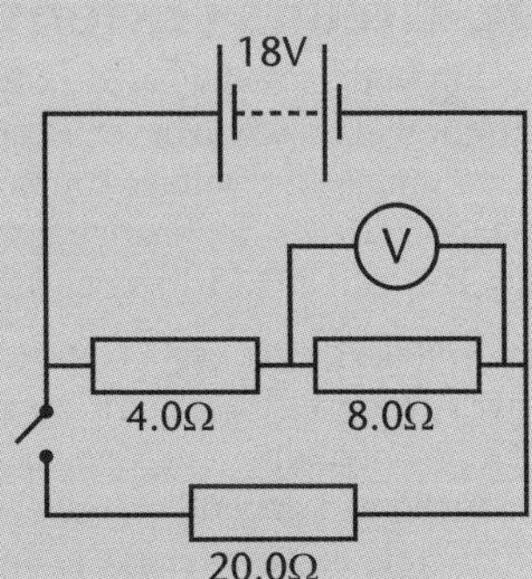

a. Use the formula $R_T = R_1 + R_2$ to calculate the total resistance of the branch that has two resistors in series.

b. Use the formula V = IR to calculate the current through the 8.0 Ω resistor.

c. What is the current through the 4.0 Ω resistor?

d. Combine information and calculate the amount of heat energy dissipated by the 4.0 Ω resistor in 13 seconds.

The switch is now closed in the circuit.

e. What is the voltage across the 20.0 Ω resistor? Give the correct SI unit with your answer.

f. If the current through the 20.0 Ω resistor is 0.9 A, select a formula and calculate the power output of the 20.0 Ω resistor. Give the correct SI unit with your answer.

g. Select a formula and calculate the energy used by the 20.0 Ω resistor in 15 seconds.

h. What is the current through the 4.0 Ω resistor now that the switch is closed?

i. The total resistance of a parallel circuit could be found using the formula $\frac{1}{R_T} = \frac{1}{R_1} + \frac{1}{R_2}$ or its alternative form $R_T = \frac{R_1 \times R_2}{R_1 + R_2}$.
Use a different method to calculate the total resistance of the circuit.

j. The students are now provided with a lamp that may be connected safely anywhere in the circuit. They are told to connect the lamp in the position that will make it glow most brightly. The circuit must otherwise remain the same and the switch must remain closed.

- **i**. Redraw the circuit diagram showing how the students will connect the lamp so that it will glow most brightly.
- **ii** Explain why the lamp will glow most brightly in this position.

(Source: NCEA Examination paper)

DC circuits (Extension)

Having understood what electric current and voltage are, we now know that an electric current will flow whenever a continuous pathway of electric conductors is connected across the terminals of a voltage source. Such a pathway is called a closed circuit and since our voltage source is not changing with time, but has a fixed magnitude (value), we call this closed circuit a DC (direct current) circuit.

The voltage source (eg a battery) is said to possess an **electromotive force** (**emf**). This emf drives or moves the electric current through the conductors. The passage of the current sets up a potential difference between the ends of the conductors. The magnitude of the current produced will depend on the electrical resistance (opposition to the flow of current) of the conductors in the circuit. The magnitude of the current decreases with the increase of the opposition to the flow of current.

There are two different types of circuit configuration (construction): series circuits and parallel circuits.

Kirchoff's Laws

In a simple circuit, and/or in complicated circuits, Kirchoff's rules can be formalised based on the concepts for voltage and currents studied in the previous section for series and parallel simple circuits. Those rules are useful in finding unknown currents, resistance, and potential differences in an electric circuit.

Kirchoff's two rules make use of the conservation of charge and conservation of energy principles. The first rule, called the ***Junction Rule***, makes use of the principle of conservation of charge. The second rule, called the ***Loop Rule***, makes use of the principle of conservation of energy.

Kirchoff's Rule 1 (The Junction Rule) can be stated as:

At any junction point, the sum of currents entering the junction must be equal to the sum of currents leaving the junction.

Mathematically it is stated as: $\sum I = 0$. That is, the algebraic sum of the currents flowing into a junction is zero. This means that whatever current enters the junction must also leave the junction, otherwise charge would accumulate at the junction. In the following figure, I_1 is entering the junction and I_2 and I_3 are leaving the junction. If the currents which are entering the junction are given positive signs, and the leaving currents are given negative signs, we will get:

$$\sum I = 0 \quad I_1 - I_2 - I_3 = 0$$

$$\therefore I_1 = I_2 + I_3$$

current entering a node equals current leaving the node.

I_2

I_1

junction

I_3

Kirchoff's Rule 2 (The Loop Rule) can be stated as:

In any closed loop, the algebraic sum of the changes in potential must be zero. Mathematically it can be expressed as: $\sum E = 0$.

This rule can be best explained using the circuit shown below. The circuit consists of a single loop **abcde**.

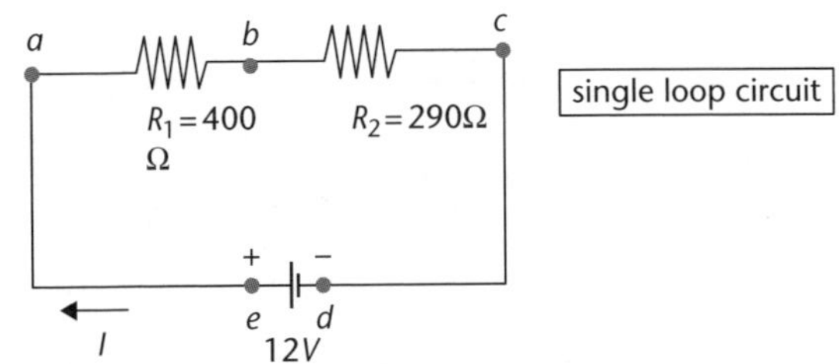

single loop circuit

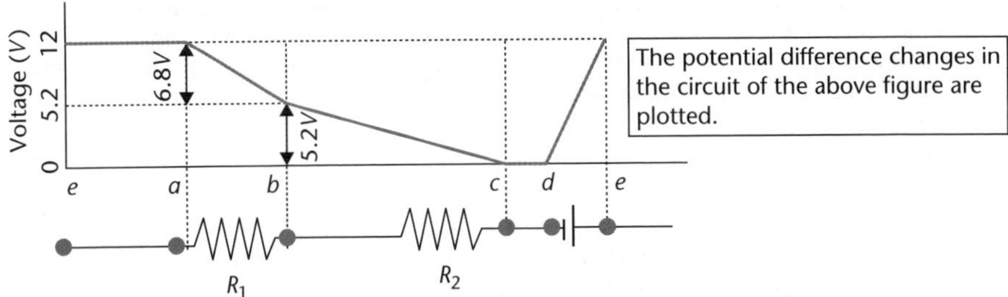

The potential difference changes in the circuit of the above figure are plotted.

Let us plot the potential difference changes as we follow the current from the positive terminal to the negative terminal. At point e the potential is $+12V$. From e to a there is no potential difference since there is no source of potential or resistance, thus there is a horizontal line in the figure above. From points a to b there is a potential decrease of $6.8V$ ($V = IR = 0.017A \times 400\Omega = 6.8V$) which is represented by the negative sloping line. Since this is a decrease in potential the potential difference is $-6.8V$.

From points b to c there is another potential drop of $-5.2V$ ($V = IR = 0.017A \times 290\Omega = 5.2V$). There is no potential difference between points a and d. From point d back to e there is an increase of potential from $0V$ to $12V$. Since it is an increase the potential difference is $+12V$. Now, summing all potential changes together, in going from point e back to e we get: $-6.8V - 5.2V + 12V = 0$. This is what Kirchoff's loop rule 2 is saying.

When applying Kirchoff's loop rule negative signs must be assigned to potential drops (going from higher to lower potentials) and positive signs must be assigned to potential increases (going from lower to higher potentials).

Example C

The following figure shows a circuit with three unknown currents I_1, I_2, and I_3 flowing. Determine the values of the unknown currents.

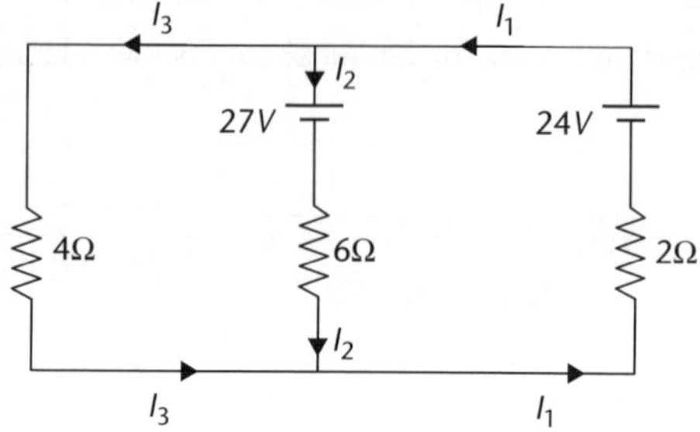

Solution:

The figure below is a redrawing of the above circuit, with circles and labelled points to use with Kirchoff's two rules.

You know from maths that if you have three unknowns, you need at least three equations to find the values of the unknowns. We are going to get one equation from Kirchoff's junction rule and two equations from Kirchoff's loop rule. In a circuit, a junction is a point where two or more currents meet. Looking at the figure below, there are only two junctions in the circuit and they are at points *B* and *E*.

Let us apply Kirchoff's junction rule to junction *B*.

Current I_1 is approaching junction *B* and currents I_2 and I_3 are leaving junction *B*.

That is: $\Sigma I = 0$

$\therefore I_1 - I_2 - I_3 = 0$ which gives $I_1 = I_2 + I_3$ (eqn 1)

Now use Kirchoff's loop rule to get two more equations.

A loop in a circuit can be thought of as a circular path taken by currents. Three loops can be identified in the circuit. Loop 1 contains points BCDEB, loop 2 is ABEFA and loop 3 is ABCDEFA. You need only two loops to give you two equations. You can choose any two loops. Let us use loop 2 and loop 3. After selecting the loops you can then decide in which direction you want to traverse the loop. Let us say that loop 1 will be traversed clockwise and loop 2 will be traversed anticlockwise as indicated in the figure below by the arrow heads of the respective circles.

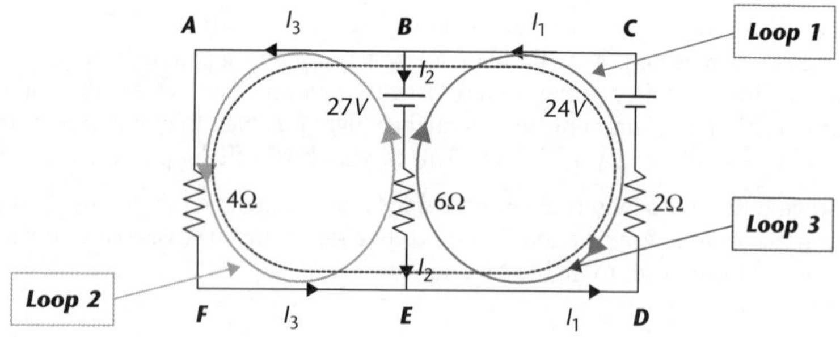

Applying Kirchoff's loop rule to loop 1 and going clockwise starting from point *C*, we get:
$\sum E = 0$.
$-24 - 2I_1 - 6I_2 + 27 = 0$,
Rearranging we get:
$-2I_1 - 6I_2 = 24 - 27$,
That is:
$-2I_1 - 6I_2 = -3$ (eqn 2)
Applying Kirchoff's loop rule to loop 2 and going anticlockwise starting from point *A*, we get:
$\sum E = 0$.
$4I_3 - 6I_2 + 27 = 0$,
Rearranging we get:
$4I_3 - 6I_2 = -27$ (eqn 3)
Now remove one unknown.
Using eqn 1, remove I_1 by expressing it in terms of I_2 and I_3 so that
$-2I_1 - 6I_2 = -3$
$-2(I_2 + I_3) - 6I_2 = -3$
$-8I_2 - 2I_3 = -3$ (eqn 4)
Solving eqns (3) and (4) simultaneously we have
$-44I_2 = -66$
$\therefore I_2 = 66/44 = 1.5$ A.
Now find I_3 using eqn4

$$I_3 = \frac{3 - 8I_2}{2} = \frac{3 - 8(1.5)}{2} = -4.5\,A$$

The negative sign means that I_3 is following in the opposite direction to the one indicated in the diagram.
Now find I_1 using eqn 1.
$I_1 = I_2 + I_3 = 1.5 + (-4.5) = -3.0$ A. Negative sign indicates opposite direction.
Therefore the values of the currents are: $I_1 = 3.0$ A, $I_2 = 1.5$ A, and $I_3 = 4.5$ A.
Now that we know the current values, we can determine potential differences across resistors, if values of resistances are given. The concept of resistance and the calculations for resistances in series and parallel will be explained in the next section.

Unit 11.5 Electricity Principles

Topic 3: Resistance

Unit 11.5 aims to cover the content on 'Electricity Principles' as set out in the Syllabus and Topic 3 aims to cover the bullet points listed under 'Resistance' on p. 20 of the Syllabus:

- Ohm's Law and resistance.
- Factors affecting the resistance of a wire.
- Resistors in series.
- Resistors in parallel.
- The resistor colour code.

Ohm's Law and resistance

George Simon Ohm (1787–1854) was a German schoolteacher who investigated the relationship between the voltage across a conductor and the current flowing through the conductor.

The following circuit repeats Ohm's experiment using modern components – a **rheostat** (variable resistor) changes the current, an ammeter measures the current through the coil of **nichrome** wire and the voltmeter measures the voltage across the coil.

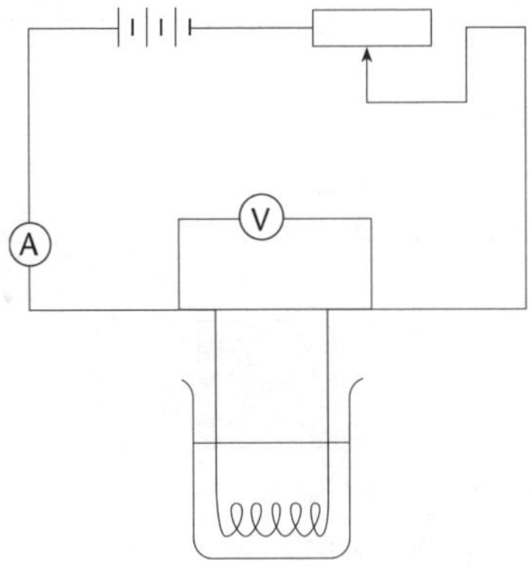

Circuit used to demonstrate Ohm's law.

The coil of nichrome wire is submerged in water to keep it at a constant **temperature**. For each different rheostat setting, the ammeter reading (current) and voltage readings are recorded. A typical set of results and a graph of voltage versus current are shown in the diagram on the next page.

Voltage (V)	Current (A)
0	0
2	0.5
4	1.0
6	1.5
8	2.0
10	2.5

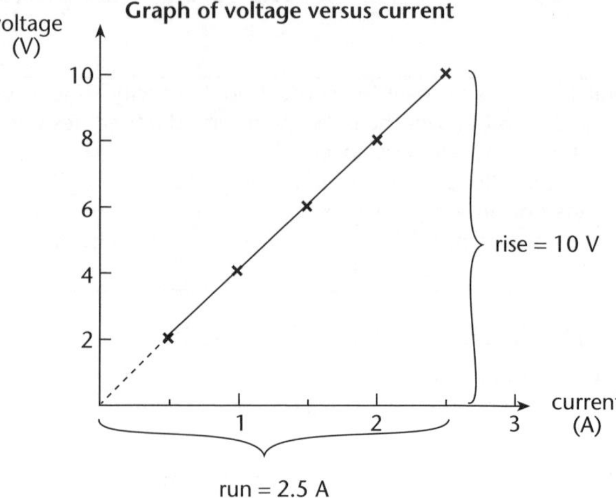

The relationship between voltage and current.

The graph shows that voltage is directly proportional to current.

The slope (gradient) of the graph is $\frac{10}{2.5} = 4\ \Omega$.

Ohm called the slope of the graph the **resistance** of the coil of wire. The SI unit of resistance is the **ohm**, Ω. The resistance of the conductor can be calculated using the following formula:

$$R = \frac{V}{I}$$

V = the potential difference across the component (V).

I = the current through the component (A).

R = the resistance of the component (**Ω**).

Example A

A lamp is connected to a 12 V supply. The current through the lamp is 1.5 A. Calculate its resistance.

Answer:

$R = \frac{V}{I} = \; = 8\ \Omega$

For an **ohmic resistor**, the slope of a *voltage versus current* graph is constant. For a **non-ohmic resistor**, such as the filament of a lamp, the slope of the voltage versus current graph is *not* constant because the resistance of the filament of the lamp increases as the filament heats up.

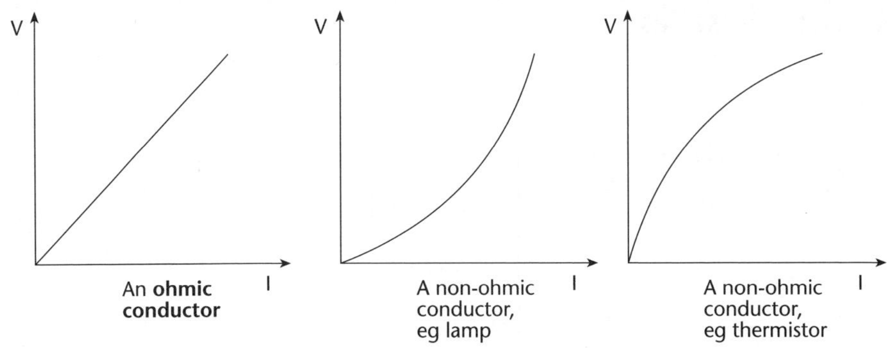

Voltage/current graphs for Ohmic and non-Ohmic conductors.

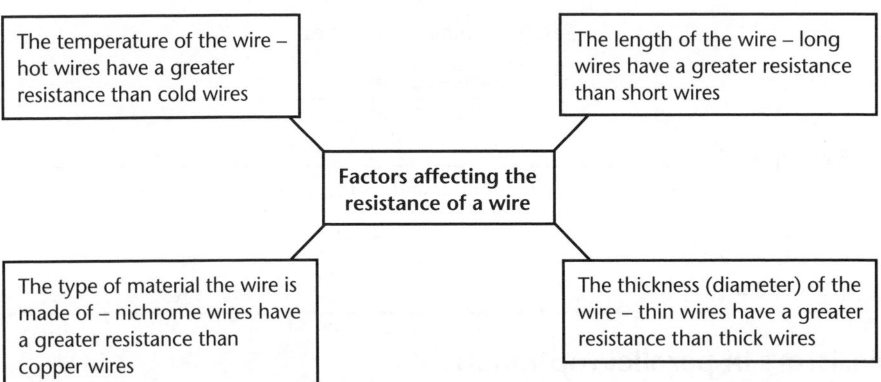

A rheostat is a variable resistor. The resistance of the rheostat is changed by a sliding contact – it increases or decreases the length of resistance wire connected into the circuit. A rheostat can be used as a dimmer to control the brightness of a lamp by changing the current in the circuit.

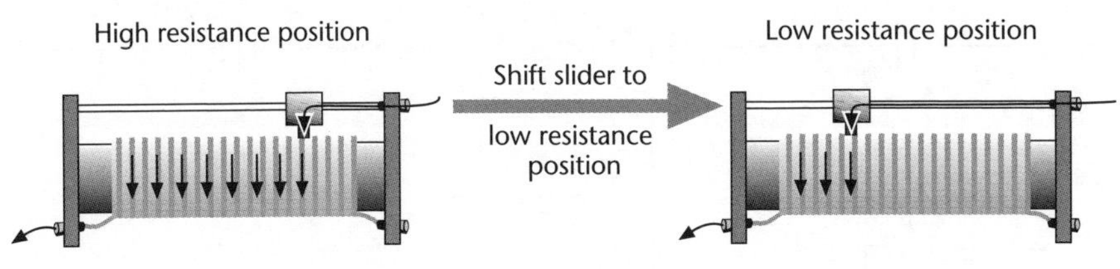

A rheostat (variable resistor).

Resistors in series

The total resistance of **resistors** connected in series is equal to the sum of their individual resistances.

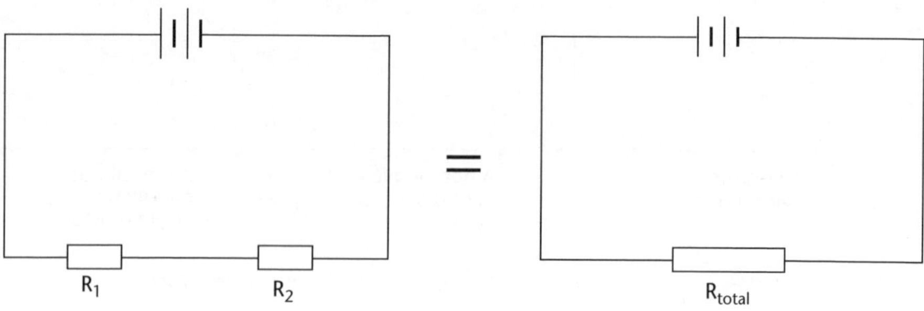

Resistors connected in series.

$$R_T = R_1 + R_2$$

Example B

What is the total resistance of 100 Ω, 1 000 Ω and 350 Ω resistors connected in series?

$R_T = R_1 + R_2 + R_3$

$R_T = 100 + 1\ 000 + 350$

$R_T = 1\ 450\ \Omega$

Resistors in parallel (optional)

The total resistance of two resistors connected in parallel can be calculated using the formula:

$$\frac{1}{R_T} = \frac{1}{R_1} + \frac{1}{R_2}$$

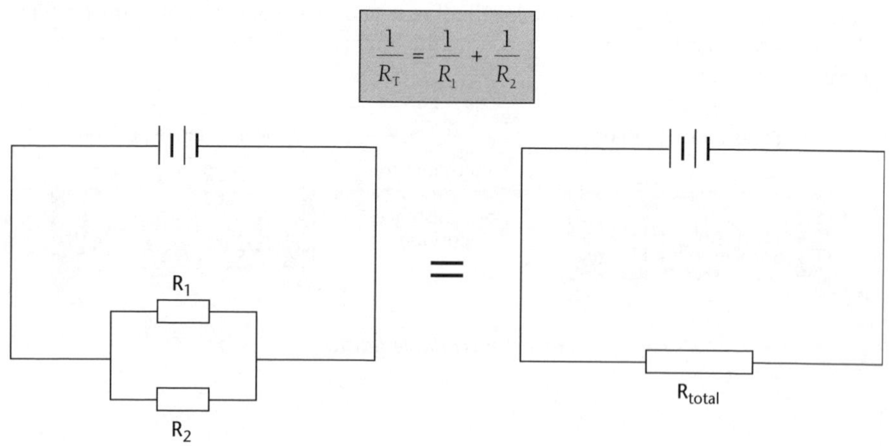

Resistors connected in parallel.

Example C

What is the total resistance of a 100 Ω and a 1 000 Ω resistor connected in parallel?

$$\frac{1}{R_T} = \frac{1}{100} + \frac{1}{1\,000} = \frac{11}{1\,000}$$

$$R_T = \frac{1\,000}{11} = 91\ \Omega$$

The resistor colour code

The value of a resistor can be determined by reading the coloured bands on the resistor. The first coloured band indicates the first digit of the value, the second coloured band indicates the second digit of the value and the third coloured band indicates the number of zeroes. The last band indicates the **tolerance** or accuracy of the value.

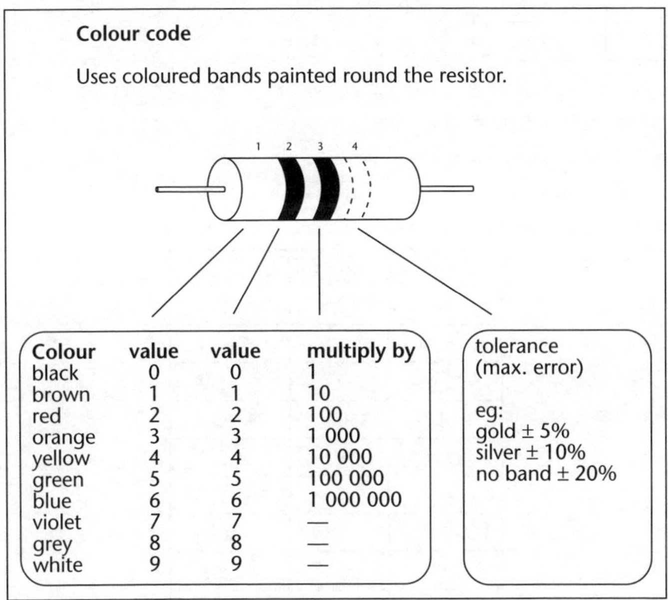

Colour	value	value	multiply by
black	0	0	1
brown	1	1	10
red	2	2	100
orange	3	3	1 000
yellow	4	4	10 000
green	5	5	100 000
blue	6	6	1 000 000
violet	7	7	—
grey	8	8	—
white	9	9	—

The colour code for resistors.

Example D

Brown, black, red = 1 0 00 = 1 000 Ω = 1 k Ω

Red, red, red = 2 2 00 = 2 200 Ω = 2.2 k Ω

Red, violet, yellow = 2 7 0 000 = 270 000 Ω = 270 k Ω

Red, red, blue = 2 2 000 000 = 22 000 000 Ω = 2.2 M Ω

Green, blue, orange, gold = 5 6 000 = 56 **Ω** (± 5%)

Unit 11.5 Activity 3A: Ohm's law and resistance

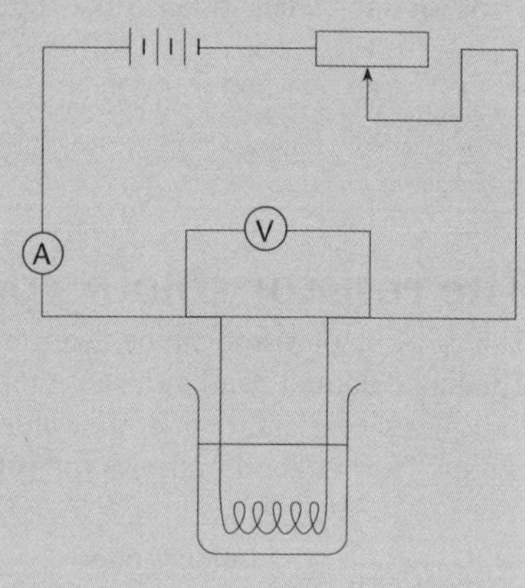

1. Jimmy and Betty set up the experiment shown in the diagram to investigate the relationship between current and voltage. They collected the following results:

Voltage (V)	Current (A)
0	0
2	0.4
4	0.8
6	1.2
8	1.6
10	2

   a. Use the data in the table to draw a graph of voltage versus current.
   b. Calculate the gradient of the graph.
   c. What is the resistance of the coil of the coil of wire?
   d. What would the current be expected to be when the voltmeter reads 12 V?

2. Calculate the quantities labelled **a** to **i** in the following table.

Current	Voltage	Resistance
a	12 V	3 Ω
5 A	6 V	b
100 mA	c	0.5 V
d	230 V	15 Ω
100 mA	e	4 Ω
4 A	f	150 Ω
1 500 mA	12 V	g
h	800 V	220 kΩ
10 A	230 V	i

3. What current will flow through a 10 Ω resistor when a potential difference of 12 V is applied across its ends?
4. What is the voltage across a 10 kΩ resistor when a current of 10 mA flows through it?
5. What is the resistance of an aquarium heater that carries a current of 5 A when connected to the 230 V mains supply?
6. What is the total resistance when a 10 Ω, 15 Ω and 5 Ω resistor are connected in series?

7. Calculate the ammeter readings in each of the following circuits.

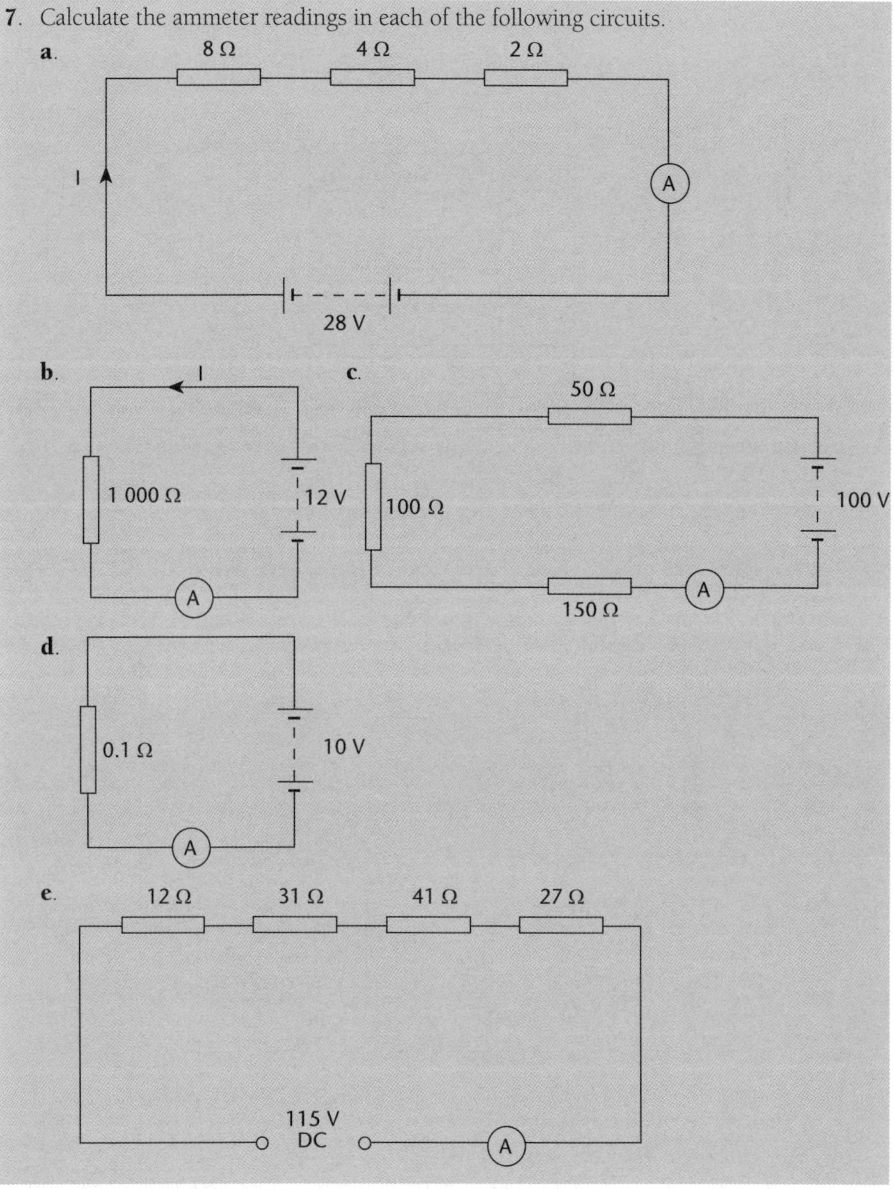

8. Naomi and John set up the following experiment to investigate the relationship between the voltage across a lamp and the current through the lamp.
They collected the following results:

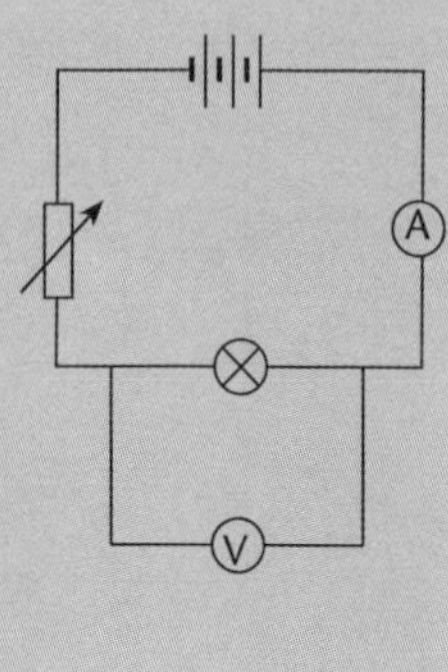

Voltage (V)	Current (A)
0.0	0.0
1.0	0.5
3.0	1.0
7.0	1.5
12.0	2.0

a. Use the data in the table to draw a graph of voltage (y-axis) versus current (x-axis).
b. Use the graph from (a) to calculate the resistance of the lamp when the current is:
i. 0.5 A. **ii**. 1.5 A
c. Explain why and how the resistance of the lamp changed.
d. Is the lamp an ohmic conductor?

9. The following circuit shows three 5 Ω resistors connected to a 10 V supply.

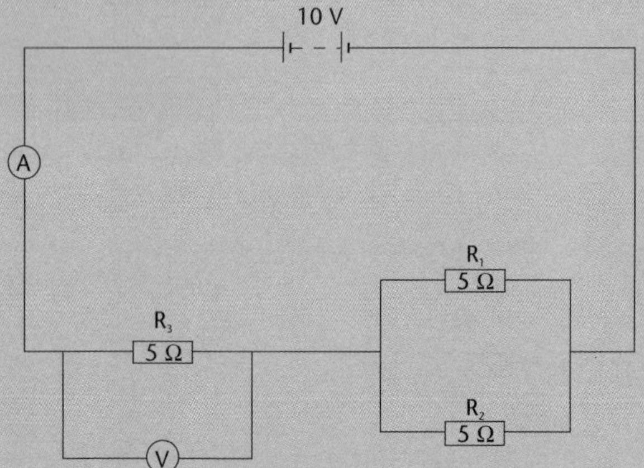

a. Calculate the total resistance of the circuit.
b. Calculate the ammeter reading.
c. Calculate the voltmeter reading.
d. Calculate the current through resistor R_2.

10. James and Serah connect a 6 Ω, 8 Ω and 10 Ω resistor in series to a DC power supply. The voltage across the 8 Ω resistor was measured as 4 V.
a. Draw a circuit diagram for this circuit.
b. What is the current through the 8 Ω resistor?
c. What is the voltage across the 10 Ω resistor?
d. What is the total voltage across all three resistors?
e. What is the voltage of the power supply?

11. A 10 Ω resistor, an ammeter and an unknown resistor are connected in series to a 15 V supply. The ammeter reads 0.5 A.

a. Draw a circuit diagram for this circuit.

b. What is the value of the unknown resistor?

c. What is the voltage across the 10 Ω resistor?

d. What is the voltage across the unknown resistor?

12. In the circuit shown, the voltage across the 4 Ω resistor is 10 V.

a. What is the current through the 4 Ω resistor?

b. What is the current through the 3 Ω resistor?

c. What is the voltage across the 3 Ω resistor?

d. What is the voltage across the battery?

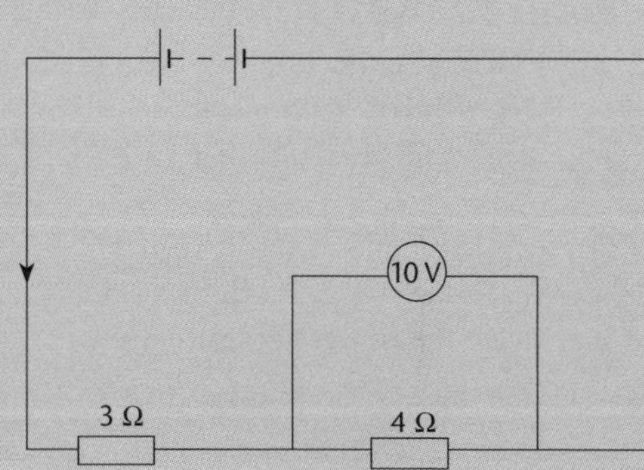

13. In the circuit shown, the voltmeter reads 12 V.

a. What is the combined resistance?

b. What is the current in the circuit?

c. What is the voltage across the 2 Ω resistor?

d. What is the voltage across the 3 Ω resistor?

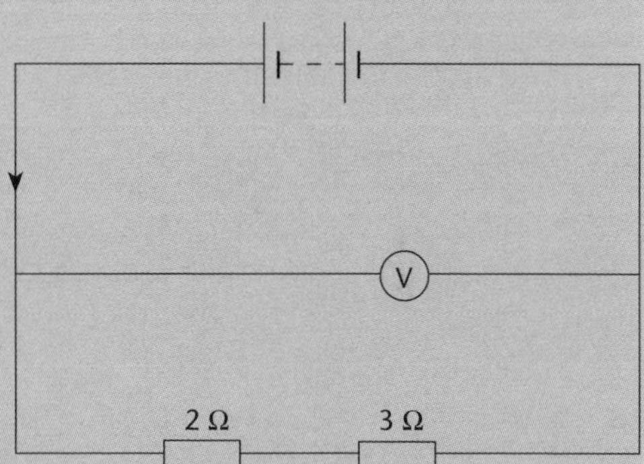

14. Calculate the current in each of the following circuits.

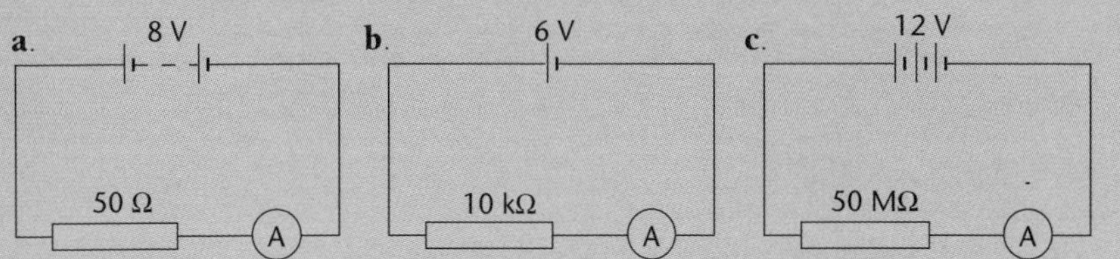

15.

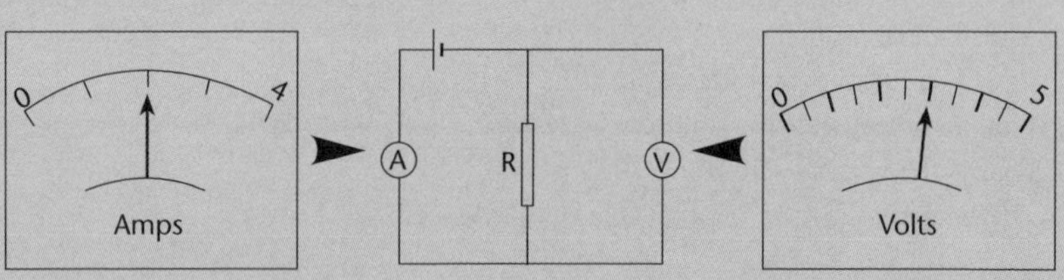

a. What is the current through the resistor?

b. What is the voltage across the resistor?

c. Calculate the resistance of R.

16. The ammeter in the circuit alongside reads 3 A.

a. What is the voltage across each resistor?

b. What is the voltage of the battery?

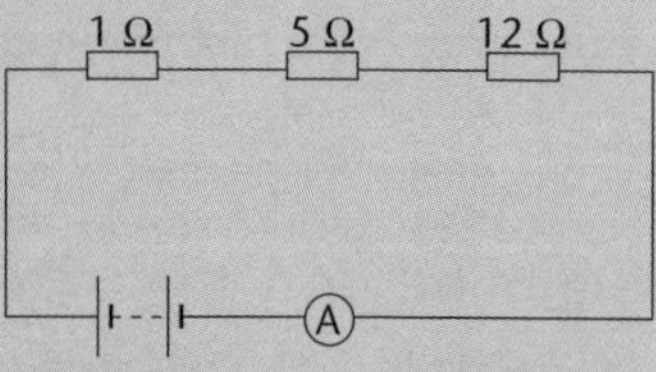

17. Calculate the resistance of each lamp in the circuits shown.

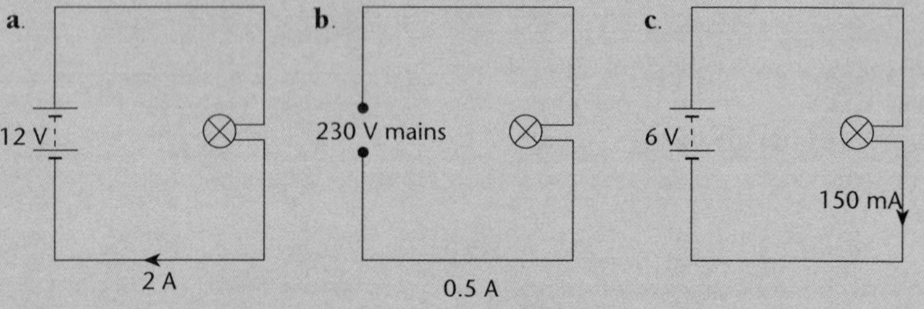

18. When the switch is closed what is the current in the 2 000 Ω resistor?

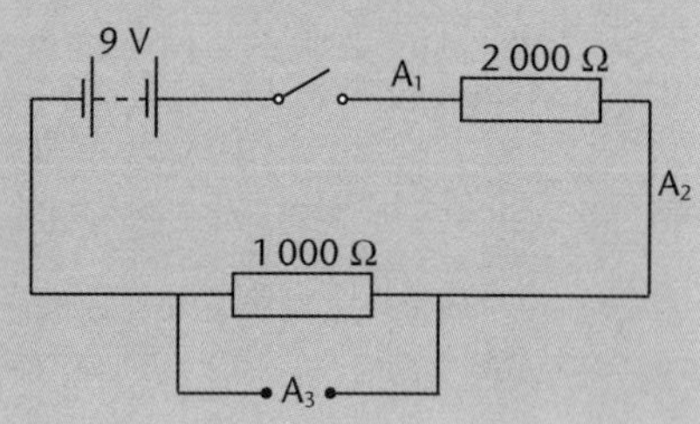

19. While investigating combinations of resistors, Anne and Betty set up the circuit shown. The reading on ammeter A_2 is 0.20 A.

a. Calculate the voltage (pd) across the 6 Ω resistor.

b. Find the current through ammeter A_3.

c. What is the reading on ammeter A_1?

d. Calculate the voltage (pd) across resistor R_1.

e. What will be the reading on the voltmeter across the cells?

f. Calculate the heat energy given out by R_2 in 1 minute.

g. In which direction does the conventional current flow in the circuit – clockwise or anticlockwise?

h. Anne and Betty add a diode to the circuit to prevent the current flowing through the 6 Ω resistor.

i. Copy the diagram and add the diode in the correct position.

ii. Calculate the total resistance of the new circuit.

20. Three resistors are connected to a 12 volt battery. What is the reading on the voltmeter?

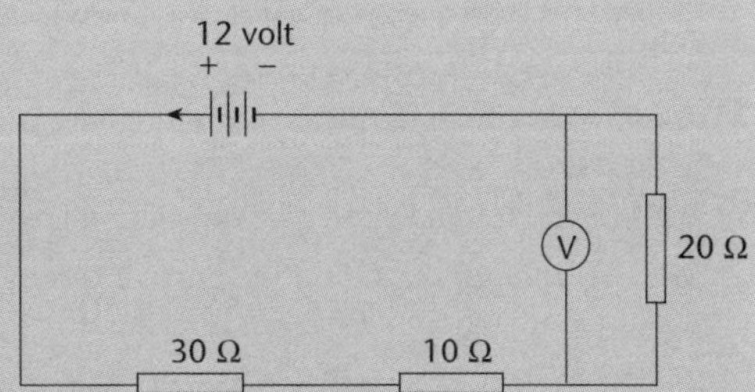

21. Calculate the total resistance of each of the following combinations of resistors.

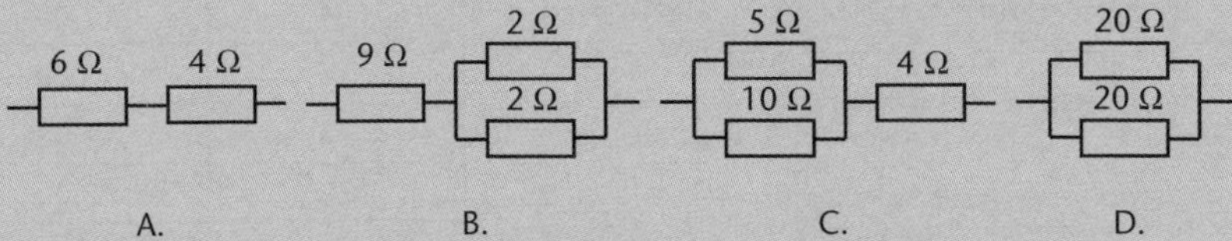

22. a. Marie's home has an outside light which is operated by a security switch. She wants to add:

- Another 240 volt light to the circuit.
- Another security switch to the circuit.

Both lights must go on when either switch is turned on.

Copy the circuit diagram shown and modify it to satisfy these conditions.

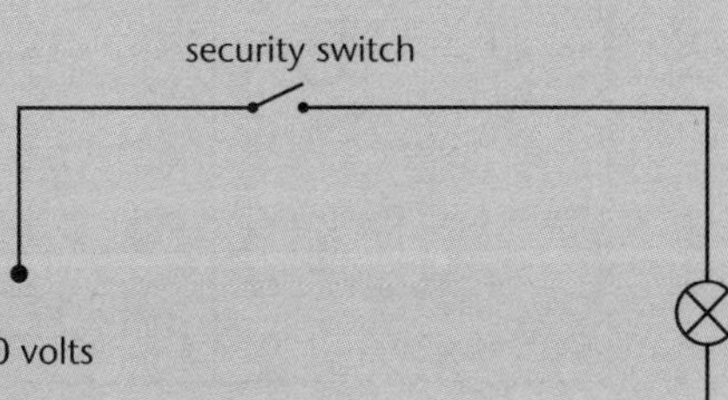

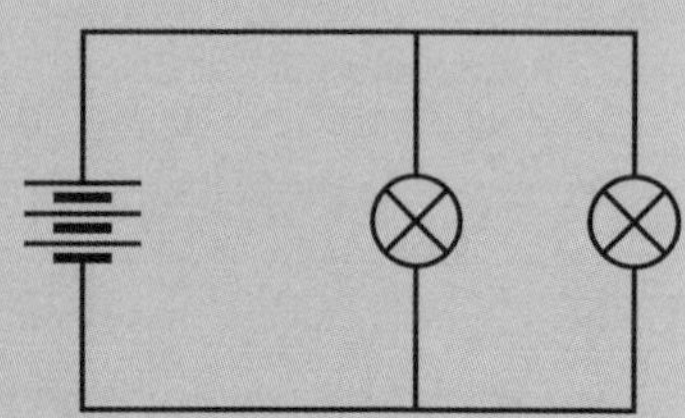

b. Tom is asked to redesign the circuit shown. He must:

- Add a switch which will turn off one light only.
- Add a variable resistor which will dim the other bulb.

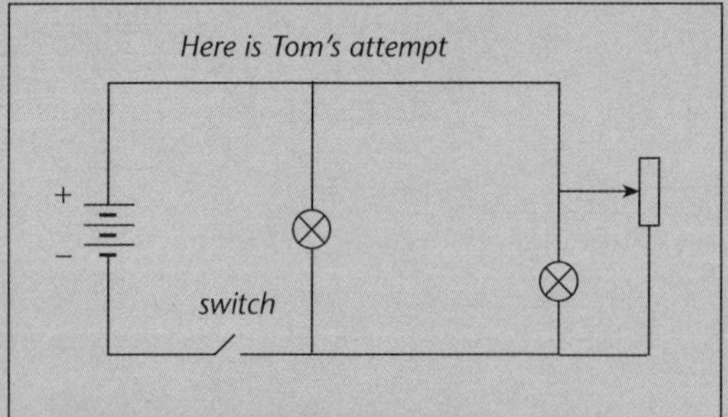

Two errors are made in Tom's circuit. Explain what is wrong with the way he connected **i** the switch **ii** the variable resistor.

iii. Draw the correct circuit diagram which Tom should have drawn.

23. Susan is looking at how the current changes when she connects different voltages to a light bulb.

a. Graph her data shown below. Put voltage on the x-axis.

b. Use your graph to find the current through the bulb when the voltage is 10 volts.

c. Alex's experiment involves connecting bulbs in series to a 12 volt battery. His bulbs are the same as the one which Susan has been testing. Use Susan's results to predict the current which Alex will measure in the two circuits **i** and **ii** below.

Sue's data	
Voltage (volts)	**Current (amps)**
0	0
2	0.70
4	1.05
6	1.40
8	1.65
12	2.00

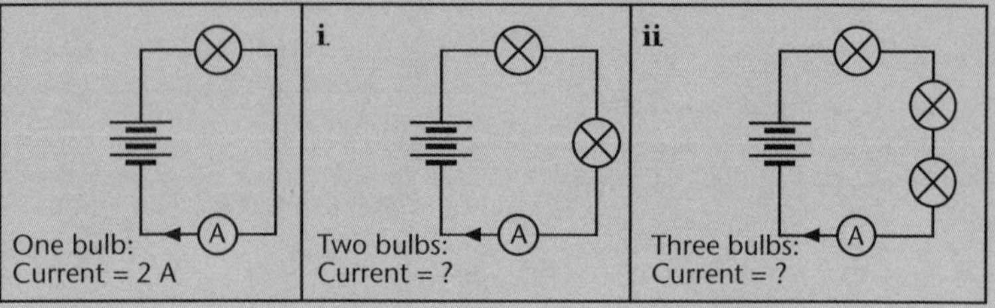

24. State the ammeter and voltmeter readings for the following (partly drawn) circuits:

a.

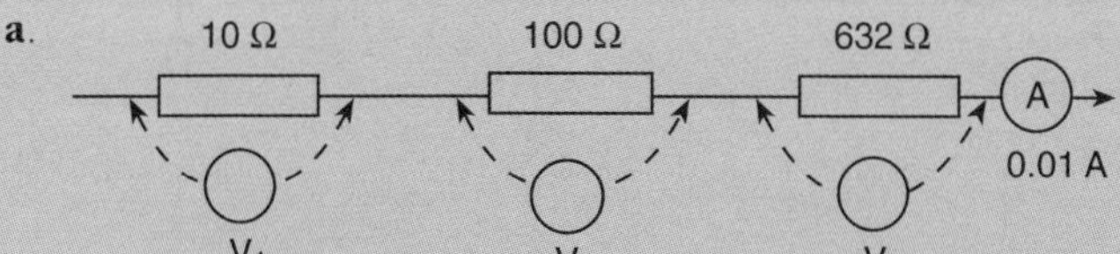

b.

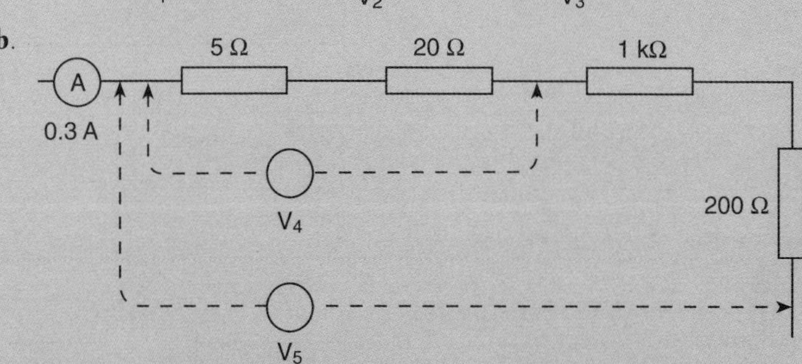

c.

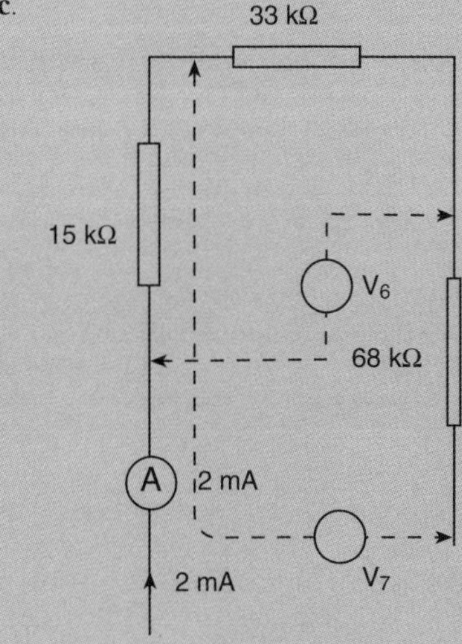

25. Three 20 Ω resistors are connected to a 12 V battery, as shown.

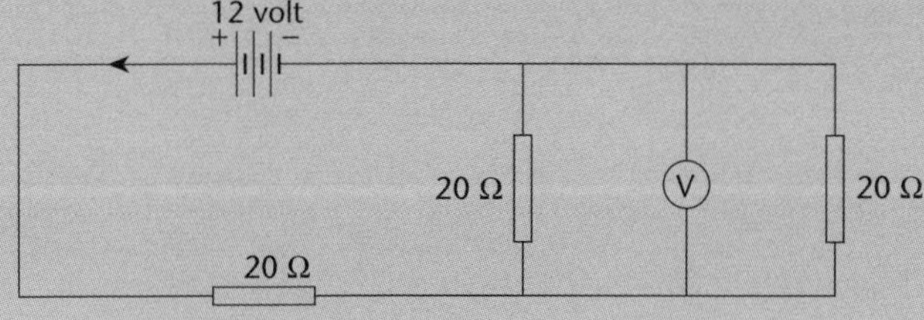

Calculate the reading on the voltmeter.

26. Calculate the current *I* in the circuit shown.

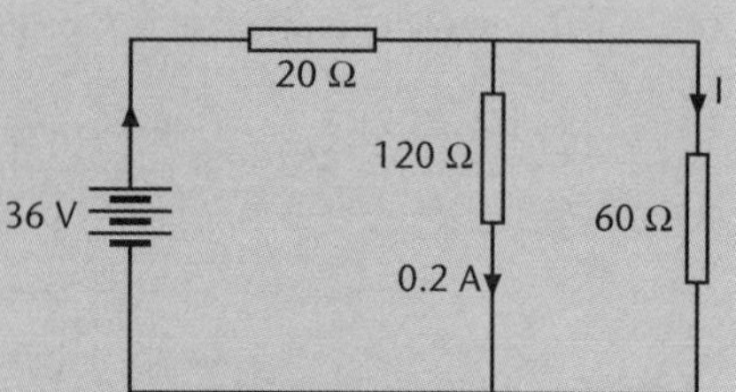

27. A heating coil that has wire 25 cm long is connected to a 10 *V* power supply.

a. Calculate the electric field formed in the coil.

b. How many coulombs of charge pass through the coil when it gives out 3 000 J of heat energy?

28. **a**. When a heater is connected to a 240 *V* household power supply, a current of 10 *A* flows. What is the resistance of the heater?

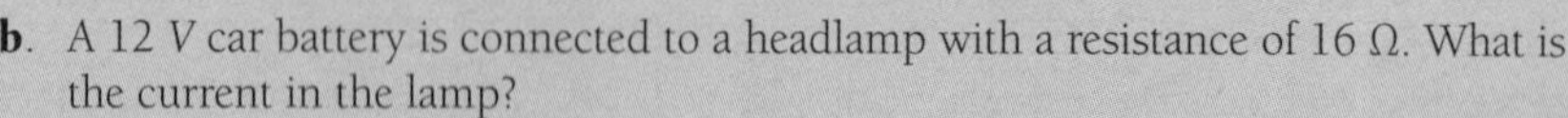

b. A 12 *V* car battery is connected to a headlamp with a resistance of 16 Ω. What is the current in the lamp?

c. A fuse is needed to protect a circuit containing a 200 *V* power supply and a 40 Ω resistor. What current rating should the fuse have?

29. **a**. A 20 Ω resistor has a power rating of 5.0 *W*. What is the maximum current which should flow in the resistor?

b. A 12 *V* car battery delivers a current of 50 *A* to operate the starter motor. What power does the battery deliver, and how much energy is given out in 10 seconds?

30. A battery is connected to the circuit shown.

The readings on the meters are:

voltmeter 5.0 V

ammeter 2.0 A.

Calculate the:

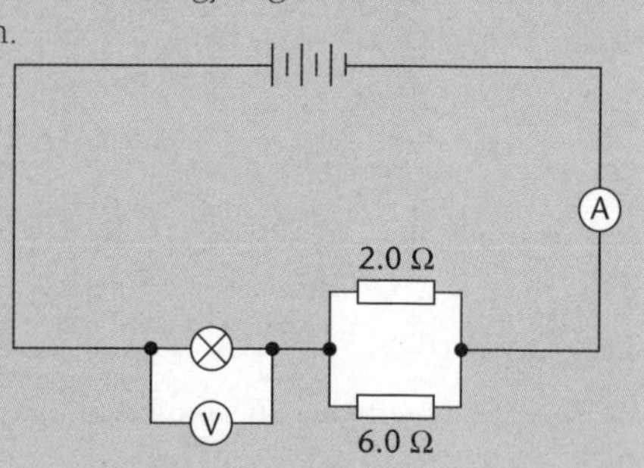

a. Resistance of the lamp.

b. Equivalent resistance of the circuit.

c. Power output of the lamp.

d. Voltage of the battery.

e. Current in the 2.0 Ω resistor.

Unit 11.5 Electricity Principles
Topic 4: Electrical power and energy

Unit 11.5 aims to cover the content on 'Electricity Principles' as set out in the Syllabus and Topic 4 aims to cover the bullet points listed under 'Electrical power and energy' on p. 20 of the Syllabus:

- Electrical power.
- The power bill.
- Fuses.

Electrical **power**, P, is the rate at which electrical energy is converted into other forms of energy such as heat and light. It is the amount of work done per second.

$$P = \frac{E}{t}$$

E = the amount of energy converted or work done (J)

t = the time taken (s)

$$P = VI$$

P = power (watts); 1 W = 1 V A = 1 J s^{-1}
V = the potential difference across the component (V)
I = the current through the component (A)

Since $V = IR$:

$$P = I^2R = \frac{V^2}{R}$$

Example A

A lamp rated at 0.5 W is connected to a 9 V battery. What is the current through the lamp?

$I = \frac{P}{V} = \frac{0.5}{9} = 0.06\ A$

Example B

How much electrical energy is converted to heat and light if a 100 W lamp is turned on for 5 minutes?

$E = Pt = 100 \times 5 \times 60 = 30\,000$ W = 30 kW

The power bill

Electricity suppliers charge consumers for the electrical energy the consumers use. The electricity meter records the total amount of electrical energy used in **kilowatt-hours**. When the previous reading is subtracted from the current reading, it provides a record of the total amount of energy used during the billing period. On electricity accounts, the kilowatt-hour is called 'the unit'.

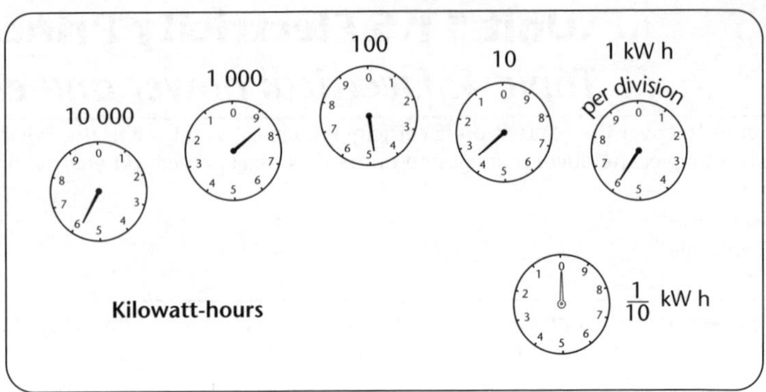

When the pointer is between figures, read the smaller figure.
When between 9 and 0, read 9. The reading above is 58 436 units.

Reading an electricity meter.

The cost of a unit of electricity varies between suppliers, but is approximately $0.12 (12 cents).

Example C

The first reading is 1 386.

The second reading is 2 586.

Calculating meter readings.

Number of units used = 2 586 – 1 586 kW h
= 1 200 kW h
Therefore, cost @ $0.12 per unit = 1 200 × $0.12
= $144.00

The following table shows typical power ratings for some common appliances:

Appliance	Typical power rating
Lamp	60–100 W
Heater	1–3 kW
Stove	8 kW
Iron	750 W
TV	200 W
Refrigerator	150 W

Power ratings for common appliances.

Example D

What is the total cost of running a fridge rated at 150 W for 1 day?
A unit (kW h) costs $0.12.

Answer:

$E = Pt = 0.150 \times 24 = 3.6$ kW h (1 day = 24 hours)
Total cost = $3.6 \times \$0.12 = \$0.43$ (43 cents)

Fuses

A **fuse** is a short length of wire made of a low melting point metal such as tin. When the current in the circuit exceeds a certain value, the fuse melts and breaks the circuit. If there wasn't a fuse in the circuit, the wiring in the circuit could overheat, melt the insulating covering and start a fire. Fires are particularly dangerous in dry, dusty and relatively inaccessible and unobserved places, such as roof spaces.

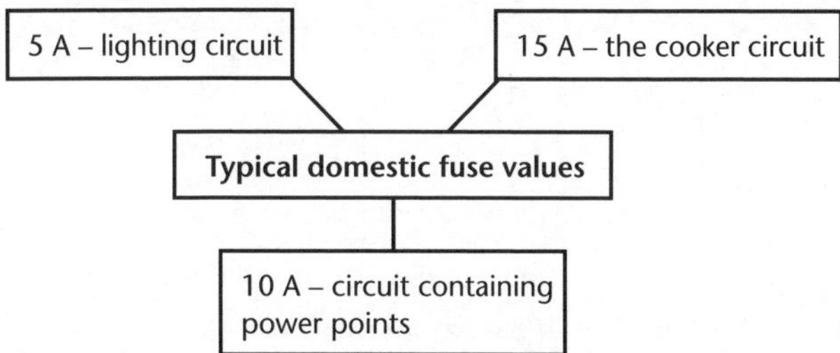

Unit 11.5 Activity 4A: Electrical power and energy

1. How much electrical energy (in joules) does a 150 W security light change in:
   a. 1 second?
   b. 8 hours?
   c. 8 hours per night for a year?
2. An electric jug is rated at 2.4 kW. What current does it draw when plugged into a 230 V power point?
3. An electric heater has three 1 kW heating elements. What current does it draw when these elements are connected in series across a 230 V house supply?

4. A light fitting for a lounge has five 60 W lamps. The cost of electricity is $0.13 per kW h. What is the total cost of running the lamps for 4.5 hours?
5. A lighting circuit of a house with a 230 V mains supply includes five 100 W and four 60 W lamps. Is a 5 A fuse sufficient to protect this circuit?
6. A 3 kW heater is switched on for 3 hours. The cost of electricity is $0.12 per unit.
   a. How many units (kW h) are used?
   b. What is the cost of using the heater for three hours?
7. An electric stove has a 3 kW oven, a 2 kW grill and four 500 W cooking elements. The stove is connected to a 230 V mains supply. The cost of electricity is $0.12 per unit.
   a. Are the cooking elements connected in series or parallel? Explain your answer.
   b. What is the total cost of using the stove for 1.5 hours if all the parts of the stove are switched on at the same time?
   c. What is the maximum current drawn by the stove?
   d. Would a 30 A fuse be suitable to protect the stove circuit? Explain your answer.
8. What is the maximum number of 150 W lamps that can be run from a 230 V mains circuit that is protected by a 5 A fuse?
9. James and Rowena set up the following circuit:

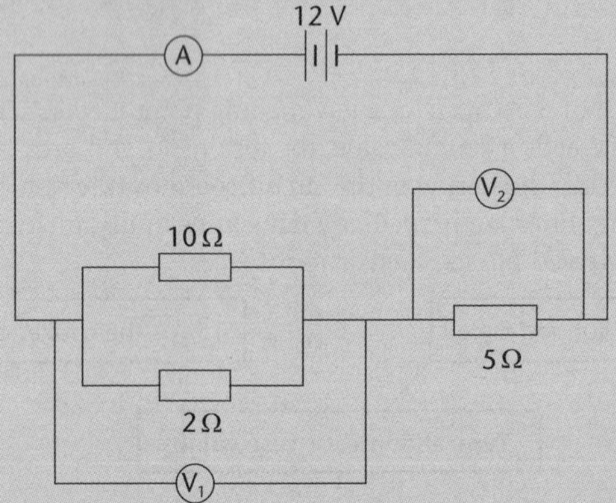

   a. Calculate the reading on the ammeter.
   b. Calculate the reading on voltmeter 1.
   c. Calculate the power developed in the 5 Ω resistor.
   d. Calculate the reading on voltmeter 2.
   e. Calculate the power developed in the 10 Ω resistor.
10. How much power should each of the following resistors be able to dissipate:
   a. 1 A, 1.0 Ω. b. 1 mA, 5 k Ω. c. 0.2 A, 60 Ω.
   d. 0.2 A, 1 000 Ω. e. 1 A, 0.7 k Ω.
11. How much power would be dissipated in the following:
   a. An 8 Ω lamp connected to a 6 V power supply.
   b. An 80 Ω heater element supplied at 230 V.
   c. A 1 000 Ω resistor with 1.5 V across it.

12. A meter reader recorded a reading of 17 084 units. The previous reading was 16 110 units. The cost per unit is 12.19 cents. Calculate the total cost of the electricity used.

13. The cost of electricity is \$0.12 per unit. Find the cost of having:

a. Five 100 Watt lamps switched on for 10 hours.

b. A 5 kilowatt electric motor run for 4 hours.

14. a. Write down the readings on the dials of the electricity meter shown. The readings were taken one month apart.

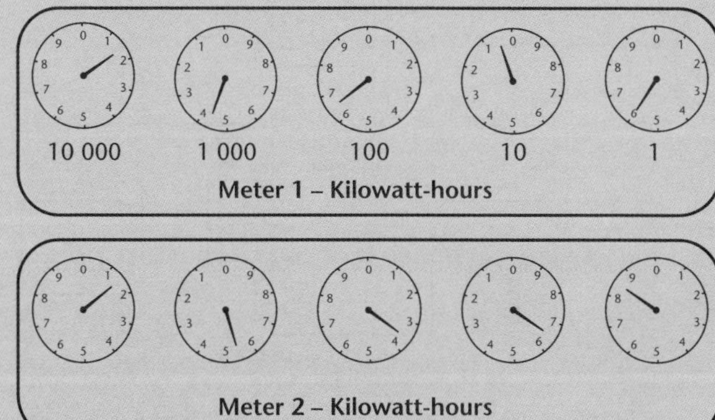

b. How much electricity was used in the month between readings?

c. How much would the power bill have been (the cost of one unit is \$0.12)?

15. A 6 V lamp has a resistance of 4 Ω. What is the resistance of another lamp which has the same power as the first when connected to 12 V?

16. Two bulbs of resistances 200 Ω and 400 Ω are connected in series across a 230 V power supply.

a. Draw a circuit diagram of the set up.

b. How much current flows through each bulb?

c. What is the power at which each bulb operates when they are in series?

17. A design engineer carries out some tests on a prototype electric jug he has designed. The element of the jug supplies 2.4 kW. The mains voltage is 230 V and the jug holds 1.5 kg of water.

a. Calculate the current through the jug element.

b. Calculate the resistance of the element.

c. How much energy does it take to raise the temperature of 1.5 kg of water from 16 °C to boiling point? The **specific heat capacity** of water is 4200 J kg^{-1} $°C^{-1}$.

d. How long does it take to raise the temperature of 1.5 kg of water from 16 °C to boiling point using this jug?

To change the design, a second heating element of resistance 12 Ω is added in series with the original element.

e. Calculate the total resistance of the combination of elements.

f. What is the new power rating of the jug?

g. i. Does the changed design take *more*, *the same*, or *less* time to heat 1.5 kg of water from 16 °C to boiling point?

ii. Calculate the time difference for the changed design to raise the temperature of 1.5 kg of water from 16 °C to boiling point compared with the original design.

h. Describe *two* design features that reduce heat loss in electric jugs.

18. The diagram shows a typical domestic water heating system.

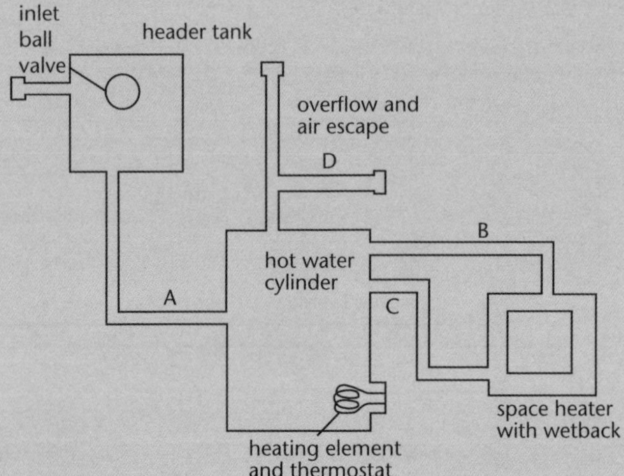

a. In the summer, the space heater is not used and the water is heated electrically. Water entering the cylinder is at a temperature of 18 °C. How much heat energy is required to raise the temperature of 180 L of water in the cylinder from 18 °C to 65 °C? The specific heat capacity of water is 4200 J kg^{-1} $°C^{-1}$. The density of water is 1 kg L^{-1}.

b. The water in the cylinder is heated by a 3 kW heating element. Calculate how many hours it takes for the temperature of the full 180 L cylinder of water to be raised from 18 °C to 65 °C.

c. The cost of 1 kW h of electricity is 11 cents. How much does it cost to heat the full 180 L cylinder of water from 18 °C to 65 °C?

d. There are six people in the family and so the cylinder frequently runs out of hot water. The cylinder takes a long time to reheat. Their electricity supplier suggests changing the 3 kW heating element to a faster, 5 kW element. Calculate how much less time it would take a 5 kW element to heat the water in the cylinder from 18 °C to 65 °C. Express your answer in hours.

Unit 11.5 Electricity Principles

Topic 5: Wheatstone bridge and potentiometer

In Topic 5 we look at the Wheatstone bridge and potentiometer, in line with the content on p. 20 of the Syllabus:

- Wheatstone bridge circuit.
- Operation principle of a Wheatstone bridge.
- How to use the Wheatstone bridge.
- Potentiometer circuit.
- Operating principle of a potentiometer.
- Using a potentiometer.

Introduction

A bridge is the name given to a special class of measuring circuits. Those circuits are mostly used for making measurements of resistance, capacitance, and inductance. Bridges are used for resistance measurements when a very accurate determination of a particular resistance value is required.

The most well known and widely used resistance bridge is the Wheatstone bridge. It is used for measuring resistance values above one ohm. Another type, the Kelvin bridge, is used for measuring resistances below one ohm.

Most commercial Wheatstone bridges are accurate to approximately 0.1 percent. Therefore, the values of resistance obtained from the bridge are far more accurate than the values obtained from the ohmmeter or the voltmeter–ammeter method.

The circuit of the Wheatstone bridge is as shown in the following figure where R_x is the resistance to be measured. The bridge works on the principle that no current will flow through the very sensitive D'Arsonval **galvanometer** connecting points *b* and *c* of the bridge circuit if there is no potential difference between them. When no current flows, the bridge is said to be balanced.

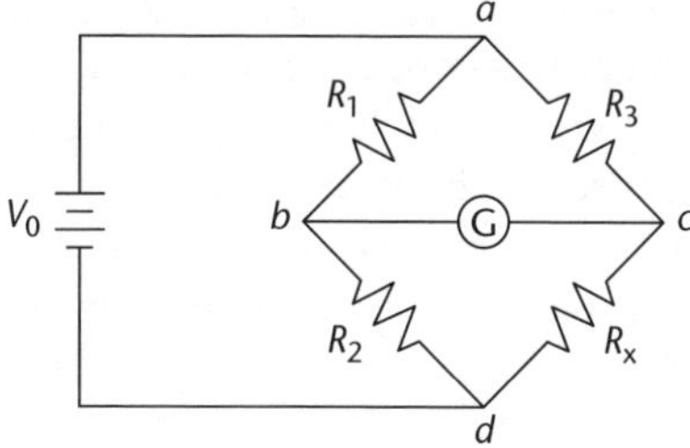

That balanced condition is achieved if the voltage V_0 is divided in path abd by resistors R_1 and R_2 in the same ratio as in the path acd by resistors R_3 and R_x. Then points b and c will be at the same potential.

Therefore, the **conduction** of no current flow through the galvanometer implies

$$(R_x/R_3) = (R_2/R_1)$$

Now, if R_x is unknown and R_1, R_2, and R_3 are known, you can find R_x from:

$$R_x = R_3\,(R_2/R_1)$$

In practical bridges, the ratio of R_2 to R_1 is usually controlled by a switch that changes this ratio by decades (factors of ten). Thus the ratio R_2/R_1 can be set to 10^{-3}, 10^{-2}, 10^{-1}, 1, 10, 10^2, and 10^3. R_3 is a continuously adjustable variable resistor. When a null is achieved, the resistance can be read directly off the dials because these dial settings correspond to the variables of the equation

$$R_x = R_3 (R_2/R_1)$$

Since the resistance value of a resistor is known to vary with frequency, resistors used in high-frequency applications should be measured at the frequency of use. When such measurements are performed, an AC source rather than a battery must be used. A number of detectors, including the **oscilloscope** or even earphones, are available to determine a null or balanced condition.

How to use the Wheatstone bridge

1. Connect the unknown resistor, R_x, to the terminals of the bridge with good, tight contacts. This will minimise contact resistance.
2. Set the scale of the galvanometer to the least sensitive setting. (A variable shunt resistor is connected across the galvanometer to allow variation of its sensitivity.) This will prevent damage to the D'Arsonval movement if the bridge is severely unbalanced.
3. Adjust the variable resistor dials until a null is reached (zero deflection of the galvanometer needle).
4. Move to a more sensitive scale setting, and null again.
5. Continue until the most sensitive scale setting is reached.
6. Calculate the resistance from:

   $$R_x = R_3 (R_2/R_1)$$

   or read directly from the dial settings.

Errors of the bridge

The possible errors that arise from using the bridge include the following.

1. Discrepancies between the true and stated resistance values in the three known branches of the bridge circuit. This error can be estimated from the resistor tolerances.
2. Changes in the known resistance values due to self-heating effects.
3. Thermal voltages in the bridge or galvanometer.
4. Balance-point error caused by lack of galvanometer sensitivity.
5. Lead and contact resistances introduced when making low-resistance measurements.

The potentiometer

When we need to perform accurate measurements of DC voltages (emf) we can use potentiometer meters, rather than deflection meters. Therefore, let us discuss the operation principle of potentiometer method for the determination of an unknown emf. The potentiometer is a null-balance instrument in which the unknown voltage is compared with an accurate reference voltage which can be adjusted until the two of them are equal.

Since no current flows at the null point, any errors due to IR (Ohm's law application to a resistance) drops in lead wires are eliminated. Such IR potential drops are always present when D'Arsonval-type meters are used to measure voltage directly. The following figures show the basic concept of the potentiometer and its operation principle.

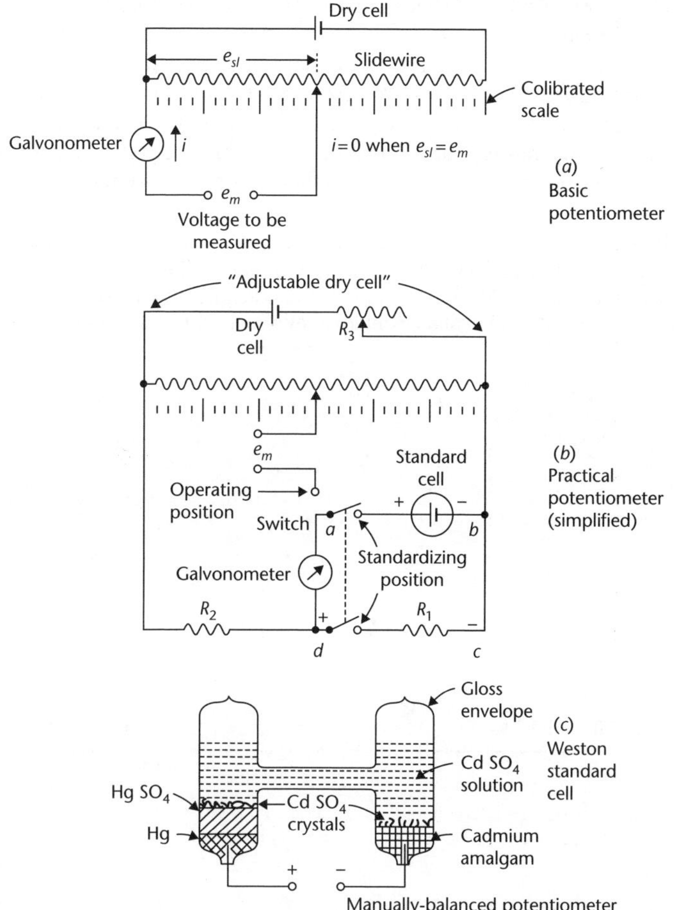

Figure a shows the basic potentiometer circuit. We see that a galvanometer (very sensitive D'Arsonval movement) is used as a null detector. It is detecting only the presence or absence of current by deflecting whenever the unknown and reference voltage are equal. However, it does not need to be calibrated since it must indicate only the presence of current, not its numerical value.

This basic circuit is not very practical since the accuracy of the reference voltage picked off the slide-wire is directly influenced by changes in the dry cell voltage. Since the dry cell supplies power to the slide-wire, its voltage is bound to gradually drop off. This problem is usually solved in a practical circuit like the one shown in figure b by the inclusion of an additional component, a standard cell.

On figure c we see a Weston cadmium saturated standard cell which is the basic working standard of voltage. Its terminal voltage is 1.018636 volts and is reproducible to the order of 0.1 to 0.6 ppm (parts per million). Its accuracy in terms of the fundamental mass, length, and time standards can be established to only about 10 ppm, however. Its temperature coefficient is –10 μv/°C; thus close temperature control must obviously be employed in the most exacting situations.

A standard cell cannot be substituted for the dry cell of figure a since its accuracy is destroyed if any appreciable current is drawn from it over a time interval. It must thus be used as an intermittent reference against which the slide-wire excitation voltage can be checked whenever desired. The unsaturated Weston cell is used in practical instruments since it is more portable. Its terminal voltage varies from one unit to another. However, its drift at constant temperature is only about –0.003 percent per year; thus it is perfectly adequate for most purposes. Its temperature coefficient is about –10 μx/°C.

Example A

A potentiometer is set up as shown in the following figure, and the balance point for the unknown e.m.f. *V* found at 74.5 cm from the left-hand end of the meter wire. If the driver cell has an e.m.f. of 1.5 volts and negligible internal resistance, find that of the unknown e.m.f.

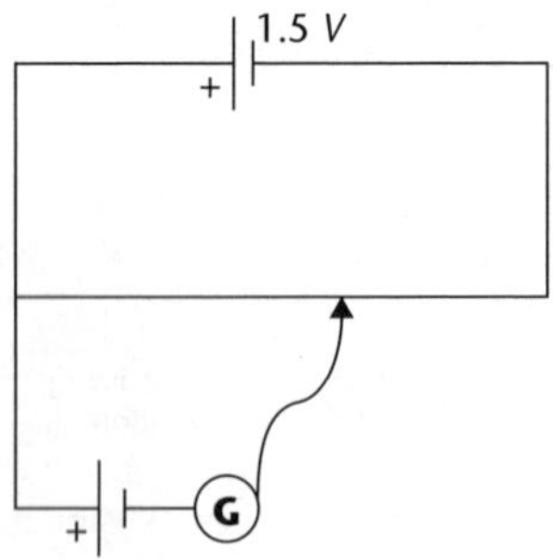

Solution:

$V = 1.5 \times 74.5/100 = 1.12\ v$

Unit 11.5 Electricity Principles
Topic 6: Alternative current (AC) circuits and effective values

Topic 6 aims to cover the content 'Alternating current (AC) circuits' on pp. 20–21 of the Syllabus:

- The alternating current (AC) voltage.
- Effective values of voltage and current.

Introduction

We will now study circuits whose power supply is an AC source that provides an alternating voltage V given by

$$V = V\text{max Sin}\omega t$$

This type of voltage is produced by a generator. These alternating voltages and currents are widely used in industry and homes as power supply for most of our appliances.

The Vmax can be controlled and changed using transformers and a wide range of electrical components like resistors, inductors and capacitors.

The major components of a generator are a coil and a magnet. If the coil rotates in a magnetic field, an alternating potential (voltage) is induced across the terminals of the coil. The generation of this alternating potential difference is graphically illustrated in the next two figures.

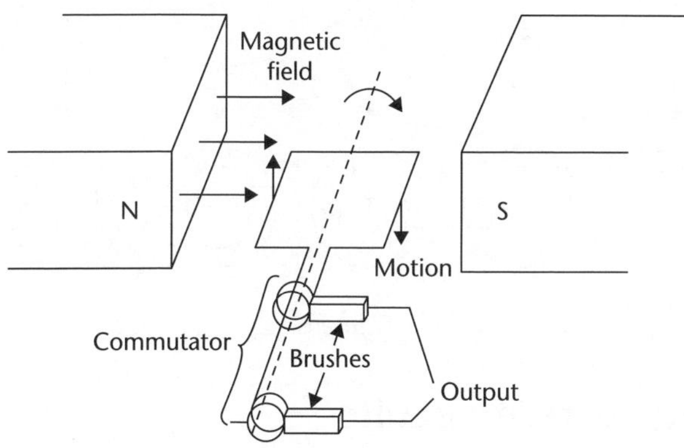

Figure A

The above figure shows that with the coil in a horizontal position, the output voltage will be at its maximum value because the sides of the coil are moving at right angles to the magnetic field. In this situation, there is no induced voltage caused by the ends of the coil since they do not cross the **magnetic field lines**.

The second figure below shows us that a quarter of a turn later the sides of the coil are moving parallel to the field lines, so in this situation the output voltage is zero.

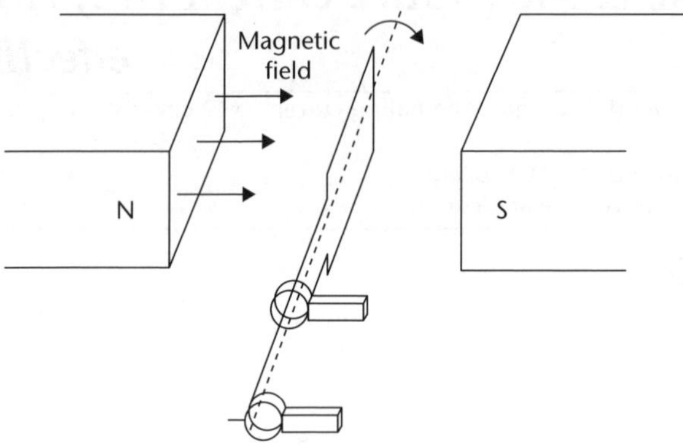

Figure B

The output voltage generated in one cycle is shown in the following figure. It is also showing the different positions of the coil.

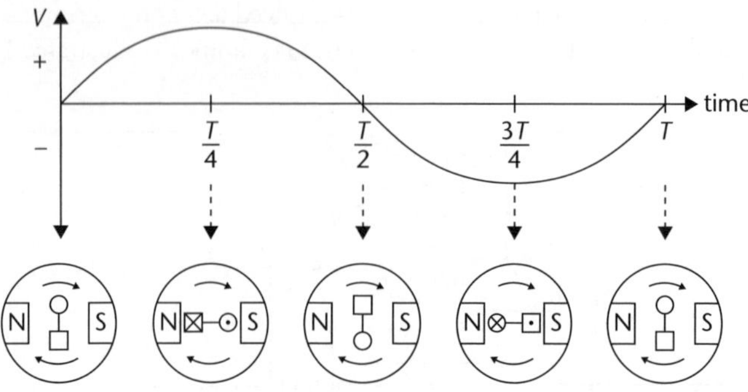

Figure C

Alternating current circuits

If an alternating emf given by

$$€ = €_m \text{Sin}\omega t$$

is applied to a circuit like the following RLC, an alternating current given by

$$i = I_m \text{Sin}(\omega t - \phi)$$

will be established in the circuit.

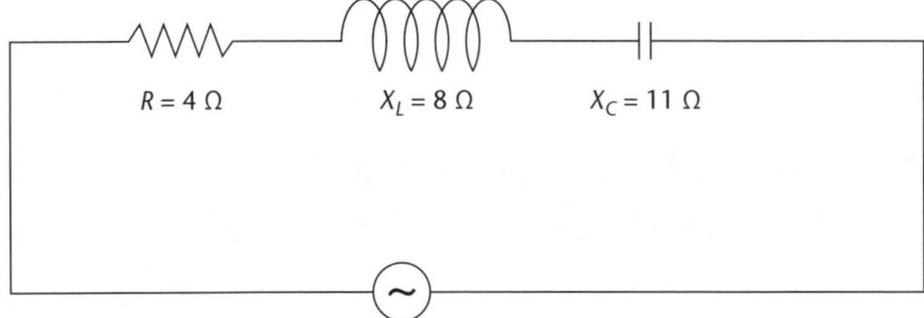

When analysing AC circuits, the basic objective is usually the following:

- Given a generator € and ω, as well as circuit elements *R*, *C*, *L*, we need to find the current *I* and ϕ (phase).

Therefore, instead of writing differential equations for the closed loops and the junction **nodes** in the circuit, and having to solve those differential equations, a geometric method using phasors will be discussed and used to analyse those AC circuits.

A phasor is a vector quantity used to represent alternating quantities. Phasors are used to depict relationships between alternating quantities in terms of so called phasor diagrams.

Rotating vectors: sine waveforms

The following figures illustrate the method to obtain a sine waveform by rotating a vector.

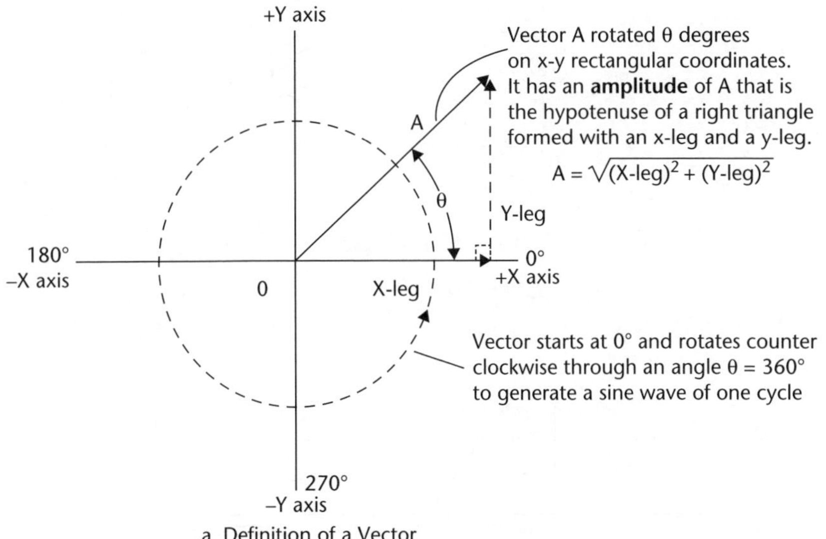

a. Definition of a Vector

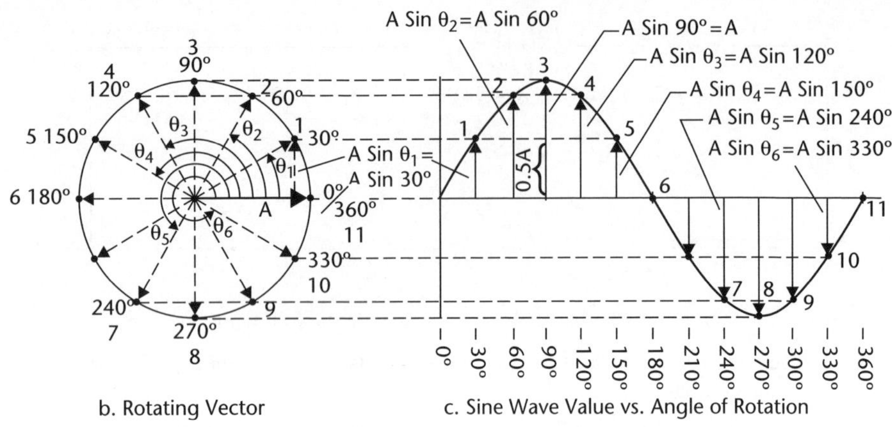

b. Rotating Vector c. Sine Wave Value vs. Angle of Rotation

(Diagrams from *Basic Communications Electronics* by Jack Hudson and Jerry Luecke. Developed and published by: Master Publishing Inc. Lincolnwood, Illinois.)

Different types of AC circuits

A resistive circuit

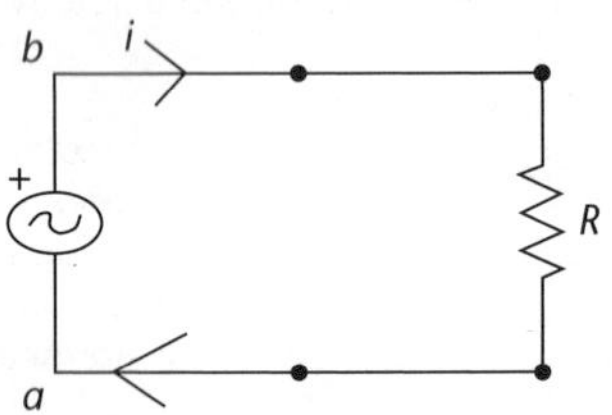

$$v_{ab} = V_R = V \cos\omega t$$

$$I = v/R = V/R \cos\omega t$$

So the current and voltage are both proportional to cosωt, i.e., they are in **phase**.

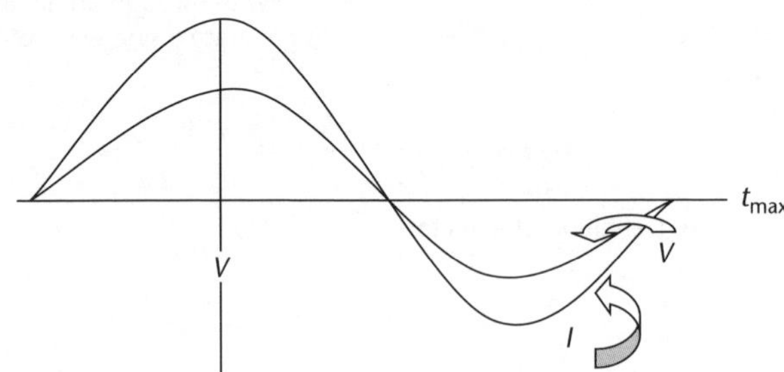

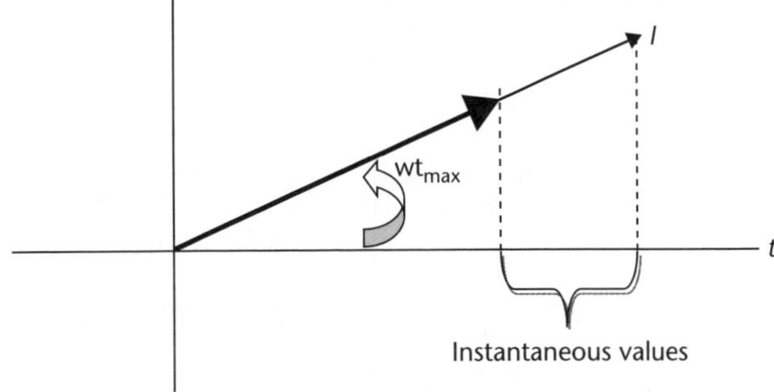

Phasors are vectors that rotate around the origin of the coordinate system.

The projections of the phasors of *V* and *I* onto the *t* axis are the instantaneous values of *V* (*v*) and *I* (*i*) with respect to t. Since *V* and *I* are in phase their phasors rotate together. Notice in the diagrams above that while the values for voltage and current rise and fall together they have different values when plotted on the same set of axes. Why? Because in this case we are dealing with pure resistive circuits, and therefore we can apply Ohm's law directly where current is directly proportional to voltage and inversely proportional to the resistance in the circuit.

A capacitive circuit

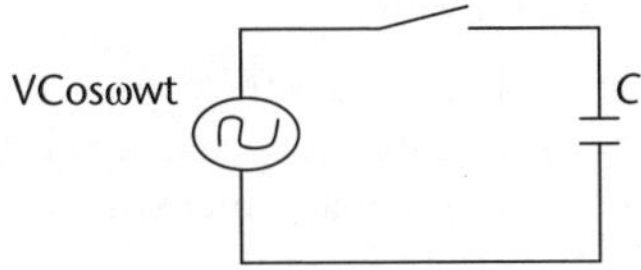

The instantaneous charge, q, on the capacitor is:

$q = Cv = CV \cos\omega t$

$i = dq/dt = -\omega CV \sin\omega t$

Compare: $v_c = V\cos\omega t$

$i_c = -\omega CV\sin\omega t$ ← ¼ cycle or 90° phase difference

$i_c = \omega CV \cos(\omega t + 90°)$

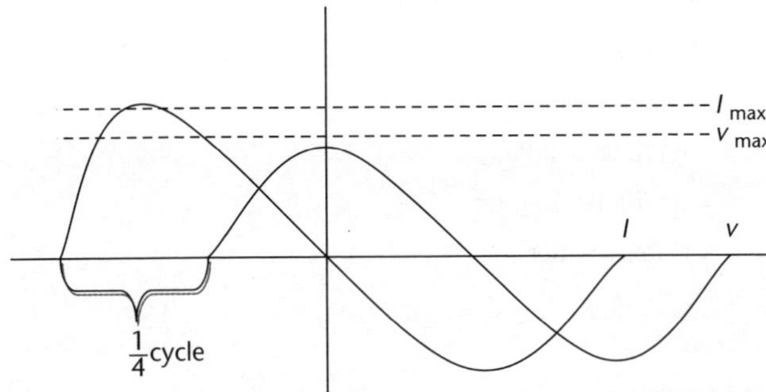

Notice that the current peaks ahead of the voltage.

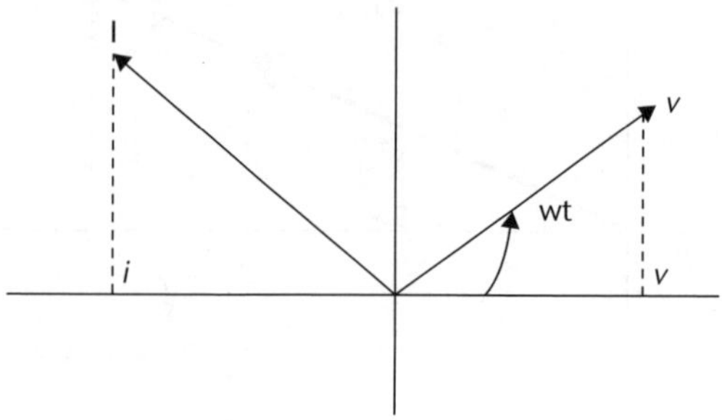

Voltage lags the current in this phasor representation.

Note: $I_{max} = \omega CV\ (1) = \omega CV$

$I = V/R$

$1/\omega C = R \longrightarrow I = C/(1/\omega) \quad X_C = 1/\omega C$

$I = V/X_C \quad X_C$ = Capacitive reactance

The units of Xc are ohms.

Capacitive reactance acts like resistance in this circuit.

Consider what is happening in a capacitive circuit as the voltage applied to the capacitor increases from zero to some maximum value.

When there is only a small amount of charge on the capacitor it readily accepts more charge and lots of current flows. As the capacitor soaks up charge, the E field between its plates increases. The potential between the plates increases; the current decreases. When Vc reaches its maximum value, the current is zero.

An inductive circuit

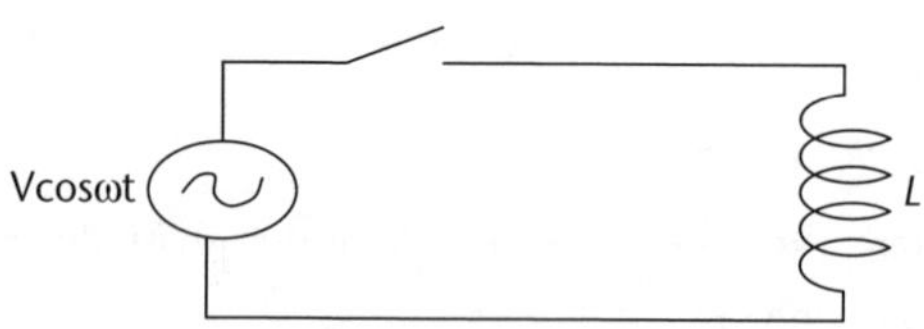

Note that the inductor has no DC resistance.

$v_L = L\,di/dt = V\cos\omega t$

$di = (V/L)\cos\omega t$

$i = \int(V/L)\cos\omega t dt$

$i = ((V/L)(1/\omega))\mathrm{Sin}\omega t + C$

$i = (V/\omega L)(\mathrm{Sin}\omega t + C$

We use the initial conditions to evaluate C

at $t = 0, i = 0 \longrightarrow C = 0$ so:

$i = (V/\omega L)(\text{Sin}\omega t)$

$I_{max} = V/\omega L$

Define $X_L = \omega L$ inductive reactance (ohms)

$I_{max} = V/X_L$

- As potential applied to the inductor rises, the magnetic flux produces a current that opposes the original current.
- The voltage across the inductor is maximum when the current is just beginning to rise, due to this tug of war.

Summary of these three basic AC circuits

Element	Symbol	Impedance	Phase of Current	Phase angle	Amplitude
Resistor	R	R	in phase with υ	0°	$V_R = IR$
Capacitor	C	X_C	leads υ_C by 90°	– 90°	$V_C = IX_C$
Inductor	L	X_L	lags υ_L by 90°	+ 90°	$V_L = IX_L$

Measuring alternating voltage

A DC voltmeter connected across an AC supply will give a measurement of zero, because, since the voltage has equal size positive and negative values, the average voltage value during a complete cycle is zero.

An oscilloscope is very useful in showing the effect of AC applied to a component.

(An oscilloscope is basically a meter that draws a graph of voltage against time for the component it is connected across. The screen that the graph is drawn on has a background grid, and adjusting the oscilloscope controls can set the scale of this grid. The 'timebase' sets the scale on the time axis, and the 'Y gain' sets the scale on the voltage axis.)

Example

An oscilloscope is connected in parallel with resistor R which is joined to an AC power supply.

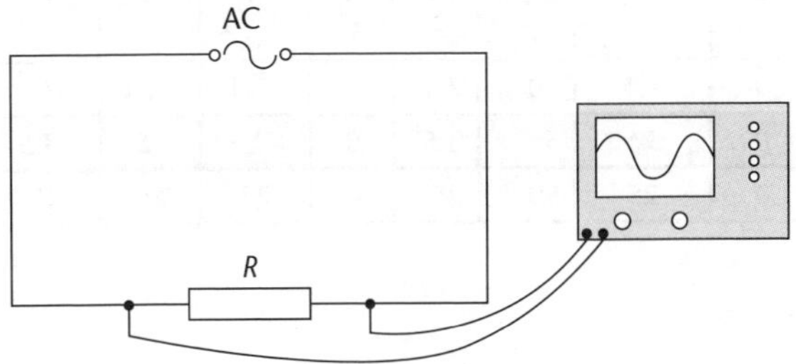

The oscilloscope controls were set to:

- Timebase setting = 2 ms.
- *Y*-gain setting = 5 *V*.

The oscilloscope display is shown.

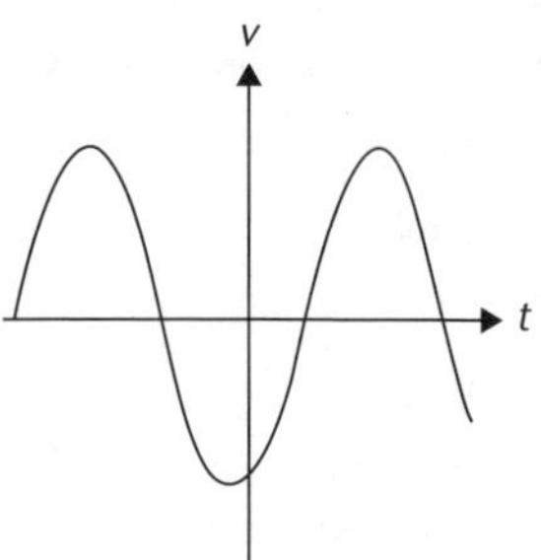

a. Calculate the maximum voltge across the resistor.

b. Calculate the frequency of the AC supply.

Solution

a. The peak voltage = 3 x 5
[height of peak x *Y*-gain setting]
= 15 *V*

b. The period *T* = 5 x 2
[length on screen of one wave x timebase setting]
= 10 ms
= 0.01 s

The frequency t

$= \frac{1}{T}$

$= \frac{1}{0.01}$

= 100 Hz

An oscilloscope is very useful, but its size prevents it from being a convenient measuring instrument.

An AC voltmeter is a more convenient meter and gives a reading that is sometimes called the DC eqivalent voltage. The DC equivalent of an AC voltage is the voltage of a DC supply that would give the same power output as the AC supply.

Power

The power in an AC circuit varies with the voltage and current.

The power output at any instant can be calculated using the formula:

$$P = VI$$

Voltage and current in an AC circuit produce a cyclic power output.

Example

The table below gives the voltage and current values in an AC circuit every $\frac{1}{8}$ of a period over one complete cycle. The power values have also been calculated.

	0	1	2	3	4	5	6	7	8	Average value
√(V)	0	7.1	10	7.1	0	–7.1	–10	–7.1	0	0
I(A)	0	3.5	5	3.5	0	–3.5	–5	–3.5	0	0
P(W)	0	25	50	25	0	25	50	25	0	25 W

The graph of *Power output against Time* is:

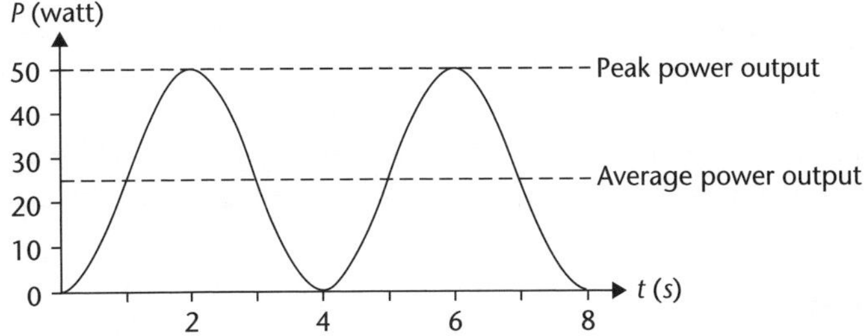

Since the graph is symmetrical, the average value of the power output is exactly half of the peak power output.

Power has two peaks during each AC cycle while V and I have a positive maximum and a negative maximum. This is because the power output is the same no matter which direction the current has:

The Shape of the AC *Power against Time* graph is similar to a sine curve. It is, in fact, a sine squared curve.

At any instant in time, $P = VI$

$$\Rightarrow P = V_{max} \sin \omega t \times I_{max} \sin \omega t$$

$$\Rightarrow P = (VI)_{max} \sin^2 \omega t$$

Average and RMS values of voltage and current

If the signals applied to a circuit are exclusively DC signals, it is relatively easy to calculate quantities as the number of amperes flowing in the circuit or the energy dissipated by the components of the circuit over a period of time. In addition, one measurement of the DC waveform at any time will reveal all that must be known about the quantity it represents.

However, the magnitudes of electrical quantities usually vary with time rather than keep constant values. If the signal is time-varying, its waveform is no longer as simple as the DC waveform shown in Figure 1 below.

Instead, the waveform has some time-varying shape. It may be **periodic** or have a random variation as shown in Figures 2 and 3.

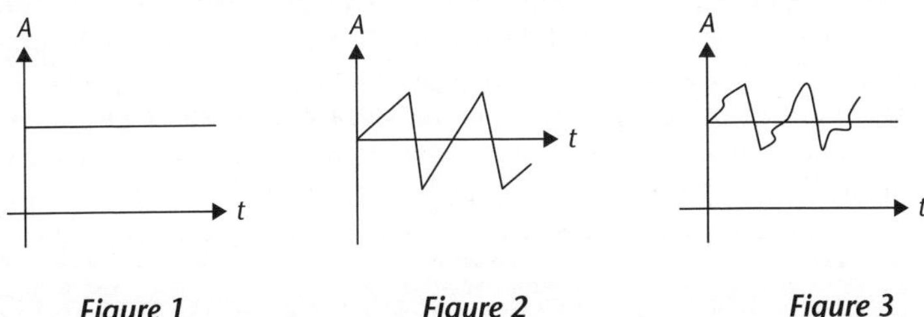

Figure 1 ***Figure 2*** ***Figure 3***

When waveforms possess time-varying shapes, it is no longer sufficient to measure the value of the quantity they represent at only one instant of time. It is not possible from one measurement to determine all that must be known about the signal.

However, if the shape of a time-varying waveform can be determined, it is possible to calculate some characteristic values of the waveform shape (such as its average value). These values can be used to compare the effectiveness of various waveforms with other waveforms, and they can also be used to predict the effects that a particular signal waveform will have on the circuit to which it is applied.

The two most commonly used characteristic values of time-varying waveforms are their average and their root-mean-square (rms) values. We will see how and why both of these values are determined.

Average value

The measuring of the average value of a waveform can best be understood if we use a current waveform as an example. The average value of a time-varying current waveform over the period, T, is the value that a dc current would have to have if it delivered an equal amount of charge in the same period, T. Mathematically, the average value of any periodic waveform is found by dividing the area under the curve of the waveform in one period, T, by the time period. We can write this as:

$$A_{av} = \frac{\text{area under the curve}}{\text{length of the period}}$$

where A_{av} is the averagve value of the waveform

Root-mean-square (rms) values

The second common characteristic value of a time-varying waveform is the root-mean-square (rms) value. In fact, the rms value is used more often than the average value to describe electrical signal waveforms. The major reason for this is that the average value of symmetrical periodic waveforms is zero. This is indeed the case, for example, for a purely sinusoidal waveform. A value of zero certainly does not provide much useful information about the properties of a signal. In contrast, the rms value of a waveform does not suffer from this limitation.

The rms value of a waveform refers to its *power delivering* capability. In connection with this interpretation, the rms value is sometimes called the **effective value**. This name is used because the rms value is equal to the value of a DC waveform which would deliver the same power if it replaced the time-varying waveform.

To determine the rms value of a waveform, we first square the magnitude of the waveform at each instant. (This makes the value of the magnitude positive even when the original waveform has negative values). Then the average (or mean) value of the squared magnitudes is found. Finally, the square root of this average value is taken to get the result. Because of the sequence of calculations that is followed, the result is given the name root-mean-square. The situation which led to an average value of zero for some waveforms is avoided because the squaring process makes the entire quantity positive before the mean is taken.

Mathematically, the rms value of a waveform can be written as:

$$A_{rns} = \sqrt{\langle f(t)^2 \rangle}$$

where the symbol ‹ › means that the average of the quantity within the brackets is taken.

For a given waveform, $f(t)$, the rms value is found by using the expression:

$$A_{rms} = \sqrt{1/T\int_0 [f(t)]^2\, dt}$$

where T is the length of one period of the waveform (in seconds).

Example A

Find the rms value of a sinusoidal waveform of maximum amplitude A_0 and period T.

Solution:

$$A_{rms} = \sqrt{1/T\int[f(t)]^2dt} = \sqrt{1/T\int A^2(\text{Sin}((2\pi/T)t)^2dt}$$
$$= \sqrt{A^2/T[t/2 - \frac{T\sin((4\pi/T)t)]^T}{4\pi}} = A_0/\sqrt{2}$$

Therefore, the rms value of sinusoidal waveforms is

$$A_{rms} = A_0/\sqrt{2} = 0.707A_0$$

When referring to sinusoidal signals, it is most common to describe them in terms of their rms values. For example, the 240 volts, 50 Hz voltage which is delivered to domestic consumers is really a sinusoidal waveform whose amplitude is about 339.463 volts and whose rms value is therefore 240 volts.

Unit 11.5 Activity 6A: AC circuits

1. If we increase the driving frequency f in the previous circuits (capacitive and inductive circuits), does the current amplitude I increase, decrease or stay the same?

2. The figure shows an emf ε_L induced in a coil. Which of the following can describe the current through the coil?

A. constant and rightward
B. constant and leftward
C. increasing and rightward
D. decreasing and rightward
E. increasing and leftward
F. decreasing and leftward

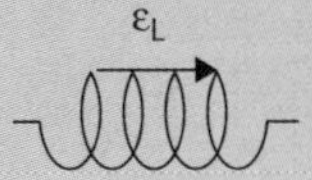

3. The following figure shows a circuit with two identical resistors and an inductor. Is the current through the central resistor more than, less than, or the same as that through the other resistor:
   a. just after the closing of switch *S*?
   b. a long time after the closing of *S*?
   c. just after the reopening of *S*?
   d. a long time after the reopening of *S*?

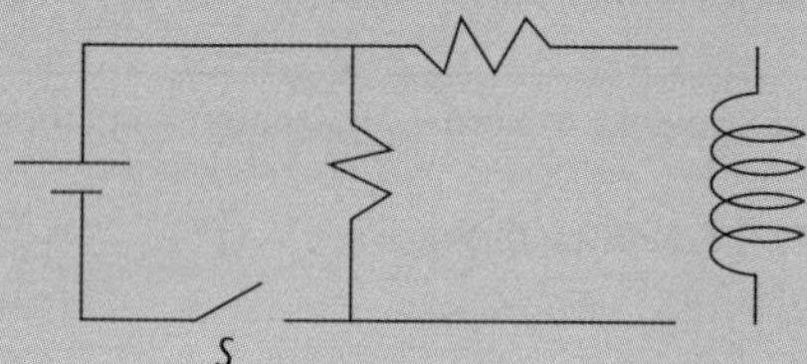

4. Suppose a 50 ohm resistor is connected, as in the following figure, to an AC generator with ε_m =30 volt. What is the amplitude of the resulting alternating current if the frequency of the emf is:
   a. 1.0 Khz?
   b. 8.0 Khz?

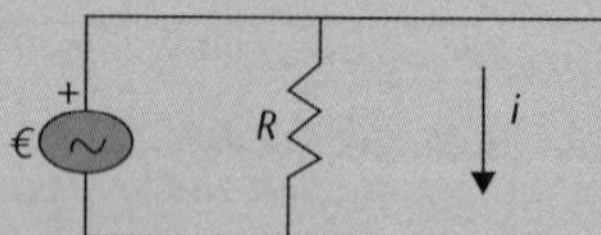

5. A 45.0 mH inductor has a reactance of 1.30 KiloOhm.
   a. What is its operating frequency?
   b. What is the capacitance of a capacitor with the same reactance at that frequency?
   c. If the frequency is doubled, what are the reactances of the inductor and capacitor?

Unit 11.6 Electronics

Topic 1: Solid state electronics—intrinsic and extrinsic semiconductors

The last Unit in the Grade11 syllabus introduces semiconductor materials. Many modern electronic devices are made from semiconductor materials. Devices such as the diode and transistor are called solid state devices because they are made of solid semiconductor materials. 'Solid state electronics' is the branch of electronics used to describe solid state devices.

Topic 1 looks at the intrinsic and extrinsic semiconductors:

- Semiconductor materials such as silicon and germanium.
- The structure of these materials.
- Intrinsic (pure) and doped semiconductors.
- The n-type and p-type semiconductors.
- The diode and transistor.

Introduction

Solid state electronics arises from the special properties of semiconducting materials like silicon (the main ingredient of sand) and germanium. Each of these has four valence electrons (electrons in the outmost shell), but they would like to have eight; therefore, a silicon or a germanium atom could link up with four of its neighbours to share electrons and change its electrical properties.

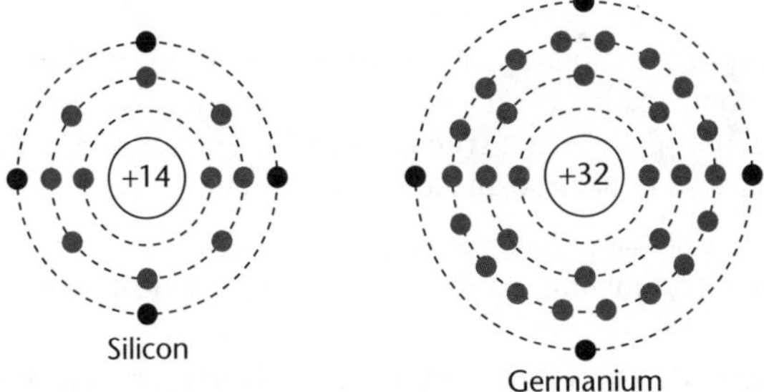

In solid state electronics, the most exciting and important electronic components are made from crystals of semiconducting materials. Ether pure silicon or germanium may be used as the *intrinsic semiconductor* material which forms the starting point for electronic device fabrication. Each has four valence electrons, but germanium will, at a given temperature, have more **free electrons** and a higher conductivity. Silicon is by far the most widely used semiconductor, partly because it can be used at higher temperatures than germanium. Depending on certain conditions, a semiconductor can act as a conductor or as an insulator.

Silicon lattice

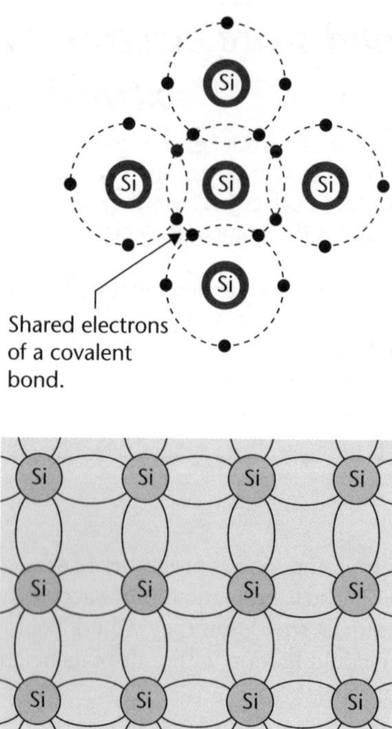

Silicon atoms form covalent bonds and can crystallise into a regular lattice by forming a cluster of silicon atoms sharing outer electrons. The actual crystal structure of silicon is a diamond lattice, but silicon is not **transparent**. This crystal is called an intrinsic semiconductor and can conduct a small amount of current.

What is important here is that a silicon atom has four electrons which it can share in covalent bonds with its neighbors, forming a regular arrangement called a crystal.

Valence electrons and semiconductor materials

In an atom, the electrons in the outermost shell are called valence electrons. They are responsible for the chemical reactions of the atom and largely determine the electrical characteristics of solid matter.

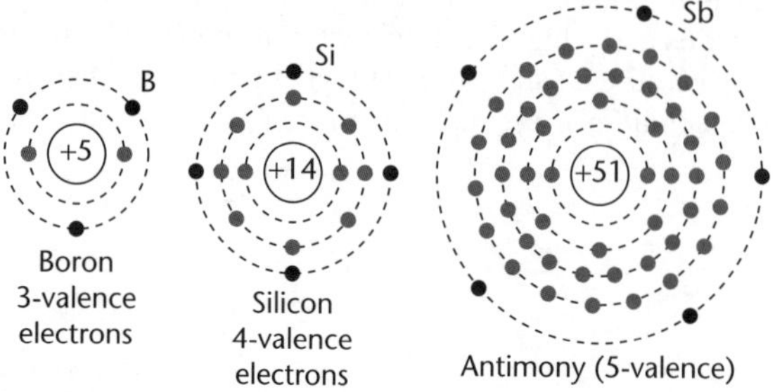

Metals are conductors because their valence electrons are free to move about inside the crystal lattice. However, the valence electrons of insulators require large amounts of extra energy if they are going to break free from their atoms; therefore, at normal temperatures insulators cannot conduct.

Semiconductors are insulators which require much less energy than normal for their electrons to break free; even at room temperature, there will be a few electrons able to carry an electric current through the material. When a valence electron leaves its atom in a metal the atom is left as an ion with a positive charge. This ion is fixed in the lattice and therefore it cannot take part in conduction.

Ions produced in semiconductors have a strange property. The ion takes advantage of the relatively weak bond between valence electron and atom and steals an electron from a neighbour. This is repeated over and over again; hence, the position of the atom that is an electron short moves inside the lattice even though the atoms themselves are stationary. This strange behaviour receives a special name: the moving shortage of an electron is called a **hole** and can be thought of as a positive charge carrier, free to move inside the semiconductor.

The value of the positive charge carried by a hole must be equal to the negative charge carried by an electron. If a hole and an electron were to meet, the electron would drop into the hole and both would cease to exist. This process is called **recombination**.

If we imagine a perfect crystal of semiconductor material with no other atoms present, then this is what is meant by an **intrinsic semiconductor**.

Intrinsic semiconductors

Even though, at certain conditions, a semiconductor can act as a conductor or as an insulator, a silicon crystal differs from an insulator because at any temperature above absolute zero there is always a probability that an electron in the lattice will be knocked from its position, leaving behind a 'hole', also called an electron deficiency position. If we apply a voltage, then the hole and the electron would contribute to a small current flow.

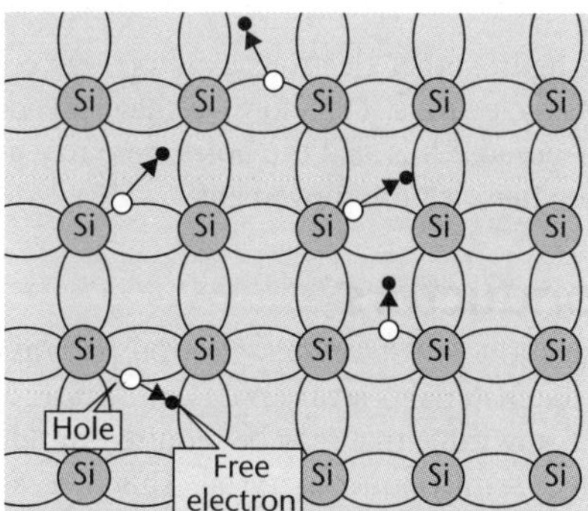

We could model the conductivity of a semiconductor in terms of the band theory of solids. This band model suggests that at normal temperatures there is always a probability that electrons can reach the conduction band and contribute to conduction of current.

The term intrinsic is used to distinguish between the properties of pure silicon and the drastically different characteristics and properties of doped n-type or p-type semiconductor material.

Intrinsic semiconductor current

As we mentioned above, both electrons and holes contribute to current flow in an intrinsic semiconductor. Electrons are attracted to the positive terminal and so will flow in one direction, and holes will flow in the opposite direction towards the negative terminal. The total current in the circuit will be 'half electron current' and half 'holes current'.

In fact, the current would be very small since at normal temperature few electron-hole pairs are formed. If however we supply energy by increasing the temperature then more and more electrons would be able to escape from their atoms and the current will increase. If we can think of the semiconductor crystal as a resistor then increasing its temperature reduces its resistance, which is the opposite of what happens in metals.

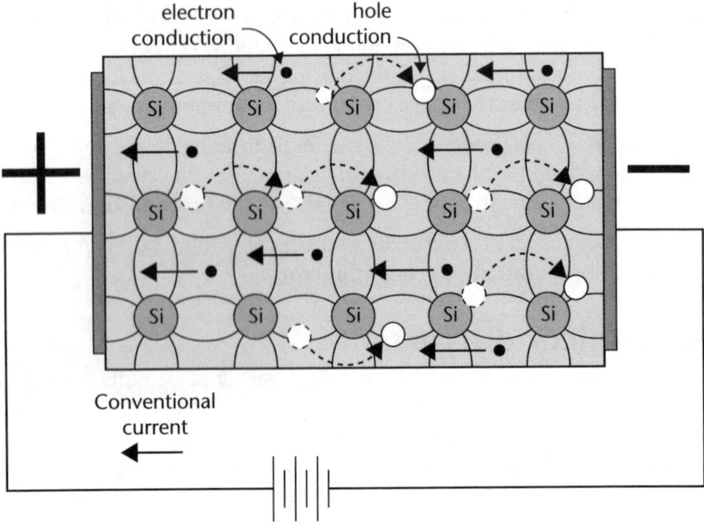

As we can see in the above figure, other electrons can hop between lattice positions to fill the vacancies left by the freed electrons. Therefore, we have an additional mechanism that is called 'holes conduction' because it is as if the holes were moving across the material in the opposite direction to the free electron movement.

Extrinsic semiconductors

Even though intrinsic semiconductors have hole-electron pairs at room temperature, they are few in number and so the material is a very poor conductor. Therefore, it is possible to increase the conductivity to any value that may be required by adding a measured quantity of an impurity to the intrinsic semiconductor crystal. This process of adding impurities to a semiconductor is called **doping**.

Adding certain types of impurities will cause either free electrons or holes to be major carriers of current. We usually add impurity elements from 3rd or 5th columns of the periodic table. Thus, for our purposes we shall consider that *conduction in semiconductors is due to two separate and independent particles' direction under the influence of an applied electric field.*

Addition of pentavalent and trivalent impurities

The addition or doping using impurity atoms with 5 valence electrons produces n-type semiconductors by contributing extra electrons, and the addition of impurity atoms with 3 valence electrons produces p-type semiconductors by producing a 'hole' or electron deficiency.

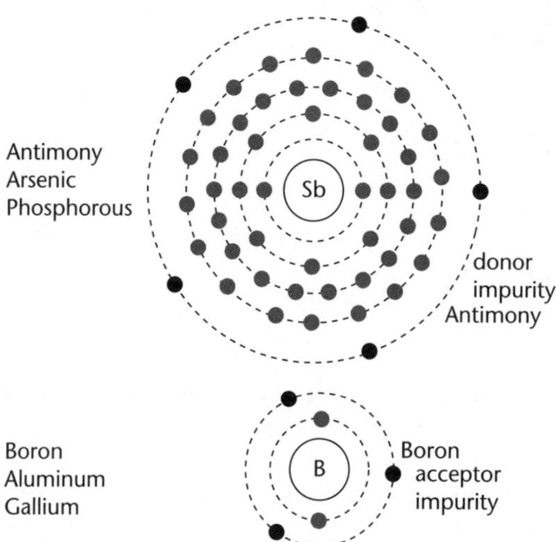

Silicon and germanium each have 4 valence electrons per atom, thus allowing them to join up with 4 other atoms within the crystal. If an impurity with 5 valence electrons is added, then one of the electrons will not be needed for bonding and will be free to move about the lattice as a conduction electron. The atom it comes from is not a semiconductor and so it leaves behind a positive ion (and not a hole). The more impurity atoms are added, the more free electrons are produced and the greater the conductivity of the material. The semiconductor is now said to be extrinsic and is described as an **n-type semiconductor**. The impurity is known as a **donor impurity**. Therefore, the main carriers of electric current in an n-type semiconductor are electrons, and at the site of each of the impurity atoms there is a fixed positive ion.

If the germanium or silicon is doped instead with an element having only three valence electrons in its atoms, an **acceptor impurity**, then the shortage of electrons produces holes in the material and negative ions. The result is a **p-type semiconductor**. The main carriers of electric current in a p-type semiconductor are holes.

P-type and n-type semiconductors

Therefore, adding pentavalent impurities such as antimony, arsenic or phosphorous to a semiconductor gives rise to free electrons, increasing the conductivity of the intrinsic semiconductor and forming an n-type semiconductor.

The addition of trivalent impurities such as boron, aluminum or gallium to an intrinsic semiconductor creates deficiencies of electrons called 'holes', forming a p-type semiconductor.

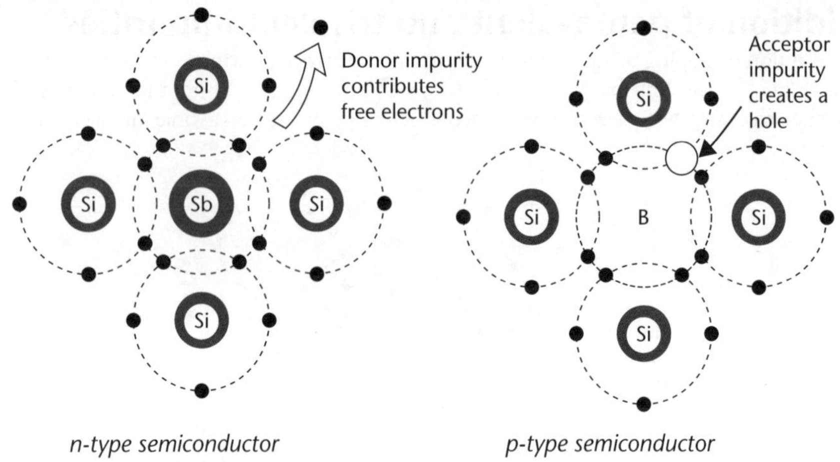

A p-n junction

When we put p-type and n-type materials in contact with each other, the junction behaves very differently than either type of material alone. Specifically, current will flow readily in one direction (forward biased) but not in the other direction (reverse biased), forming what we know as the basic **diode**. This non-reversing behavior arises from the nature of the charge transport process or current mechanism in the two types of materials.

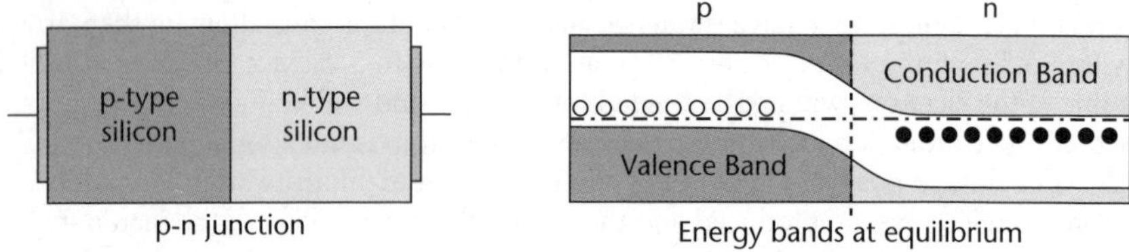

In the above figure, the open circles on the left side of the junction represent 'holes' or deficiencies of electrons in the lattice which act like positive charge carriers. The solid circles represent the available electrons from the n-type dopant. Near the junction, electrons diffuse across it to combine with holes, giving rise to a 'depletion region'.

Depletion region

When we form a p-n junction, holes from the p-type region diffuse into the n-type region and leave behind negative acceptor ions which have no corresponding holes. Thus, the p-side of the junction has more negative charges. Similarly, the n-region near the junction becomes lined with fixed positive donor ions, and has more positive charges. Only the fixed ions remain near the junction, negative on the p-side and positive on the n-side. *There are no charge carriers in this region.*

If a hole on the p-side tries to diffuse to the n-side, the positive ions near the junction will repel it. Likewise, if an electron from the n-side tries to diffuse to the p-side, the negative ions near the junction will repel it. This will stop further diffusion of charges, unless it is helped by adding a forward bias on the junction.

The region near the junction that contains these fixed ionic charges and no current carriers is called the **space-charge region** because of the lack of current carriers in this region, and it is referred to as the **depletion region**. An internal electric field is thus created. Under the action of this electric field, the minority charge carriers are *drifted* across the junction; therefore, current flow in a p-n junction is due to these two mechanisms' diffusion and drifting of charge carriers.

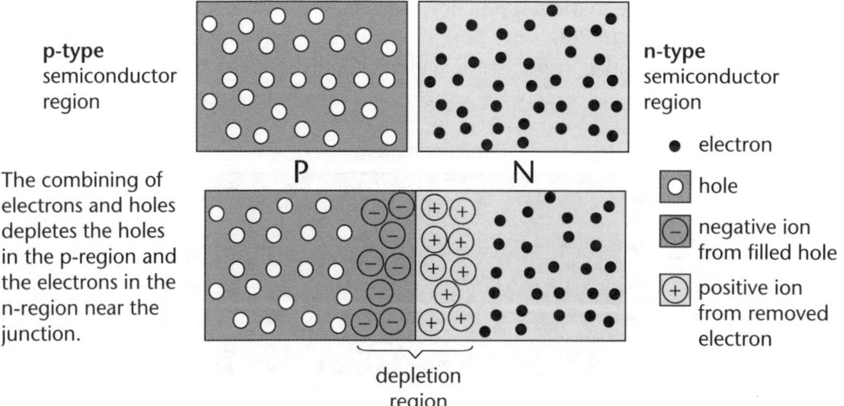

Forward biased p-n junction

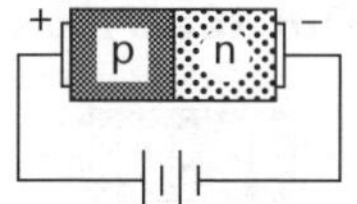

The figure above shows the appropriate way for forward biasing the p-n junction that drives holes to the junction from the p-type material and electrons from the n-type material. At the junction the holes and electrons combine for a continuous electric current to be maintained.

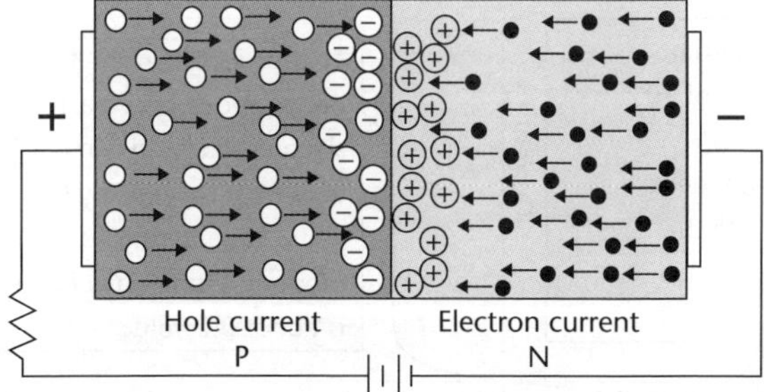

Reverse biased p-n junction

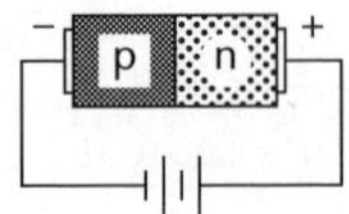

Applying a reverse voltage to the p-n junction with the polarity as shown in this figure causes a transient current to flow, since both electrons and holes are pulled away from the junction. When the potential created at the widened depletion region equals the applied voltage, the current will cease except for a small thermal effect current.

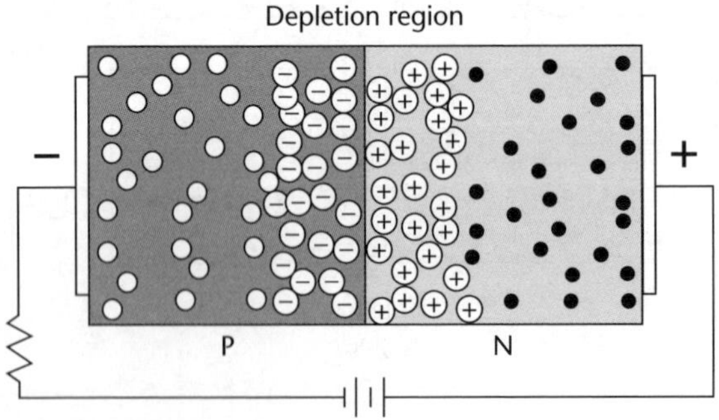

The p-n junction diode: characteristics

This behavior of the p-n junction as conducting current in the forward direction but not in the reverse direction is therefore a basic characteristic used to manufacture a device called a **diode**, to be used for rectification in the design and fabrication of DC power supplies.

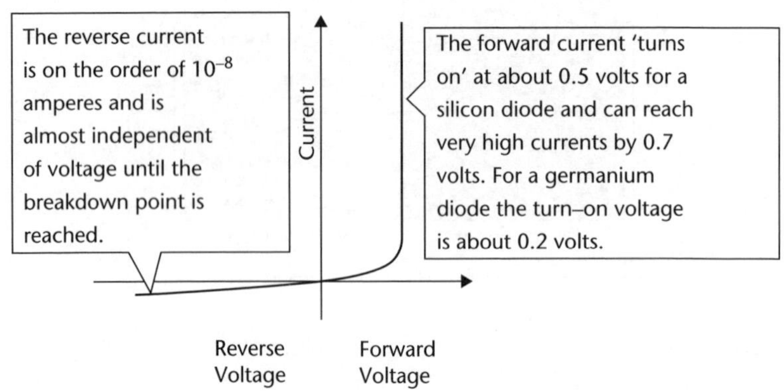

Unit 11.6 Electronics

Topic 2: Solid state devices—diode and transistor

Topic 2 continues the introduction to electronics by looking at the diode and transistor, both solid state devices. In particular, the Topic covers:

- Use of the diode as a rectifier and as a switch.
- NPN and PNP symbols for diode and transistor.
- Diode and transistor operating principles, including I-V and biasing.
- The diode as a rectifier and as a switch.
- The transistor as a switch and as an amplifier (common-emitter, common-base, common collector).

Diode as a rectifier

As we have said, a diode is a device that allows current to flow in one direction only. We also know that electric current flows in the opposite direction to the flow of electrons. Thus: *current in a diode flows from the anode to the cathode, providing the anode is positive with respect to the cathode*. If the polarity is reversed, the diode would be reverse biased, and no current would flow.

The term **bias** is used to describe the DC conditions. If a diode is always forward biased by the DC conditions, it is always conducting and can be viewed as a closed switch. Otherwise it can be thought of as an open switch (reverse biased).

Diodes are used in rectifier circuits, in which an AC voltage is converted to a DC voltage. This voltage conversion process is called **rectification**, and the diodes used to do it are called rectifiers. As we know, the mains power (supplied by PNG Power) is AC; because most electronic circuits need DC, there is a rectifier in virtually all mains-powered electronic circuits. The simplest circuit is the **half-wave rectifier** and this is shown below.

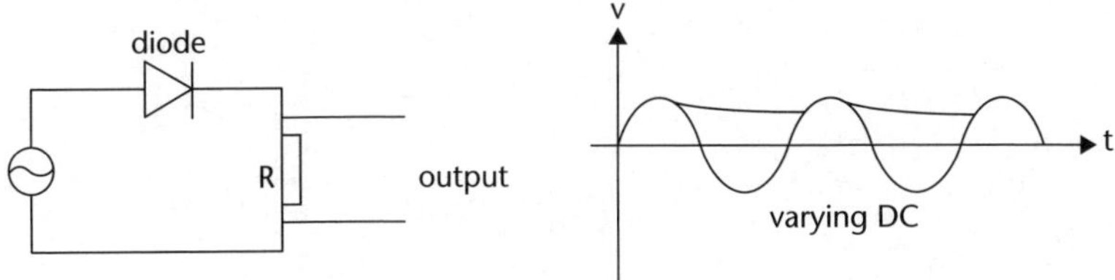

Half-wave rectification

In half-wave rectification, the diode lets the forward parts of the alternating current pass through, and blocks the backward part (the negative side of the AC voltage); therefore, the current through the resistor flows one way only and becomes a sort of jerky form of DC. This pulsating current from the rectifier can be made smooth by using a capacitor to collect the charge during the surges and release it when the current from the rectifier falls. Smoothing the current this way, we now get a current that looks like a steady DC from a battery.

One way of illustrating the behavior of the diode as a switch is in the above process of generating DC from AC, as shown in the following figures:

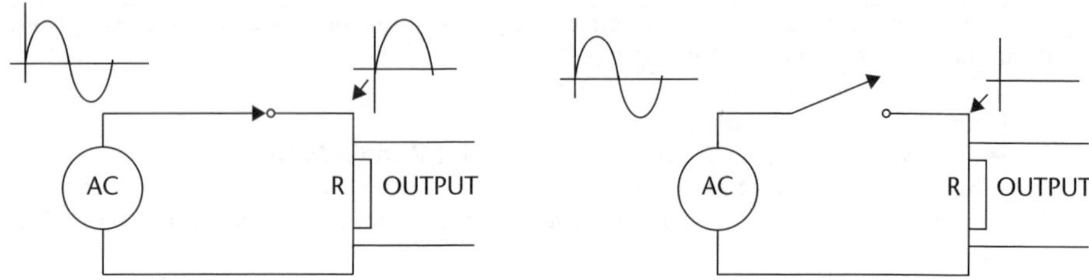

In both figures above, the diode has been replaced with a switch. For the positive half cycle of the AC supply, as shown on the left, the switch representing the diode is closed because the anode is positive. The AC supply is therefore connected to the load resistor, and current flows.

When the AC supply reverses polarity, the switch opens (shown in the figure on the right) because the anode is negative (reverse bias). The voltage across the load resistor falls to zero, and no current flows.

An important point to note is that for the connections shown in the circuit, the DC output is positive. Why is this? As we have being saying, the polarity at the cathode should be negative; in fact, the anode is more positive (by at least 0.6 volt) than the cathode. Think of the diode as a switch, but with a 0.6 volt drop across when it is closed (like a very small resistor).

Unit 11.6 Activity 2A: Diodes

1. Construct each of the following circuits and measure the voltage across the diode(s):

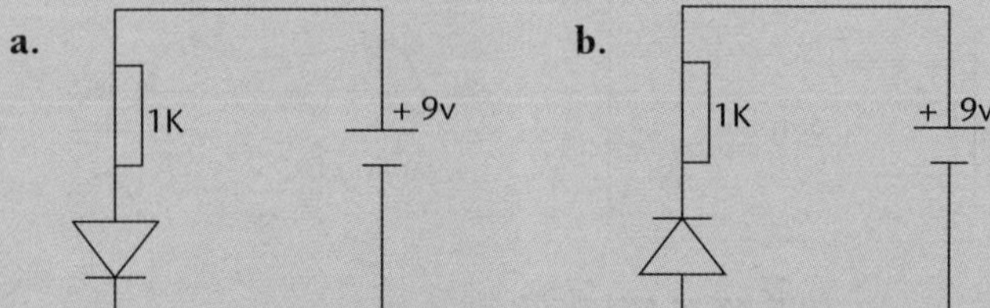

2. Explain what will happen if you connect a diode across the battery as shown:

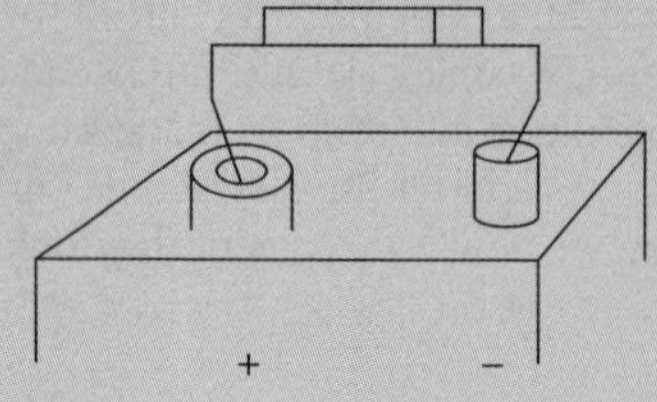

3. If you connect the following circuit, what is the voltage across the two diodes? Are they in forward bias or reverse bias?

1K
+ 9v

Other useful diodes

The LED: light-emitting diode

The LED is an electronic component. The equivalent in electrics is the torch globe, so let us see how they differ from each other. The diagram below shows the similarity between a torch using a globe and one using an LED. First, note the need for a resistor in the LED circuit. The reason for this will be explained.

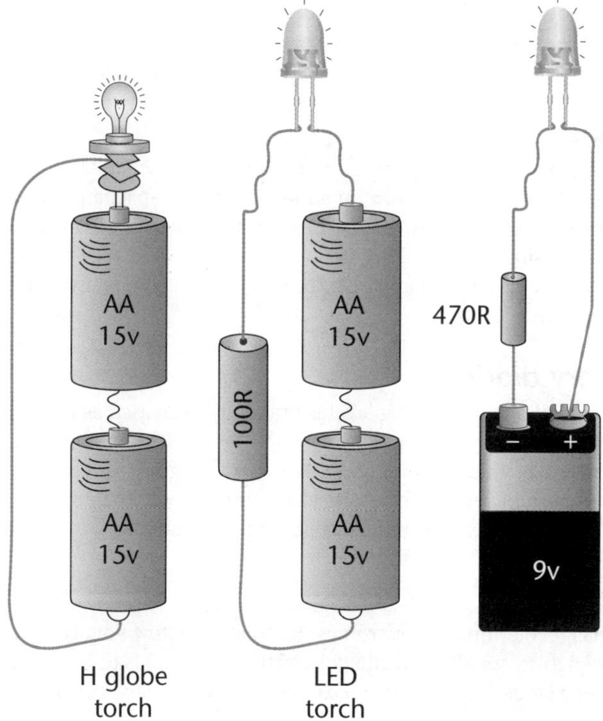

The LED torch does not give the same **illumination** as a normal torch and would only be suitable as a key-hole light. The LED is an electronic component because the light it gives off is produced by a crystal and not by the heating of a wire (in fact, almost no heat is released by an LED).

It is a very efficient light producer and although it does not give off as much light as a torch globe, it requires considerably less current. For the same output, high efficiency LEDs consume about 1/10th less power.

Study the LED and observe some of the characteristics that make this component so important in electronics. Most important is its polarity sensitivity. You must connect correctly within a circuit; if it is connected incorrectly, the crystal will not emit light.

You can verify this by building the following circuit. Connect a resistor of about 470 ohms to the negative of a 9 v battery and connect an LED across the two terminals as shown in the diagram above.

When you connect the LED in one direction (the correct way), it will illuminate. If you reverse the connection, it will not illuminate.

You must not omit the resistor. It is a current-limiting resistor and prevents the LED from 'burning out'. With this basic experiment you have established that an LED will only work when connected correctly, whereas a torch globe will operate in either direction.

Another difference you can observe is the colour of the light emitted by an LED. The crystal of the LED determines the colour of the emitted light; it is generated electronically and depends on the structure of the crystal used, so the light exhibits a specific color. This is another major difference between the LED and the torch globe/bulb.

The PIN diode

The PIN diode has heavily doped p-type and n-type regions separated by an intrinsic region. When reverse biased it acts like an almost constant capacitance, and when forward biased it behaves as a variable resistor. The forward resistance of the intrinsic region decreases with increasing current. Since its forward resistance can be changed by varying the bias, it can be used as a modulating device for AC signals. It is used in microwave switching applications.

Step-recovery diode

In the step-recovery diode the doping level is gradually decreased as the junction is approached. This reduces the switching time since the smaller amount of stored charge near the junction can be released more rapidly when changing from forward to reverse bias. The forward current can also be established more rapidly than in the ordinary junction diode. This diode is used in fast-switching applications.

Transistors

The transistor has revolutionised electronics. Its small size, low voltage, low power and cheapness have enabled smaller and smaller electronic products to be created. Computers can be created on a single chip and video cameras can be no larger than a postage stamp.

These products use many thousands of transistors in integrated circuits and you cannot see the individual transistors. We will be studying a single transistor. We will do this without any mathematics or formulas, and will explain its operation with a number of diagrams, including a water valve equivalent model, that will show how the transistor operates and the concept of how it is considered as an amplifier. We will start by showing how it looks like and its symbol(s).

A junction transistor is a sandwich type, composed of material of either a p-type semiconductor between two slices of n-type (NPN transistor), or the reverse (PNP transistor). Transistors can be made of germanium or silicon, but the most common device is the silicon NPN transistor.

The junction transistor

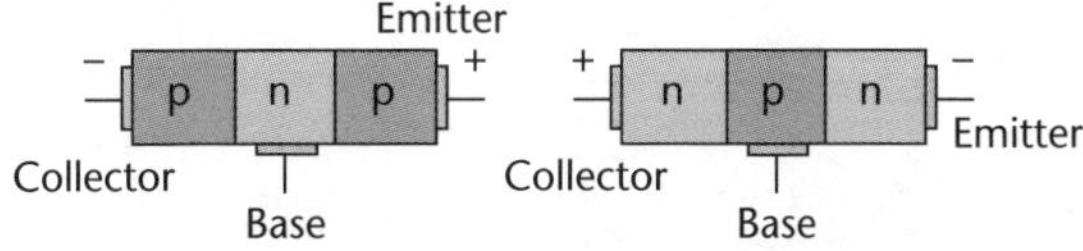

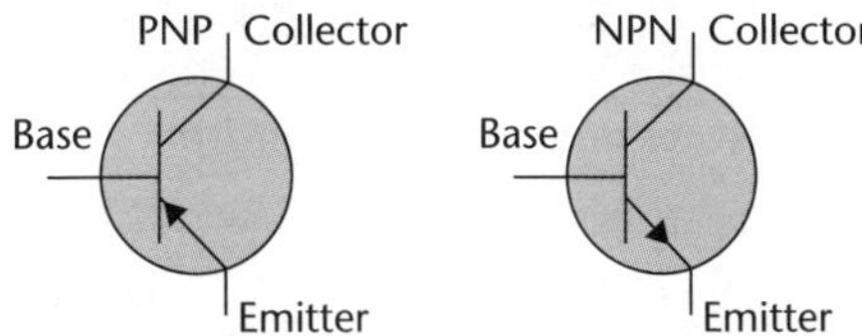

A bipolar junction transistor has three regions of doped semiconductors. A small current in the center region or **base region** can be used to control a larger current flowing between the end regions (**emitter** and **collector**). The device can be characterised as a current amplifier, having many applications for amplification and switching.

NPN transistor as a current amplifier

In an NPN transistor, the collector current I_C is proportional to the base current I_B according to the relationship $I_C = \beta I_B$; or, more precisely, it is proportional to the base-emitter voltage V_{BE}. The smaller base current controls the larger collector current, achieving current amplification.

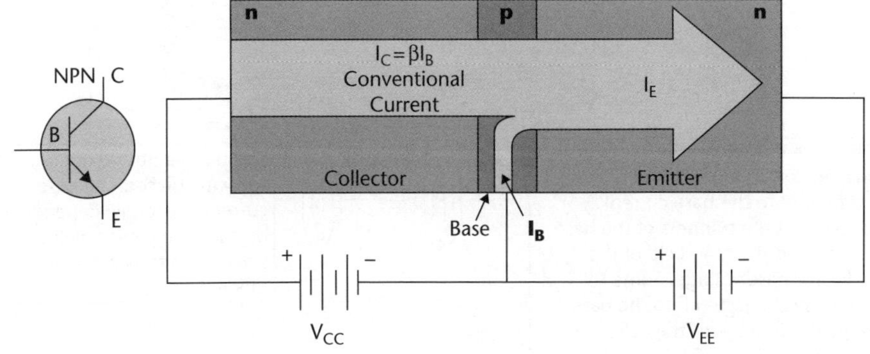

The effect of the transistor is amplification. To understand the transistor and amplification, the analogy of a water valve can be helpful. This is illustrated in the figure below and it works as follows: one drop of water entering the base will be amplified to allow 100 drops of water to flow through the collector path, an amplification of 100.

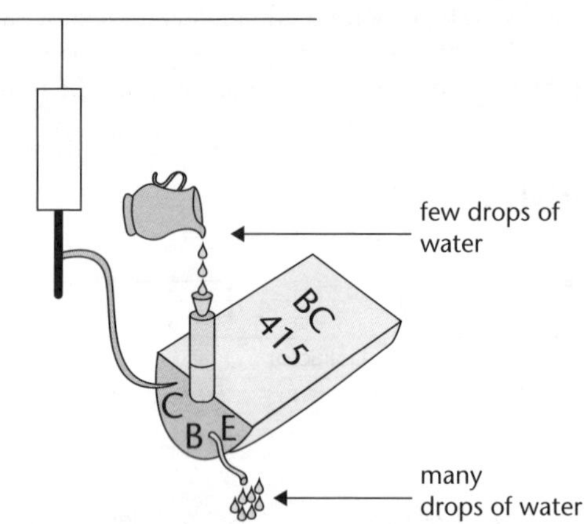

Water analogy for a transistor

In terms of an electric circuit, the smaller current in the base acts as a water valve controlling the larger current from collector to emitter. A 'signal' in the form of a variation in the base current is reproduced as a larger variation in the collector-to-emitter current, producing an amplification of that signal.

Transistor structure

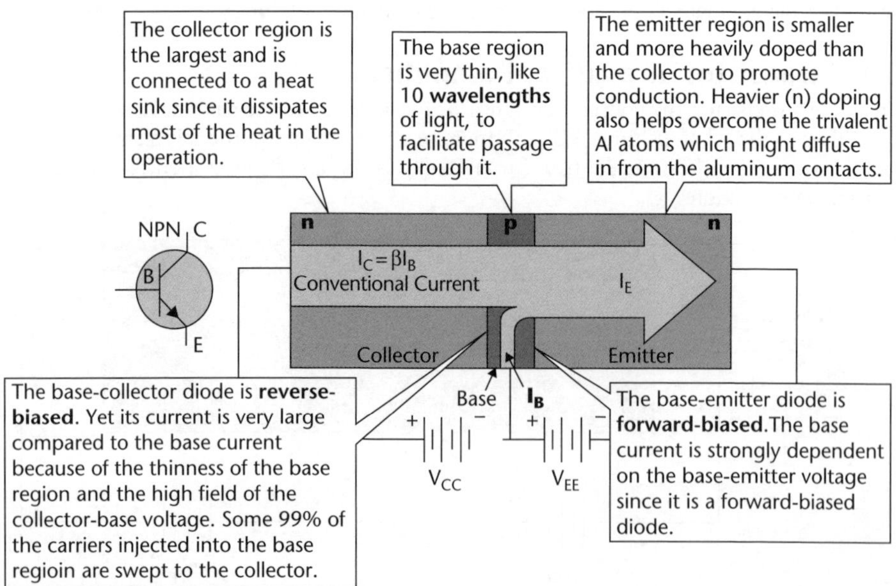

PNP transistor as current amplifier

The larger collector current I_C is proportional to the base current I_B according to the relationship $I_C = \beta I_B$; or, more precisely, it is proportional to the base-emitter voltage V_{BE}. The smaller base current controls the larger collector current, achieving current amplification.

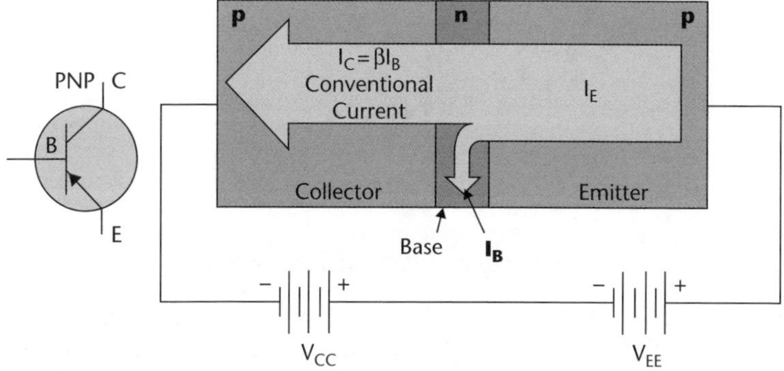

Transistor operation—an example

The water analogy for the transistor mentioned above is not entirely true as water represents conventional current flow in that model, whereas the transistor operates on electron flow (electrons flow in the opposite direction).

However, if we replace the water with a small current flow (such as 1 mA) into the base, we can expect a flow of 100mA in the collector emitter circuit (assuming the transistor has a gain of 100 and we use a globe of 6v/100 mA type). This is sufficient current to illuminate a small globe.

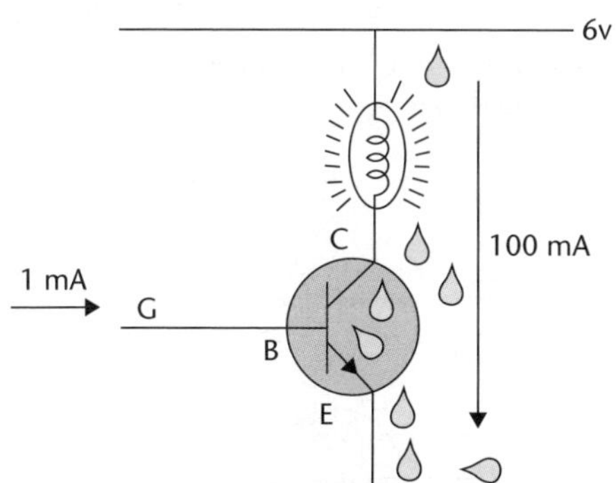

Transistor with a gain of 100

The figure above shows the lamp illuminating, but how do we get 1mA to flow into the base?

The water analogy in the figure below shows a fine stream of water flowing into the base and since the transistor has a gain of 100, the flow through the collector-emitter path will be 100 times greater.

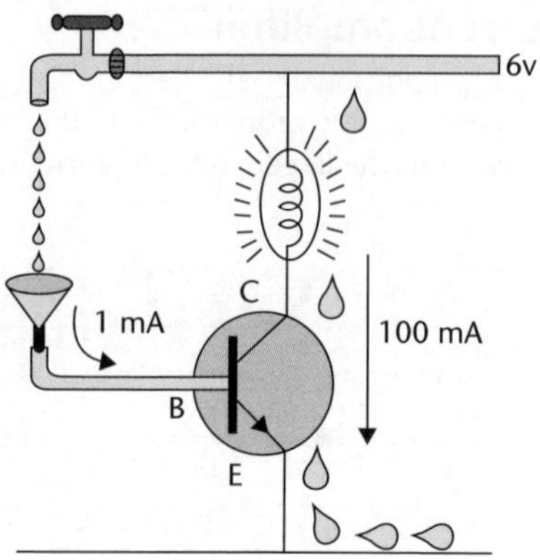

'1mA of water' flowing into the base

Again, the water analogy shows that a transistor requires current at its input for a flow to occur between collector and emitter. The transistor is like a tap or regulator on a large water pipe, depending on the size of the transistor, so the maximum flow is determined. The transistors we are talking about can handle up to about 500 mA so that if 5mA were to flow into the base, the transistor would allow 500mA to flow through the collector-emitter circuit, provided the globe is a 6v/500mA type.

If the tap is turned off, the transistor will cease to conduct and the globe will go out. If only 0.5mA flows into the base, the collector current will be 50mA and the globe will illuminate very dimly.

Basic terms explained

In our water model, the transistor is an NPN type. The tap is providing base current and the collector load is the small globe. The gain in the example above is 100.

In place of the tap we can use a resistor of approximately 6 kohm to provide 1mA into the base of the transistor, as illustrated.

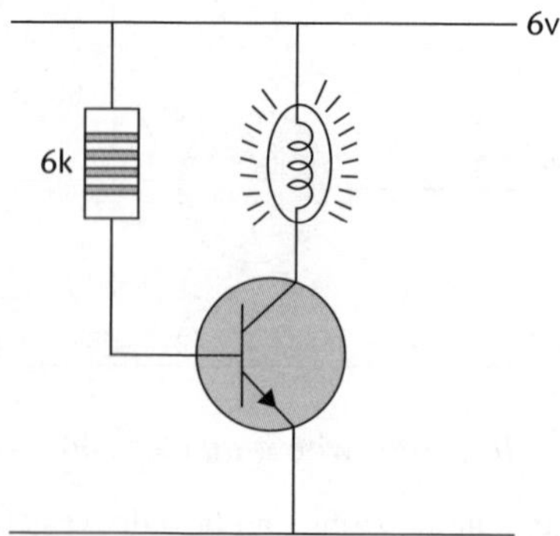

Transistor is turned ON via the 6k

What the transistor is doing is this: It is reducing the 6 kohm by a factor of 100 (the gain of the transistor) to create an effective resistance of 60 ohms between collector and emitter.

The transistor has been converted into a 60 kohm resistor and the globe will light up with the same brightness. In other words, the transistor is acting exactly like a 60 ohm resistor, as illustrated by the following figure.

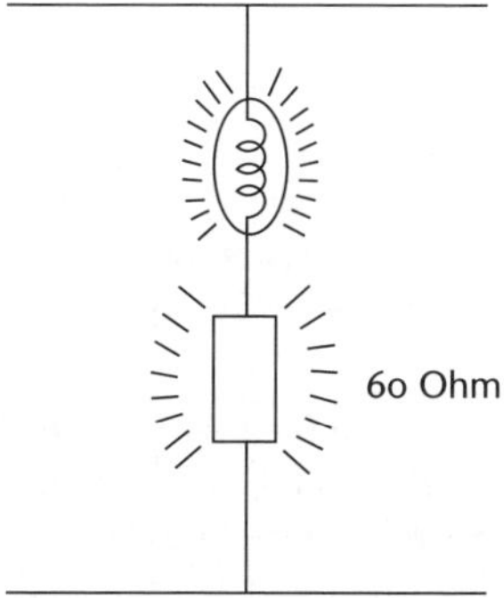

The transistor has been converted into a 60 ohm resistor

Characteristics and modes of operation of the transistor

A transistor in a circuit will be in one of three conditions or modes of operation:

1. Cut off (no collector current) – useful for switch operation.
2. In the active region (some collector current, more than a few tenths of a volt above the emitter) – useful for amplifier applications.
3. In saturation (collector a few tenths of a volt above emitter) – large current useful for 'switch on' applications.

Collector current

Normal transistor action results in a collector-to-emitter current which is about 99% of the total current. The usual symbols used to express the transistor current relationships are shown.

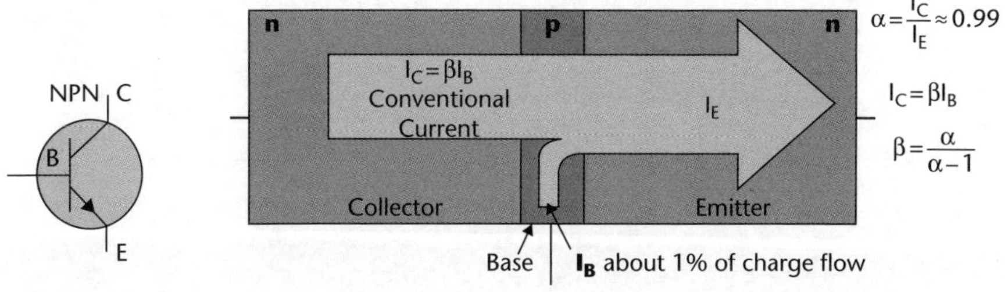

The proportionality β can take values in the range 20 to 200 and is not a constant, even for a given transistor. It increases for larger emitter currents because the larger number of electrons injected into the base exceeds the available holes for recombination so the fraction which recombine to produce base current declines even further.

Any circuit that depends on a specific value of the current gain β is a bad circuit because that value varies for a given transistor as well as between different transistors of the same type.

The transistor as a switch

In many electronic systems, transistors are used as controlled switches. A digital computer, for example, will use several thousand transistor switches. The speed with which the switches operate is of paramount importance. Let us consider the time response of a simple transistor switch.

The transistor can be operated as a switch by designing the associated circuit (biasing the transistor) so that the transistor is either in the cut off region or in the saturation region of operation. When the transistor is in cut off, no collector current flows and the switch is open. When the transistor is in saturation, maximum collector current will be flowing and the switch is said to be closed. The switch is then controlled by the current applied to the base.

Such a transistor switch would be as shown in the following figure. The response of the transistor switch to a pulse-wave form input voltage v_i is also shown.

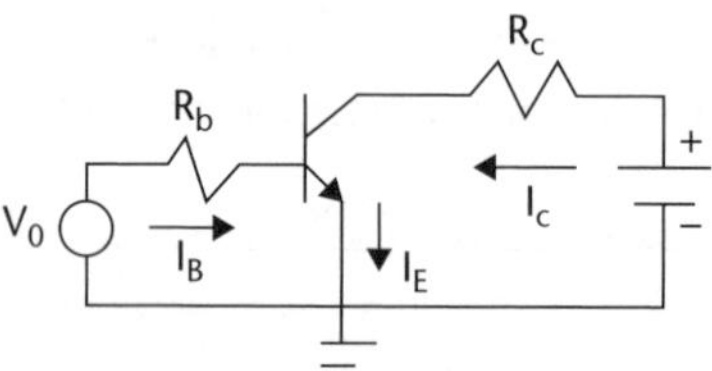

Transistor as a switch: circuit configuration

Unit 11.6 Activity 2B: The transistor

When the transistor turns ON, what is the current flow through the load resistor?

6v
1K
NPN
C
B
E

The transistor as an amplifier

Having previously described the mechanism of current flow in the junction transistor and how the current amplification is achieved, let us now present the different basic configurations of transistor amplifiers for the NPN type of transistor.

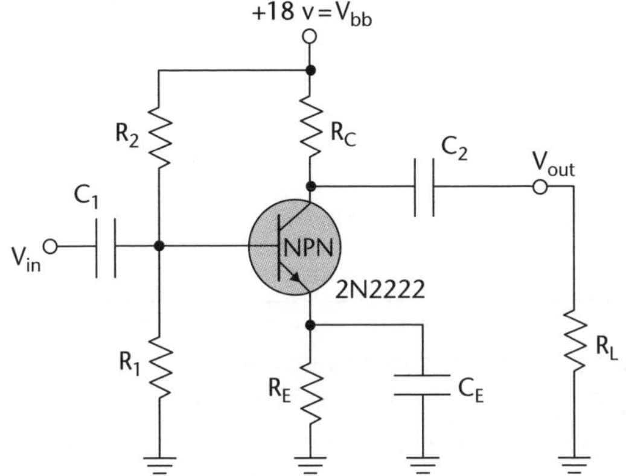

NPN common emitter amplifier

NPN common collector amplifier

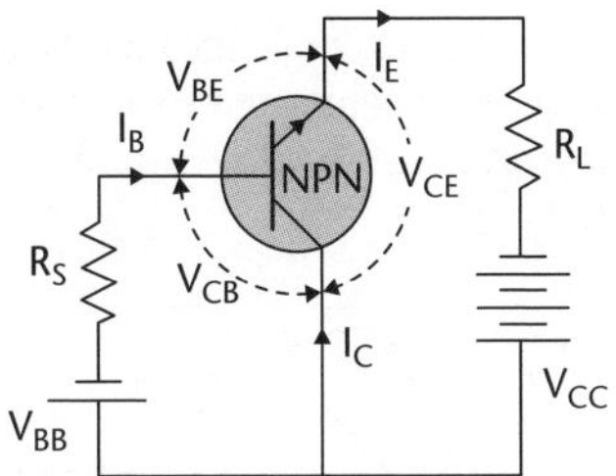

The common collector amplifier, often called an emitter follower since its output is taken from the emitter resistor, is useful as an impedance matching device since its input impedance is much higher than its output impedance. For this reason, it is also termed a 'buffer' and is used in digital circuits with basic gates.

Emitter follower discussion

This common collector transistor amplifier is commonly called an emitter follower. The voltage gain of an emitter follower is slightly less than unity since the emitter voltage is constrained at the diode drop of about 0.6 volts below the base. Its function is not voltage gain but current or power gain and impedance matching. Its input impedance is higher than its output impedance so that a signal source does not have to work so hard.

It is generally used as an impedance transformer in the input and output circuits of amplifier systems. When placed in the input circuit, its high input impedance reduces the loading of the source. When placed in the output circuit, it serves to isolate the preceding stage of the amplifier from the load and, in addition, provides a low output impedance.

NPN common base amplifier

The circuit for the common-base amplifier is illustrated by the following figure.

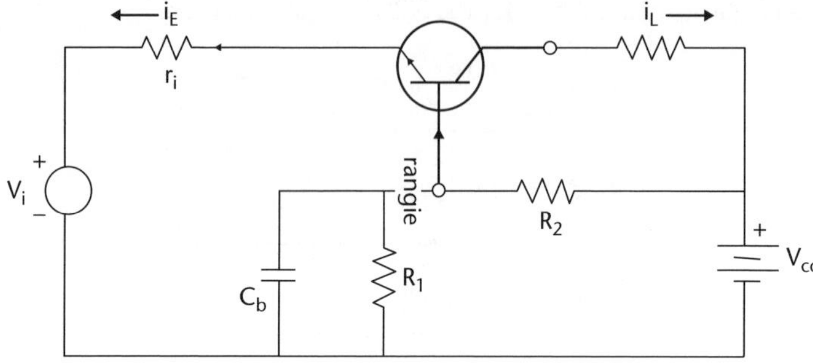

This configuration does not provide current gain but does provide some voltage gain. It also has properties which make it useful at high frequencies.

From our previous example above on the operation of the transistor and the analogy to a water flow, we expect you to understand the operation of transistors and the concept of amplification.

Transistors are ideal for audio work. They are small, low cost, low-power devices that can produce amazing results. They have been used for such applications as pocket radios and hearing aids, where the amplitude of the signal must be increased hundreds of times with a number of stages of amplification. The example in the following figure is not complete but shows you how an NPN transistor is connected to a power supply with the collector connected to the positive of the source via a load resistor and the emitter to the negative (zero volts) of the power source.

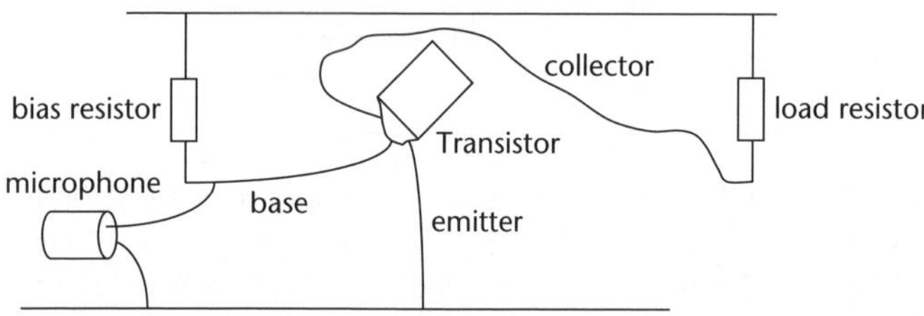

As you can see, the signal is fed to the base and appears on the collector in an amplified form. If you connect a microphone (as shown) to the base, the output of the microphone will be about 10 mV when you whistle on it. The transistor will amplify this; if it amplifies it 100 times, the collector will produce a waveform of 100×10 mV = 1000 mV = 1V.

The following figure shows the approximate biasing voltages for an audio amplifier:

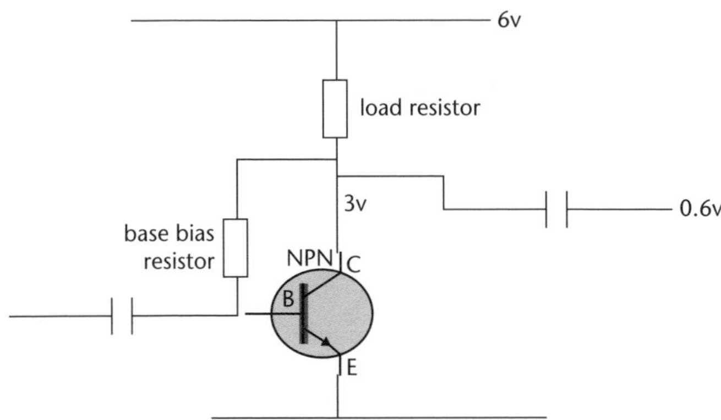

By selecting the correct value for the load resistor and bias resistor, the transistor will operate with the collector voltage at half the value of the power source. At this point we say it will provide the maximum amplification. The amplifier in this configuration is the common emitter amplifier, as the emitter is common to both the input and the output signal. It is a self-biasing stage and the equilibrium point is obtained by the bias resistor turning on the transistor and causing the collector voltage to drop. The voltage across the base bias resistor reduces and thus the base current fails. This makes the transistor turn OFF slightly and the collector voltage rises.

After a short time an equilibrium point is reached called the **bias point**, and if the component values are chosen correctly, the collector will be at the mid-point of the power source.

Unit 11.6 Activity 2C: Transistors

1. What is the name given to a transistor that has its load connected to the emitter lead?
2. What is another name for a fully conducting transistor?
3. What is the value of base voltage when a transistor begins to turn ON?
4. When a transistor is fully conducting, what is usually the voltage between collector-emitter?
5. What is another name for a transistor that is switched OFF?
6. What is the maximum current through the load resistor?

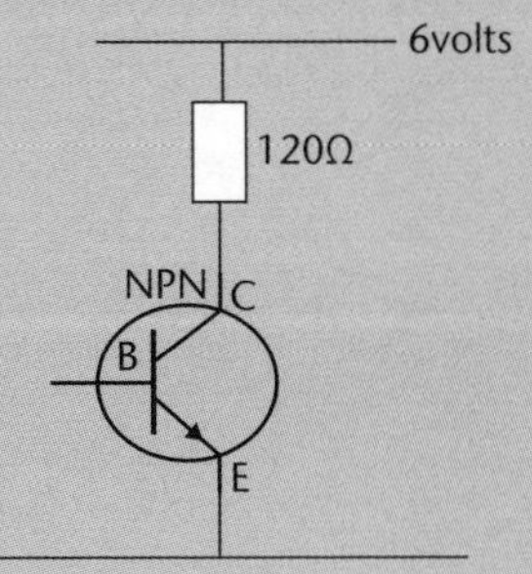

Unit 11.6 Electronics

Topic 3: Digital electronics

Digital electronics is introduced in Topic 3 and its operational principles of logic gates as function blocks are covered. The Topic looks at:

- Digital signal levels – 'high' and 'low' states.
- Logic gates and electronic symbols AND, OR, NOT, NOR, NAND, XOR and combinations.
- 2 inputs AND gate and 2 input OR gate.
- The output of each of AND, OR, NOT, NAND, XOR.
- Basic uses (or applications) of logic gates.

Introduction

The development of the transistor practically replaced the old vacuum tubes you may have heard about. The transistor can be used as either an analog or a digital device, but the current trend is toward integrating thousands or even millions of transistors and similar operating devices into digital integrated circuits (ICs) that can have capabilities far beyond those of discrete analog transistor circuits.

Analog circuitry is still used for especial applications but digital electronics is quickly taking over most of the trends in electrical and electronics design, because digital circuits or logic circuits usually provide the best options for speed, efficiency, accuracy, size, and cost.

The concept of digital

Let us consider the voltage level at the output of a transistor. Suppose the input voltage is about 12 volts. Biased as an analog circuit, the output can vary from 0 to 12 volts or any voltage level in between. However, suppose the transistor is wired or used as a switch. This device would then have only two conduction estates: conducting or not conducting. The transistor then is said to be either 'ON' or 'OFF'.

In digital circuitry, ON and OFF are the only two possible states at which a given point can exist. We can also express those two estates as 'high' or 'low', ' true' or 'false', or 1 or 0 (hence the term digital).

A digital circuit is one that expresses a voltage or current value as digits: 1 or 0. This concept is in many ways simple to understand and manipulate compared to analog circuits.

The numbering system that uses only the digits 1 and 0 is the binary number system, so the first step is to understand the binary and other number systems and how to manipulate and make conversions among those number systems. Therefore, let us briefly review the decimal number system first and explain in general terms the binary and the octal number systems, as well as how to perform number conversions in those systems.

Decimal number system

This is the most common and familiar number system. We represent decimal numbers with the digits 0, 1, 2, 3, 4, 5, 6, 7, 8, and 9. In a multi-digit decimal number, each digit has a value ten times the value of the next digit to its immediate right. In other words, every position can be expressed by the base of the system (10) raised to some power. Therefore, the form of every decimal number can be expressed as:

$$\ldots 10^4\ 10^3\ 10^2\ 10^1\ 10^0\ 10^{-1}\ 10^{-2}\ 10^{-3} \ldots$$

In the decimal system, the most significant digit (the one with the largest power of ten) is always at the left, and the least significant digit is at the right. This convention is applicable to any number system. Other number systems can be represented in the same way. In each case, the weights of the various positions are powers of a particular number. This number is the base (also called radix) of that number system and equals the number of digits used in that system (eg decimal = 10, binary = 2, octal = 8, hexadecimal = 16, etc).

Binary number system

The binary number system is the most useful in digital circuits because there are only two digits, 0 and 1. These binary digits or also called bits are used to represent the states of switches, relays, transistors, etc.

Because the binary number system uses two digits, it has a radix of two, and the position weights are powers of two. Counting in the binary number system is performed much the same way as in the decimal number system. With decimal numbers, a count proceeds as 0, 1, … 9, until a single digit can no longer represent a number. Then a second digit is added and the count continues as: 10, 11, … 20,… and so on. Adding of digits is continued for higher counts so that there are always enough digits to represent any number.

The binary counts progress in a similar manner. Counting from zero to four in binary, the counts would be:

	Binary				Decimal
			0	=	0
			1	=	1
		1	0	=	2
		1	1	=	3
	1	0	0	=	4
Power of two	2^2	2^1	2^0		

As you can see, every time the two digits 1 and 0 in one position are exhausted (counted as high as they can go), a 1 is added to the left, and all digits to the right are made 0, and the count continues. The column to the far right represents the ‘ones’ column, or 2^0. The next columns to the left equal 2^1 and 2^2 respectively, as explained previously. Binary numbers are therefore made up of 0s and 1s only – the first two numbers from the decimal numbering system. Binary numbers can be used in the same way as decimal numbers. They can be added and subtracted, relatively easily and, of course, converted to a decimal number.

Addition of binary numbers

The procedure for adding numbers in binary is similar to adding in decimal except that the binary sum is made up only of 1s and 0s. As in the decimal system, when the sum exceeds the radix, you must carry a 1 to the next highest place. There are four basic rules for adding binary digits:

Rule 1: 0 + 0 = 0

Rule 2: 0 + 1 = 1

Rule 3: 1 + 0 = 1

Rule 4: 1 + 1 = 10 (0 with a carry of 1)

You should notice that in Rule 4, the 1 in the sum is carried to the next column as in regular decimal addition.

Here are some examples of binary addition with the decimal equivalent shown for reference.

0101	5	1111	15	110	6
+0010	+2	+1100	+12	+100	+4
0111	7	11011	27	1010	10

Binary-to-decimal conversion

We can convert binary numbers to decimal numbers in an easy way. The method for performing this conversion consists of adding together all the positions' weights where a 1 appears, and ignoring all positions containing a zero.

The whole procedure can be thought of as a two-step process. First, determine the position weights and write the proper position weights above the binary number to be converted. Second, add the position weight containing 1s to obtain the resultant decimal number. The following are some easy examples to understand.

Powers of two:	2^5	2^4	2^3	2^2	2^1	2^0
Decimal weights:	32	16	8	4	2	1
Example 1:	1	1	0	1	0	1 = 32 + 16 + 4 + 1 = 53
Example 2:	1	0	1	0	0	1 = 32 + 8 + 1 = 41

Usually, long binary expressions are broken up into groups of bits to aid in readability. These groups of 8 bits are called bytes. Therefore, a byte is a sequence of adjacent bits treated as a unit. Groups of greater numbers of bytes form a word.

Decimal-to-binary conversion

There is more than one method or technique to perform this conversion. Let us review the process that allows us to convert any number containing an integer part and a fractional part (eg 1.25). Convert the integer part first, then convert the fractional or decimal part.

We convert the integer part by successively dividing the number by two. Each time there is a remainder of zero, write down a zero, and each time a 1 remains, write down a 1.

To convert the fractional part, we repeatedly multiply by two until the decimal/fractional part of the number becomes zero or until sufficient accuracy is obtained.

This method works well for fractional numbers because you do not need to know the negative powers of two to perform the conversion.

Let us illustrate the method by converting the number 53.3125, as follows:

Division	Remainder	
$53 \div 2 = 26$	1	least significant digit/bit (LSB)
$26 \div 2 = 13$	0	
$13 \div 2 = 6$	1	
$6 \div 2 = 3$	0	
$3 \div 2 = 1$	1	
$1 \div 2 = 0$	1	most significant bit (MSB)

Result: $53_{10} = (1\ 1\ 0\ 1\ 0\ 1)_2$

This method does not require us to write down the powers of two. To convert the fractional part, we repeatedly multiply by two, as follows:

Multiplication	Whole number part	
$0.3125 \times 2 = 0.6250$	0	Most significant bit
$0.6250 \times 2 = 1.2500$	1	
$0.2500 \times 2 = 0.5000$	0	
$0.5000 \times 2 = 1.000$	1	Least significant bit
$0.000 \times 2 = 0$		

Result: $0.3125_{10} = (0.\ 0\ 1\ 0\ 1)_2$

Our final result for $(53.3125)_{10} = (1\ 1\ 0\ 1\ 0\ 1\ .\ 0\ 1\ 0\ 1)_2$

Octal number system

In the octal number system, the radix is 8, so it uses 8 digits: 0, 1, 2, 3, 4, 5, 6, 7. The position weights in the system are powers of 8, which also happen to be every third power of two. The first six powers of 8 are:

$8^0 = 1$

$8^1 = 8$

$8^2 = 64$

$8^3 = 512$

$8^4 = 4096$

$8^5 = 32768$

This octal system is frequently used in digital circuits because it can easily be converted to binary. Also, there are significantly fewer digits in a given octal number than in the corresponding binary number, so it is easier to work with the shorter octal numbers.

Unit 11.6 Activity 3A: Binary numbers

1. Add the following pairs of binary numbers:
   a. 0011100 and 0111010
   b. 101011 and 010101
2. Find the decimal equivalent of the binary number 101110_2.
3. Convert 87_{10} to its binary equivalent.
4. Convert the following binary to decimal: 01010011.
5. Convert 83_{10} to a binary equivalent:
6. Add the following 3 binary numbers: 11011; 01101, and 00110.
7. Represent the decimal numbers 46 and 83 into their octal number equivalent.

Basic logic circuits

Logic or digital circuits were originally developed to process data inside digital computers, but their use has spread into all branches of electronics.

Inside the digital computer, information is expressed in terms of binary numbers; that is, numbers consisting of ones and zeros. Those binary states are represented by low and high voltages. The information is transferred and processed inside the computer according to rules built into the computer in the form of circuits we call logic gates. We call them gates because these circuits can stop information or allow it to pass; we use the word logic because they do so according to rules we can describe by logic statements using the words AND, OR and NOT.

Logic gates

While each logical element or condition must always have a logic value of either 0 or 1, we also need to have ways to combine different logical signals or conditions to provide a logical result.

For example, consider the logical statement: 'If I move the switch on the wall up, the light will turn on.' At first glance, this seems to be a correct statement. However, if we look at a few other factors, we realise that there is more to it than this. In this example, a more complete statement would be: 'If I move the switch on the wall up *and* the light bulb is good *and* the power is on, the light will turn on.'

If we look at these two statements as logical expressions and use logical terminology, we can reduce the first statement to:

Light = Switch

This means nothing more than that the light will follow the action of the switch, so that when the switch is up/on/true/1 the light will also be on/true/1. Conversely, if the switch is down/off/false/0 the light will also be off/false/0.

Looking at the second version of the statement, we have a slightly more complex expression:

Light = Switch *and* Bulb *and* Power

Normally, we use symbols rather than words to designate the *and* function that we are using to combine the separate variables of Switch, Bulb, and Power in this expression. The symbol normally used is a dot, which is the same symbol used for multiplication in some mathematical expressions. Using this symbol, our three-variable expression above becomes:

Light = Switch • Bulb • Power

When we deal with logical circuits (as in computers) we not only need to deal with logical functions; we also need some special symbols to denote these functions in a logical diagram. There are three fundamental logical operations from which all other functions, no matter how complex, can be derived. These functions are named AND (as in our above example), OR and NOT. Each of these has a specific symbol and a clearly defined behaviour.

The AND gate

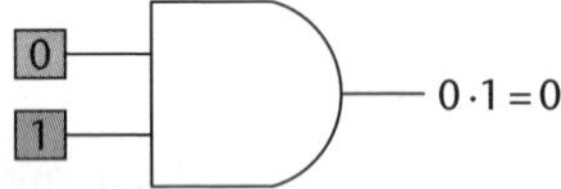

The AND gate implements the AND function. With the gate shown above, both inputs must have logic 1 signals applied to them in order for the output to be a logic 1. With either input at logic 0, the output will be held to logic 0.

This behaviour can be conveniently summed up by a special type of table called a truth table. Other ways in which the AND function can be described are the logic symbol and the Venn diagram. Both of these representations are shown in the following figure:

A	B	F
0	0	0
0	1	0
1	0	0
1	1	1

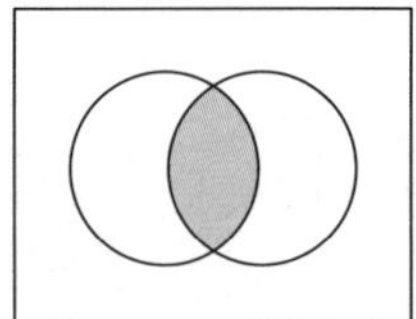

There is no limit to the number of inputs that may be applied to an AND function, so there is no functional limit to the number of inputs an AND gate may have. However, for practical reasons, commercial AND gates are most commonly manufactured with 2, 3 or 4 inputs. A standard integrated circuit (IC) package contains 14 or 16 pins, for practical size and handling. A standard 14-pin package can contain four 2-input gates, three 3-input gates, or two 4-input gates, and still have room for two pins for power supply connections.

The OR gate

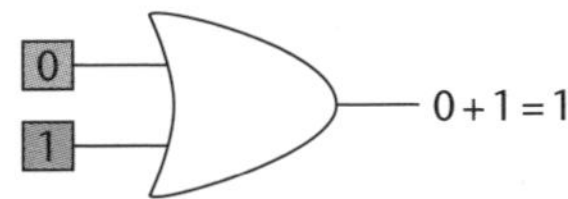

The OR gate could be described as the reverse of the AND gate. The OR function, like its verbal counterpart, allows the output to be true (logic 1) if any one or more of its inputs are true.

In symbols, the OR function is designated with a plus sign (+). In logical diagrams, the symbol in the above figure designates the OR gate.

Let us illustrate the OR function with the following statement: 'Paul can go to the cinema if either his mother OR his father will give him the money.' If we denote going to the cinema as F, mother giving him money as A and father giving him money as B, then we can represent the OR function with a truth table and a Venn diagram, as follows:

A	B	F
0	0	0
0	1	1
1	0	1
1	1	1

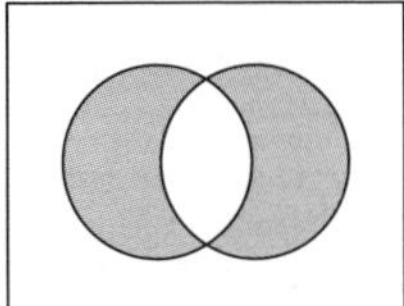

As with the AND function, the OR function can have any number of inputs. However, practical commercial OR gates are mostly limited to 2, 3, and 4 inputs, as with AND gates.

The NOT gate

As we previously mentioned, the basic logic gates are based in three words: AND, OR and NOT. NOT is used in statements such as 'Paul will NOT go to the cinema if Bill AND Fred are going'. The word NOT reverses the normal meaning of the AND condition. The truth table for this NOT AND or NAND condition is the AND truth table with the 1s changed to 0s and vice-versa in the F column, as shown in the following figure:

A	B	F
0	0	1
0	1	1
1	0	1
1	1	0

NAND

A	B	F
0	0	1
0	1	0
1	0	0
1	1	0

NOR

Statements like 'John will NOT drive the car if his father OR mother drive it' give rise to a condition of NOT OR, or NOR. This has a truth table like the one for OR with its F column complemented, as shown in the above figure for NOR.

As you can see, to perform the NOT function electronically, a circuit is required that will give a (1) output for a (0) input and vice-versa. We do not have to invent such a circuit; it is already a basic circuit, an inverter (a transistor used as a switch). In a logic circuit, the symbol for the inverter is the following:

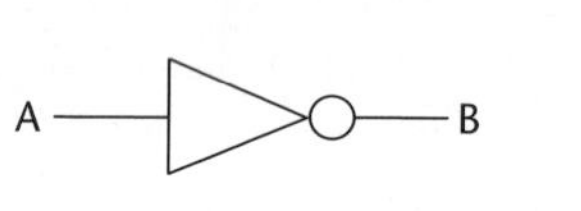

A	B
0	1
1	0

Example

The following figure is a representation of the three basic circuits, AND, OR and NOT, and their truth tables. Truth tables show the conditions of the output for all input conditions. A true condition is represented by a closed switch and a false condition (logic 0) by an open switch.

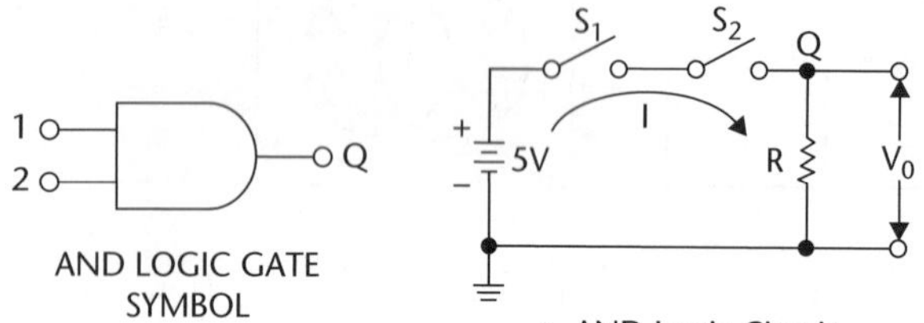

AND LOGIC GATE SYMBOL

S_1 S_2 Q I + 5V − R V_O

TRUTH TABLE

INPUTS		OUTPUTS V_O	
S_1	S_2	Q	VOLTS
0	0	0	0
0	1	0	0
1	0	0	0
1	1	1	+5

a. AND Logic Circuit

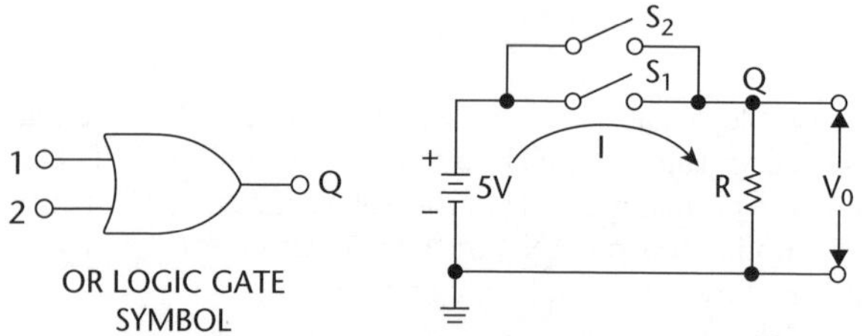

OR LOGIC GATE SYMBOL

S_2 S_1 Q I + 5V − R V_O

TRUTH TABLE

INPUTS		OUTPUTS V_O	
S_1	S_2	Q	VOLTS
0	0	0	0
1	0	1	+5
0	1	1	+5
1	1	1	+5

b. OR Logic Circuit

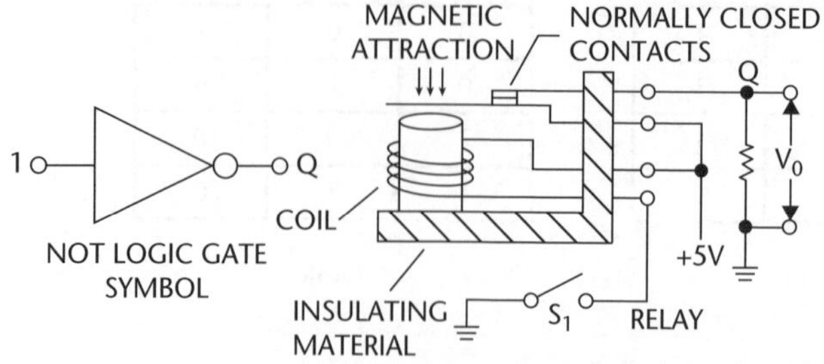

NOT LOGIC GATE SYMBOL

MAGNETIC ATTRACTION

NORMALLY CLOSED CONTACTS

Q V_O COIL +5V INSULATING MATERIAL S_1 RELAY

TRUTH TABLE

INPUTS		OUTPUTS V_O	
S_1		Q	VOLTS
OPEN	0	1	+5
CLOSED	1	0	0

c. NOT Logic Circuit (Inverter)

Electronic logic levels

To make an electronic logic gate you must represent a 1 or a 0 by an electrical quantity such as a voltage. The following figure helps you to see how this can be done. Four voltage levels are defined:

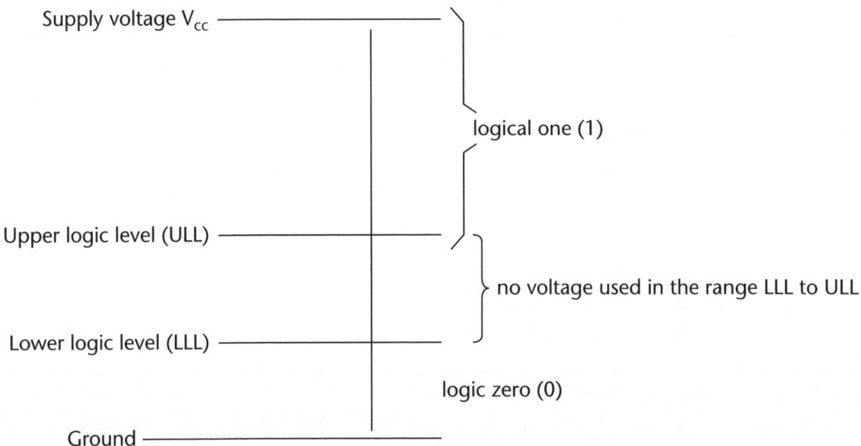

Ground (zero, earth): usually the negative power supply voltage.

Lower logic level: a voltage just above ground, often as big as 0.7 volts.

Upper logic level: a higher voltage, often $V_{cc}/2$.

Supply voltage V_{cc}: usually the positive supply rail.

Any voltage in the range zero to LLL is taken as logic zero (0). A voltage level between the ULL and V_{cc} is represented by a logic one (1). Circuits must be designed so that there are no voltages at input or output in the range LLL to ULL. If voltage is used in this range, the operation of the circuit will be uncertain.

When we define our logic system with these considerations in mind, it is said to be using positive logic. If the supply leads were reversed so that V_{cc} was negative with respect to ground, we would be using negative logic. Either system is equally good, but we must not mix them.

Basic use of a logic gate

The following diagram shows you the use of a logic gate. The recording device will start recording IF and only IF the 'record' AND 'play' buttons have been pressed simultaneously.

In most practical applications, you will need combinations of different gates in your circuit. Most of the time, each of the input sensors in your circuit or system forms part of a **potential divider**, as in a transistor switch, for example; and the small output current controls or switches ON and OFF lamps, motors, and other devices via **relays**.

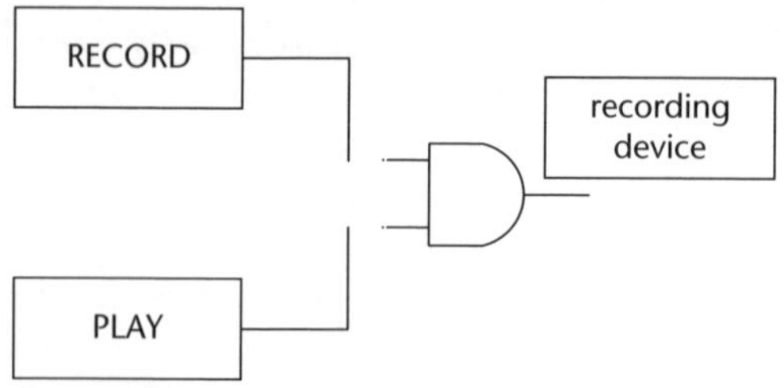

Unit 11.6 Activity 3B: Truth table

1. Write a truth table for this combination of gates, showing all the possible states of A, B, C, and D.
2. What must the states of inputs A and B be for the output to be HIGH?

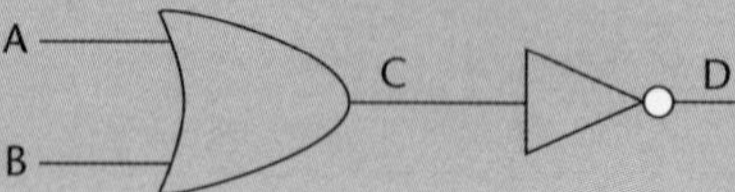

Derived logical functions and gates

While the three basic functions AND, OR, and NOT are sufficient to accomplish all possible logical functions and operations, some combinations are used so commonly that they have been given names and logic symbols of their own.

The NAND gate

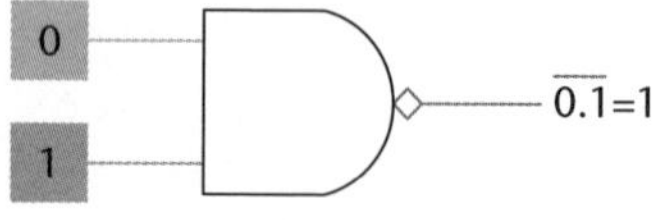

The NAND gate implements the NAND function, which is exactly inverted from the AND function you already examined. With the gate shown above, both inputs must have logic 1 signals applied to them in order for the output to be a logic 0. With either input at logic 0, the output will be held to logic 1.

The circle at the output of the NAND gate denotes the logical inversion, just as it did at the output of the inverter. Also in the figure above, note that the over bar is a solid bar over both input values at once. This shows that it is the AND function itself that is inverted, rather than each separate input.

As with AND, there is no limit to the number of inputs that may be applied to a NAND function, so there is no functional limit to the number of inputs a NAND gate may have. However, for practical reasons, commercial NAND gates are most commonly manufactured with 2, 3, or 4 inputs, to fit in a 14-pin or 16-pin package.

The NOR gate

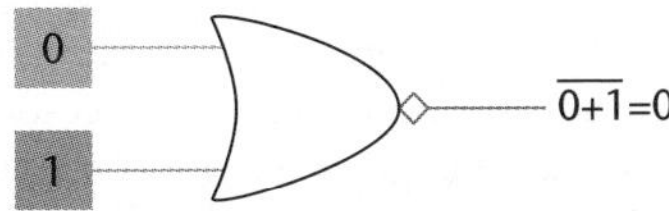

The NOR gate is an OR gate with the output inverted. Where the OR gate allows the output to be true (logic 1) if any one or more of its inputs are true, the NOR gate inverts this and forces the output to logic 0 when any input is true.

In symbols, the NOR function is designated with a plus sign (+), with an over bar over the entire expression to indicate the inversion. In logical diagrams, the figure above designates the NOR gate. As expected, this is an OR gate with a circle to designate the inversion.

The NOR function can have any number of inputs, but practical commercial NOR gates are mostly limited to 2, 3, and 4 inputs, as with other gates in this class, to fit in standard IC packages.

The Exclusive-OR, or XOR, gate

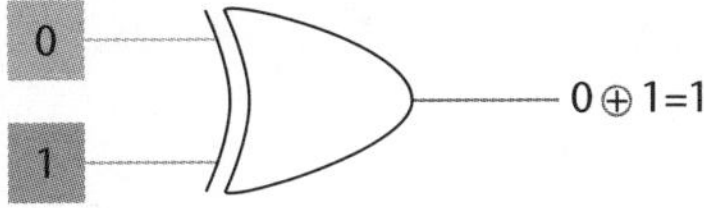

The Exclusive-OR, or XOR, function is an interesting and useful variation on the basic OR function. Verbally, it can be stated as, 'Either A or B, but not both'. The XOR gate produces a logic 1 output only if its two inputs are *different*. If the inputs are the same, the output is a logic 0.

The XOR symbol is a variation of the standard OR symbol. It consists of a plus (+) sign with a circle around it. The logic symbol, as shown here, is a variation on the standard OR symbol.

Unlike standard OR/NOR and AND/NAND functions, the XOR function always has exactly two inputs, and commercially manufactured XOR gates are the same. Four XOR gates fit in a standard 14-pin IC package.

Using a combination of different logic gates

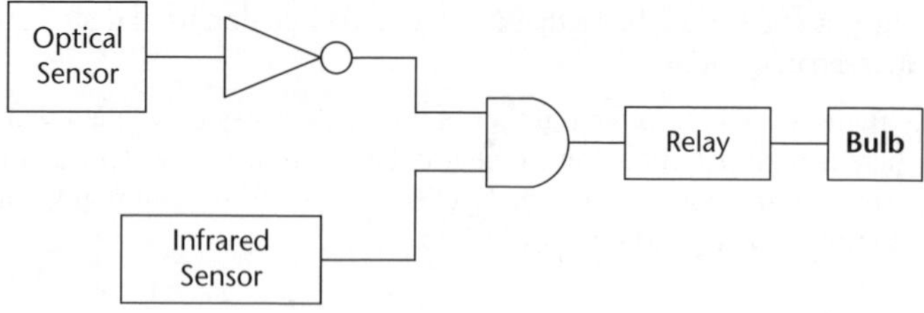

The above diagram shows you how sensors and logic gates can be connected so that, in this specific example, if it is dark and someone approaches the bulb will come ON automatically (infrared sensor detects the body heat). The reason to have the relay is because the AND gate cannot provide enough power for the lamp, so it switches ON a relay instead, which switches ON a separate circuit with the bulb on it.

Logic gates are the building blocks of combinational logic, in which the state of the outputs is directly determined by the state of the inputs. That is, change the inputs and the outputs will change automatically. There is a lot more to say about logic gates, but this has been a basic introduction to digital electronics.

Answers for many questions include a 'Marking Guide':

- **A** ('Achievement', meaning 'satisfactory achievement').
- **B** ('Merit', meaning 'high achievement').
- **E** ('Excellence', meaning 'very high achievement').

The Marking Guide has been made by the authors and the publishers and is not an official guide but we hope it will be a help to students who are striving for the best possible results.

Preliminary Unit

Investigating and research in physics

Preliminary Unit Activity A: Forces on a current-carrying wire (page 3)

1. a.

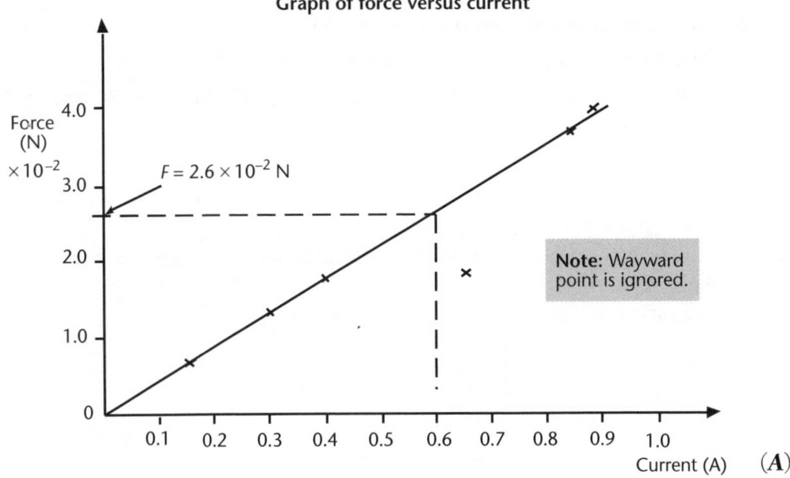

(**A**)

b. Gradient = $\frac{\text{rise}}{\text{run}} = \frac{3.80 \times 10^{-2}}{0.9}$ = 0.04 N A^{-1} (**A**)

c. Length of wire in the field; magnetic field strength. (**A**)

d. Length of wire in the field. (**A**)

e. Magnetic field strength and current. (**A**)

f.
- Reading uncertainties in the ammeter and force meter readings.
- Friction in the hinges. (**E**)

2. **a**. Number of windings on coil; number of iron rods; size of paper clips. (***M***)

b.

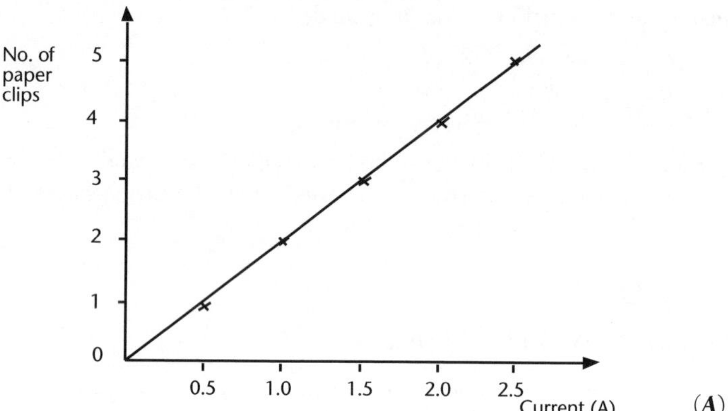

(***A***)

c. **i**. The number of turns. (***A***)

ii. The number of rods and the current. (***M***)

iii. Number of paperclips versus number of turns. (***M***)

iv. Number of turns (*independent variable is always plotted on the x- axis*). (***M***)

3. **a**. **i**. length (***A***)

ii. thickness (***A***)

iii. type of material resistance wire will increase – this will cause the resistance of the wire to increase. (***A***)

b. Yes, the results support her hypothesis because the graph has a constant positive gradient. (***E***)

4. **a**. **i**. volume **ii**. temperature **iii**. pressure (***A***)

b.

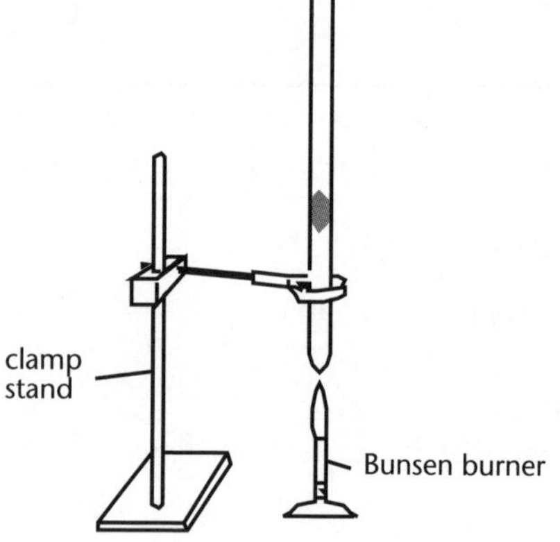

(***A***)

c.

Graph of length versus temperature

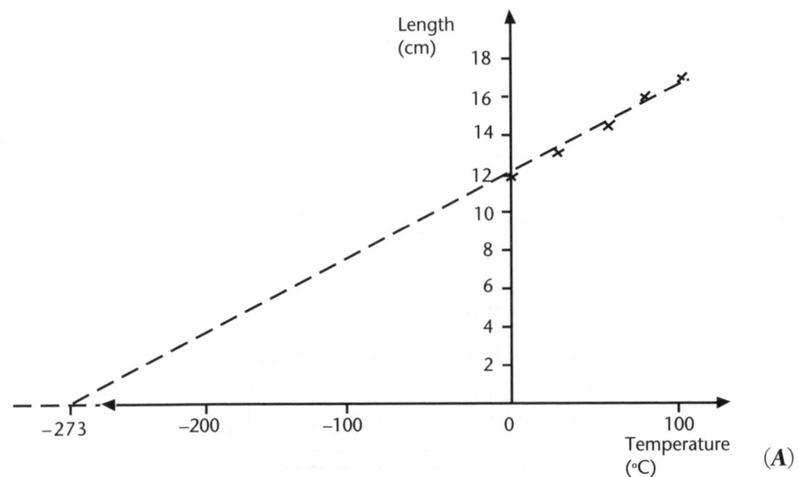

(*A*)

d. −273 °C

e. Set up the apparatus outside, record the length of mercury and use the graph to find the temperature. (***M***)

Unit 11.1 Measurement

Topic 1: Quantities and units

Unit 11.1 Activity 1A: Units and measurement (page 10)

1.

Quantity	SI Unit	
	Name	**Unit**
Velocity	metre per second	$m\ s^{-1}$
Mass	kilogram	kg
Voltage	volt	V
Charge	coulomb	C
Heat energy	joule	J
Pressure	pascal	Pa
Magnetic field	tesla	T
Energy	joule	J
Density	gram per cubic centimetre	$g\ cm^{-3}$

(***A***)

2. **a**. 1.8 m **b**. 360 s **c**. 0.25 A **d**. 10×10^6 J **e**. 1 200 kg (***A***)

3. **a**. 0.33 A **b**. 35.276 kg **c**. 1.7×10^6 J **d**. 5 300 m **e**. 0.0174 g
f. 7 980 s **g**. 13.9 $m\ s^{-1}$ **h**. $2.7 \times 10^{-3} \times 10^6 = 2.7 \times 10^3 = 2\ 700\ kg\ m^{-3}$ (***A***)

4. A Newtons B joules C watts D m s^{-2} (***A***)

5. 18 m (***A***)

6. **a**. Meter A measures current, 0.45 A. (***M***) **b**. Meter B measures voltage, 4.5 V. (***M***)

7. **a**. For 2.2 kg the position is 64 + 2 × 3 = 70 cm. (Slider slides 3 cm for every 0.1 kg.) (***A***)

b. Mass of kumara = 1.0 + 6 × 0.1 = 1.6 kg. (Position of sliders for kumara = 52 cm.) (***A***)

c. Maximum position of slider = 100 cm; mass corresponding to slider at 100 cm = 3 + 0.2 = 3.2 kg. (***M***)

d. Heavier masses could be measured but for very light masses the beam would be too sensitive for balancing. (***E***)

Unit 11.1 Activity 1B: Graphing (page 12)

1. **a**.

Graph of length versus temperature

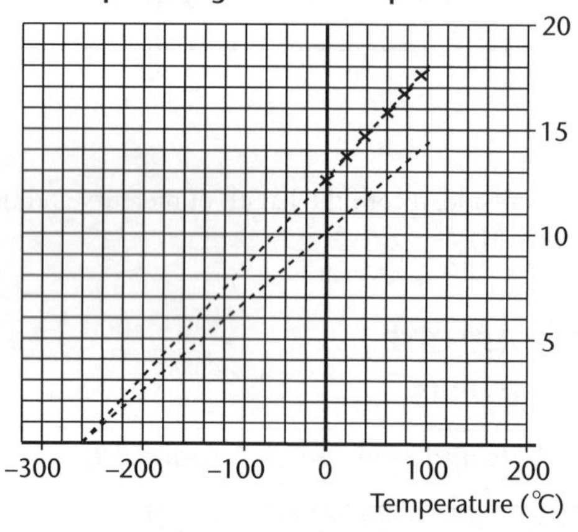

(A)

b. –273 °C **c**. –25 °C **d**. *See graph.*

2. **a**.

Graph of speed versus time

speed ($m\ s^{-1}$)
70
60
50
40
30
20
10
2
4
6
8
10
12
Time (s)

(***A***)

3. a.

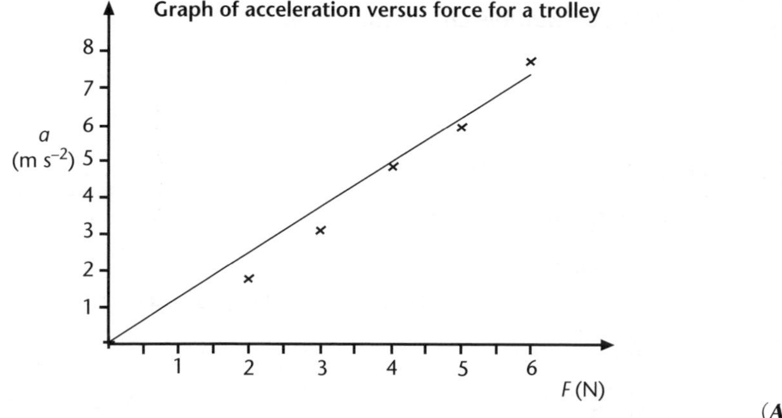

(***A***)

b. Slope = $\frac{7.5 - 1.5}{6 - 2} = \frac{6}{4} = 1.5 \text{ kg}^{-1}$. (***M***)

c. Mass is the inverse of the slope. (***E***)

4.

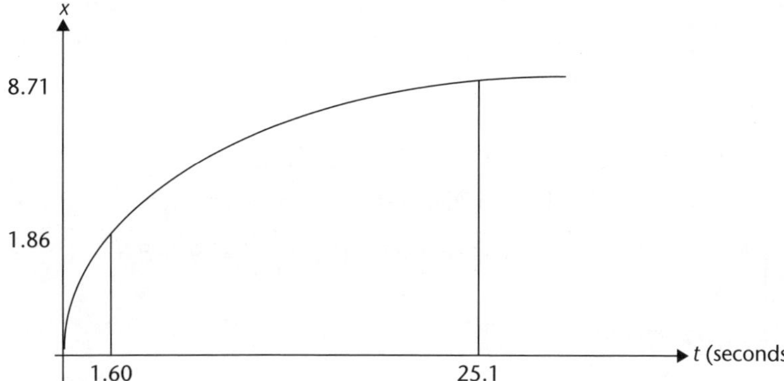

5.

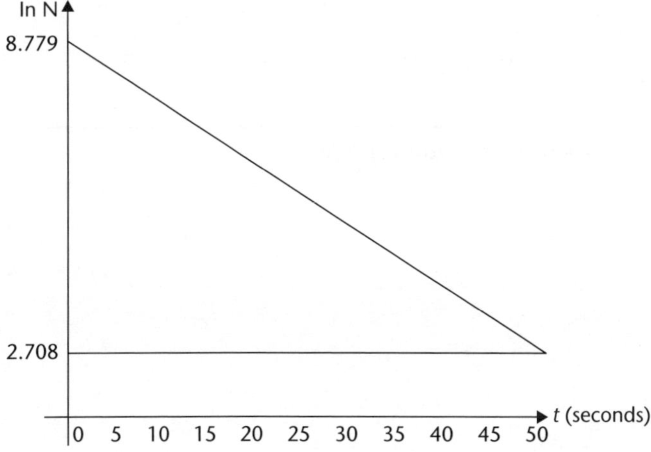

Unit 11.1 Activity 1C: Multiple choice (page 14)

1. E. 192

2. B. 2.6 μm

Topic 2: Vectors and scalars

Unit 11.1 Activity 2A: Vectors (page 17)

1. Velocity, acceleration, force.

2. A scalar B scalar C vector D scalar. (***A***)

3. **a**.

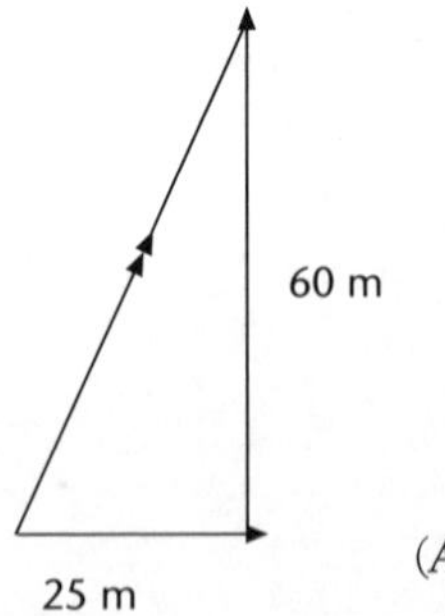

(***A***)

b. Displacement = $\sqrt{25^2 + 60^2}$ = 65 m (*See ↗ on graph in a.*). (**M**)

4. **a**.

2.5

resultant

θ

1.5

(**M**)

b. Resultant = $\sqrt{2.5^2 + 1.5^2}$ = 2.9 m s^{-1};

angle to bank = $\tan^{-1}\left(\frac{2.5}{1.5}\right)$ = 59°. (**E**)

Unit 11.1 Activity 2B: Scientific notation (page 18)

1. 1.25×10^5

2. 9.32×10^{-4}

Unit 11.1 Activity 2C: Scientific notation, errors and units in measurements (page 20)

6. **a**. 6.3 Megaseconds
   b. 56 Megahertz

Topic 3: Measuring instruments

Unit 11.1 Activity 3A: Taking measurements (page 25)

1. **a**. 7.34 cm (***A***) **b**. 2.72 mm (***A***) **c**. 5.060 or 5.065 cm (***A***)
2. **a**. +0.01 mm (***A***)
   b. The zero error should be subtracted from the readings.
   The thickness of 100 page thicknesses is 10.30 – 0.01 = 10.29 mm. The thickness of one page is thus $\frac{10.29}{100}$ = 0.1029 mm. Adjusting for the zero correction has had almost no effect on the final measurement. However, measuring just one thickness gives 0.10 – 0.01 = 0.09 mm, which is not very accurate, since it has only 1 sig fig. The zero error is significant since the measurement has changed by 10%. (***E***)
3. The multimeter reading is 8.206 kΩ. This is $\frac{8.206 - 8.2}{8.2} \times \frac{100}{1}$ = 0.07% higher than 8.2 kΩ and so is well within the ±5% range. (***M***)
4. **a**. Used carefully, a wooden metre ruler should be able to measure the length to the nearest millimetre. Because ruler is thick, care must be taken to avoid parallax error. (***M***)
   b. **i**. The ruler will be used 10 times in taking the measurement. If, at best, a mark is made each time where the ruler's scale ends and the ruler is carefully placed end-on-end with that mark, the total measurement will have an accuracy of 10 × 1 mm = 1 cm. It is more probable that the measurement will be accurate to between 2 cm and 5 cm. (***M***)
   ii. Advantage: One measuring scale, so do not have to take partial measurements. This would give more accuracy. (***M***)
   Disadvantage: Tape can sag and needs to be held tightly or supported at various points. A sagging tape would tend to indicate a larger distance than actual distance. (***M***)
5. The mass bounces regularly, so it would be more accurate to take multiple measurements and divide by the number of bounces than simply to measure the time for one bounce. A good number to time would be between 10 and 20 bounces. These bounces have to be accurately counted at the same time as operating the stopwatch. Pick a particular point in the motion (usually the centre or top or bottom), and start and stop the stopwatch in the same manner as the mass reaches that point. This reduces timing errors in stopping and starting. (***E***)
6. 99.19 g The zero error is negative since mass has to be added to bring the balance reading to zero. So, subtracting a negative error gives 98.76 – (–0.43) = 99.19 g. (***M***)

Topic 5: Error analysis

Unit 11.1 Activity 5A: Combining measurements (page 37)

1. a. 14.1 m^2 (**A**) b. 24.5 cm (**A**)
   c. 0.3 cm Using subtraction of measurements rule. (**A**)
2. a. 64 cm^3 (**A**) b. 180 cm^2 (2 sig figs) (**A**)
3. 95.6 cm^3 (**A**)

Topic 6: Graphs

Unit 11.1 Activity 6A: Graphs (page 45)

1. a. $d = 3.5t$ metres (**A**) b. $v = 0.5t$ m s^{-1} (**A**)
2. a. 3.3 V (**A**) b. 23.5 A (**A**)
   c. slope = $\frac{25.0 - 0}{8.5 - 0}$ = 2.9 A V^{-1}. The mathematical relationship is: $I = 2.9\ V$ amperes. (**M**)
3. a. Graph of extension against mass

   extension (cm): 0, 5, 10, 15, 20, 25; mass (g): 100, 200, 300, 400, 500 (**A**)

   b. 100.0 g (from graph) (**A**)
   c. slope = $\frac{25.0 - 0}{500.0 - 0} = \frac{25.0}{500.0}$ = 0.0500 cm g^{-1} (**A**)
4.

Speed (km h^{-1})	20	30	40	50	60	70	80	90	100
Speed (m s^{-1})	**5.6**	**8.3**	**11.1**	**13.9**	**16.7**	**19.4**	**22.2**	**25.0**	**27.8**
Air resistance (N)	370	833	1 481	2 315	3 333	4 537	5 926	7 500	9 259
A graph of air resistance against speed curves upwards like an x^2 graph, so square the speed values and plot a graph of air resistance against speed squared.									
Speed squared (m^2 s^{-2})	30.9	69.4	123.5	192.9	277.8	378.1	493.8	625.0	771.6
Air resistance (N)	370	833	1 481	2 315	3 333	4 537	5 926	7 500	9 259

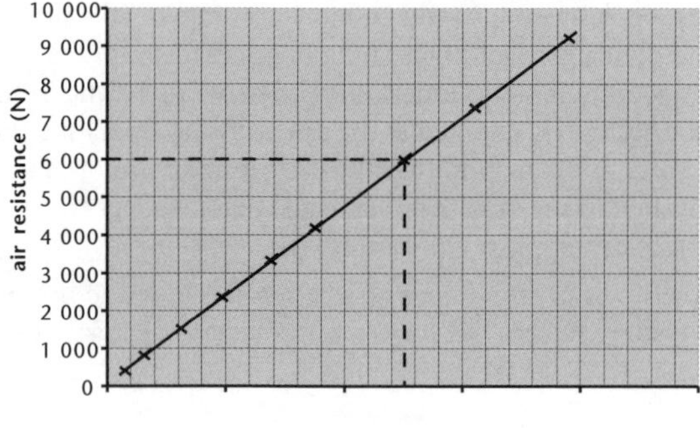

This graph is linear. Slope = $\frac{6\,000 - 0}{500 - 0}$ = 12 N m^{-2} s^{2}.
The relationship is: air resistance = 12 × speed^2 (newtons). (***E***)

5. A graph of resistance against diameter curves rapidly like an x^{-2} graph, so calculate the reciprocal squared of the diameter values and plot a graph of resistance against 1/diameter^2.

1/(diameter squared) (mm^{-2})	25.00	4.00	1.00	0.40	0.30
Resistance (ohms)	37.50	6.00	1.50	0.67	0.38

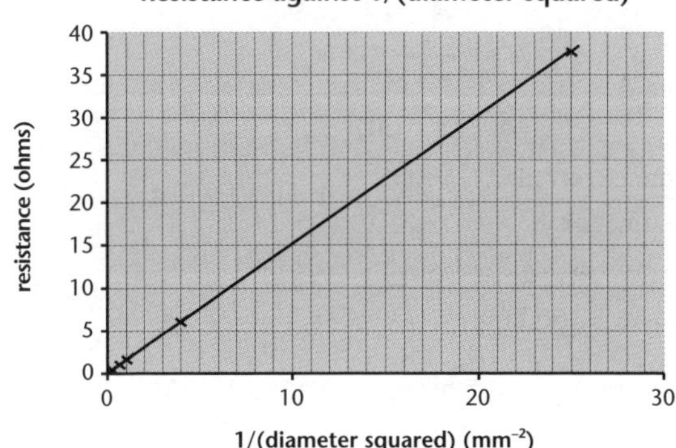

This graph is linear. Slope = $\frac{30.00 - 0}{20.00 - 0}$ = 1.5 ohms mm^2.
The relationship is: resistance = 1.5 × 1/diameter^2 (ohms). (***E***)

Unit 11.2 Motion (Kinematics)

Topic 1: Characteristics of motion – introduction

Unit 11.2 Activity 1A: Motion (page 49)

1. 1.4 cm (***A***)

2. **a**. 2 m s^{-1} **b**. 3 km h^{-1} **c**. 880 m **d**. 50 m **e**. 4 s
f. 2 s

3. **a**. 8 000 m **b**. 0.9 m **c**. 3 h **d**. 1 800 s (***A***)

4. **a**. 2 m s^{-2} **b**. 4 km h^{-2} **c**. 880 m s^{-1} **d**. 50 m s^{-1} **e**. 4 s
f. 2 s (***A***)

5. 20 s (***A***)

6. 20 m s^{-1} (***A***)

7. (***M***)

8. (***M***)

9. **a**. Total distance travelled = 0.65 + 0.5
= 1.15 m (***A***)

b. Average speed = $\frac{1.15}{15}$ = 0.08 m s^{-1} (***A***)

c.

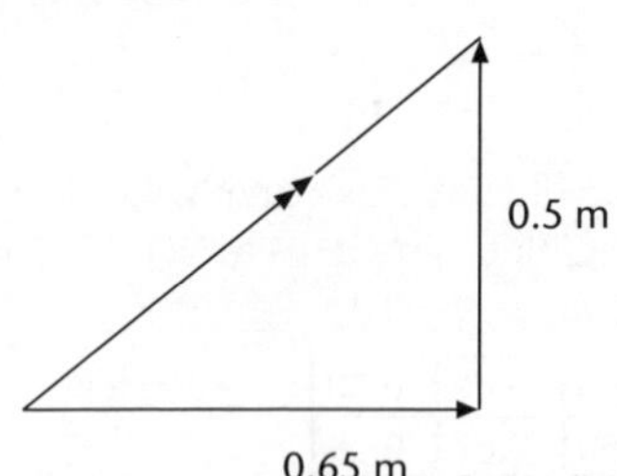

(***A***)

d. Displacement $= \sqrt{0.65^2 + 0.5^2}$

$= 0.82$ m (***A***)

e. Average velocity $= \dfrac{0.82}{15} = 0.05 \text{ m s}^{-1}$ (***M***)

10. **a**. Distance from A to C $= \sqrt{400^2 + 80^2}$

$= 408$ m (***M***)

b. Speed $= \dfrac{\text{distance travelled}}{\text{time taken}} = \dfrac{408}{2} = 204 \text{ km h}^{-1}$ (***A***)

c. 40 km h^{-1} South; 200 km h^{-1} West.

$V_{plane} = 200 - 40 = 160 \text{ km h}^{-1}$

$v = \dfrac{d}{t}; t = \dfrac{d}{v} = \dfrac{80}{100} = 0.5$ hours = 30 minutes (***M***)

Topic 2: Motion – extension

Unit 11.2 Activity 2A: Motion and velocity (page 55)

a. positive

b. negative

c. negative

d. negative

Topic 3: Vectors and relative motion

Unit 11.2 Activity 3A: Vector calculations (page 64)

1. 3 (***A***)

2. **a**. 130° (***A***) **b**. 056° (***A***) **c**. 240° (***A***) **d**. 350° (***A***)

3. **a**. → (***A***) **b**. ↓ (***A***) **4**.

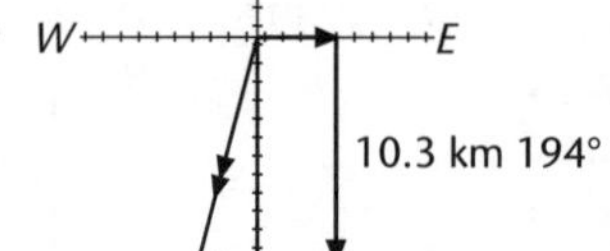

(***A***)

5. (*A*)

6. W, S, 5 km 217° (*A*)

7. 180° 12 m, 5 m, 7 m (*A*)

8. 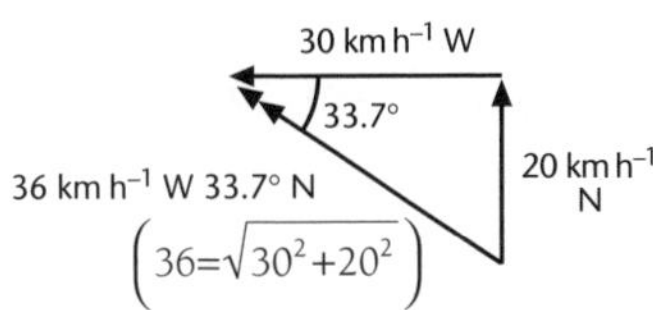

(*M*)

9. **a**. 3.0 (*A*)
   b. 4.0 (*A*)
   c. 5.0 (*A*)

10. 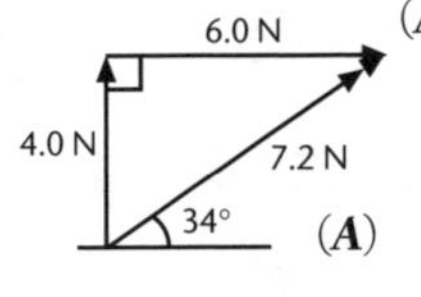

(*A*) (*A*)

11. ← (*M*)

12. 9 m s^{-1} away from the crossbar. (*M*)

13. 40 km h^{-1} W ($-\underset{\sim}{v}_i$), 20 km h^{-1} N ($\underset{\sim}{v}_f$), 44.7 km h^{-1}, 63.4°

($\Delta\underset{\sim}{v}$) = 44.7 km h^{-1} 63.4° W of N (*M*)

14. ← (*A*)

Unit 11.2 Activity 3B: Relative velocity (page 69)

1. **a**. 0.80 km (*A*) **b**. 5.0 km h^{-1} (*A*) **c**.

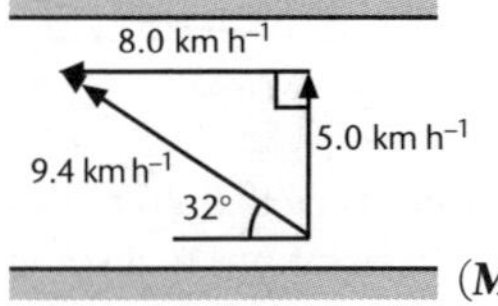

(*M*)

v_{boat} = 9.4 km h^{-1} downstream at 32° to riverbank. (*E*)

2. **a**. 600 km h^{-1} S, 603 km h^{-1}, 5.7°, 60 km h^{-1} E (*M*)

   b. v_{jet} = 603 km h^{-1} 5.7° E of S (*M*)

3. **a**. 50 m 53° to river bank downstream. (*M*)
   b. 2.5 m s^{-1} (*M*)
   c. 1.5 m s^{-1} $\frac{30\text{ m}}{20\text{ s}}$ (*M*)
   d. 2.0 m s^{-1} $\frac{40\text{ m}}{20\text{ s}}$ (*M*)
   e.

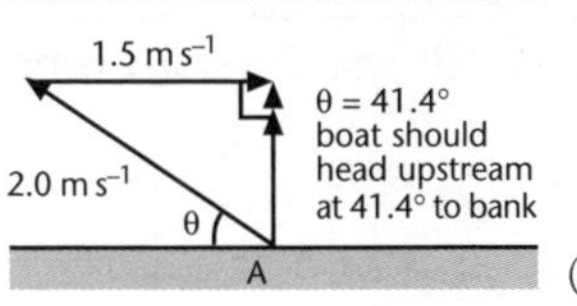

(*E*)

4. 1.3 m s^{-1}

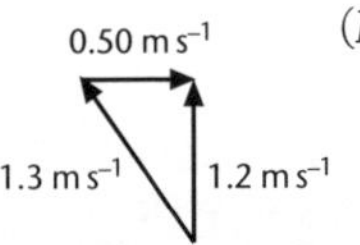

(*M*)

5. 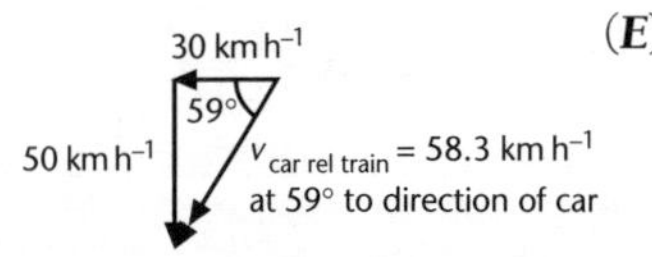

(*E*)

6. 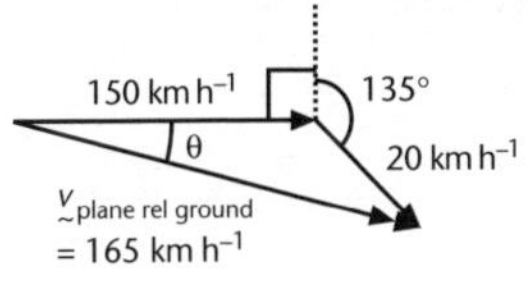

$v_{plane\ rel\ ground}$ = 165 km h^{-1} 4.9° S of E (*E*)

Topic 4: Graphs of motion

Unit 11.2 Activity 4A: Distance versus time graphs (page 72)

1. Constant speed, stopped, constant speed, stopped. (**M**)
2. C (***A***)
3. Acceleration, constant speed, at rest. (**M**)
4. **a**. After approximately 12 s (when the distances travelled by Rex and Ruth are equal). (**M**)
   b. Ruth's average speed = $\frac{100}{25}$ = 4 m s^{-1} (***A***)
   c. Rex's speed after 17 s = slope of d-t graph at t = 17 s = 0 m s^{-1}. (**M**)
   d. Ruth is still travelling at 4 m s^{-1}. (***A***)
5. **a**.

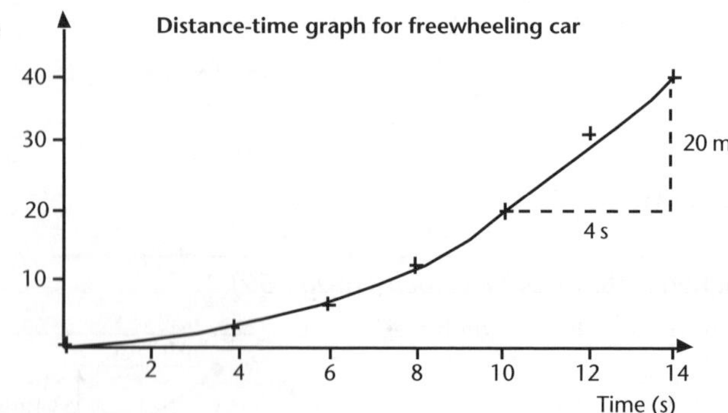

   b. After it has travelled 20 m the car stops accelerating and the graph changes from a curve (acceleration) to a straight line (constant speed). (**M**)
   c. **i**. At t = 5 s the distance travelled is 7.5 m. (***A***)
   ii. At t = 11 s the speed is equal to the slope of the graph:

$$= \frac{\text{rise}}{\text{run}} = \frac{20 \text{ m}}{4 \text{ s}} = 5 \text{ m s}^{-1}$$ (**M**)

Unit 11.2 Activity 4B: Speed versus time graphs (page 75)

1.

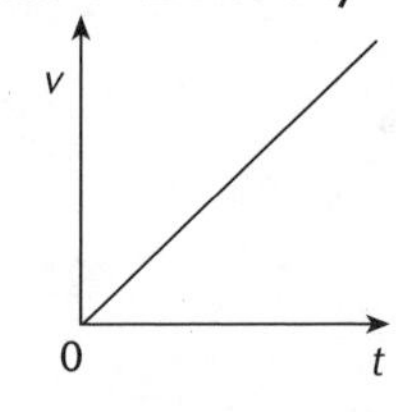

(***A***)

2. 1 m s^{-2} (***A***)
3. Total distance travelled by the car = area under the velocity-time graph

$$= \frac{1}{2} \times 5 \times 5 + 5 \times 5 + 2 \times \frac{1}{2} \times 2.5 \times 5 + \frac{1}{2} \times 5 \times 3$$
$$= 12.5 + 25 + 12.5 + 7.5$$
$$= 57.5 \text{ m}$$ (**M**)

4. Total displacement $= \frac{1}{2} \times 5 \times 5 + 5 \times 5 + \frac{1}{2} \times 2.5 \times 5 - \frac{1}{2} \times 5 \times 3$

$= 12.5 + 25 - 7.5$

$= 30$ m (***M***)

5. Average speed $= \dfrac{\text{total distance travelled}}{\text{total time taken}}$

$= \dfrac{\frac{1}{2} \times 5 \times 5 + 5 \times 5 + \frac{1}{2} \times 2.5 \times 5}{12.5}$

$= \dfrac{12.5 + 25 + 6.25}{12.5}$

$= 3.5$ m s^{-1} (***M***)

6. Acceleration over Section C = gradient of graphs over Section C

$= \dfrac{-5-5}{5} = -2$ m s^{-2}.

A negative slope implies acceleration. (***M***)

7.

(***M***)

8.

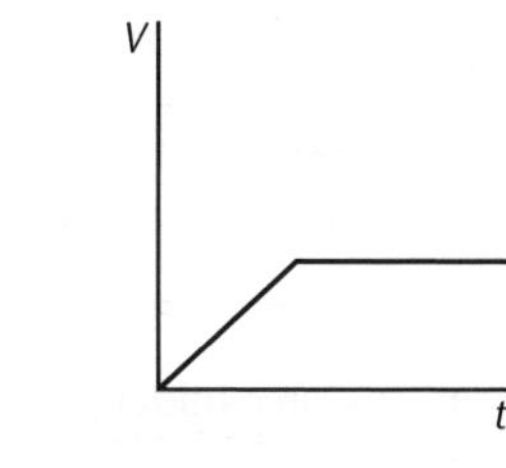

(***A***)

9. Total distance travelled = area under the graph

$= \frac{1}{2} \times 10 \times 10 + 5 \times 10 + \frac{1}{2} \times 5 \times 10$

$= 50 + 50 + 25$

$= 125$ m (***M***)

10.

(***M***)

11. Average speed $= \dfrac{\text{total distance travelled}}{\text{total time taken}}$

$= \dfrac{\text{area under speed–time graph}}{\text{total time taken}}$

$= \dfrac{2 \times 8 + 16 \times 8}{24}$

$= 6$ m s^{-1} (***M***)

12. a. Acceleration = slope of graph = $\frac{40}{15}$ = 2.7 m s^{-2} (***A***)

b. Acceleration = slope of graph = $-\frac{40}{7.5}$ = –5.3 m s^{-2} (***A***)

c. Total distance = area under graph

$= \frac{1}{2} \times 15 \times 40 + 10 \times 40 + \frac{1}{2} \times 7.5 \times 40$

$= 300 + 400 + 150$

= 850 m (***M***)

13. a.

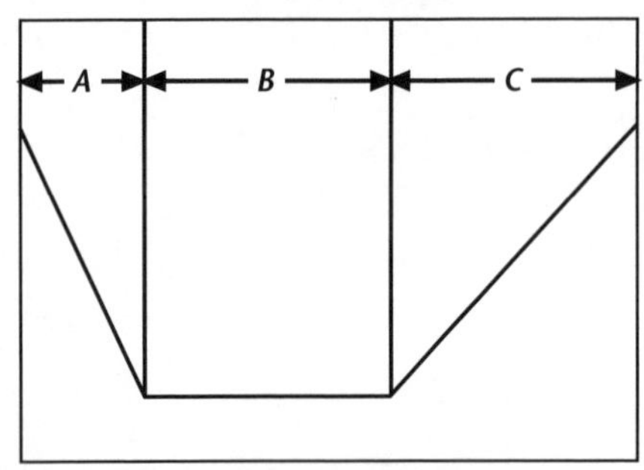

(***A***)

b. Mere's acceleration = slope of graph over section A

$= \frac{8}{20}$ = 0.4 m s^{-2} (***A***)

c. Total distance travelled = area under graph

$= \frac{1}{2} \times 20 \times 8 + 40 \times 8 + \frac{1}{2} \times 40 \times 8$

$= 80 + 320 + 160$

= 560 m (***M***)

d. Average speed = $\frac{\text{total distance}}{\text{time taken}} = \frac{560 \text{ m}}{100 \text{ s}}$ = 5.6 m s^{-1} (***M***)

14. a. Speed at t = 5 s = 22 m s^{-1}

Speed at t = 10 s = 30 m s^{-1} (***A***)

b. For the first 10 s the car accelerates; for the remaining 20 s the car travels at a constant speed of 30 m s^{-1}. (***A***)

c. Acceleration *decreases* because the slope of the graph decreases. (***M***)

d. Total distance travelled = area under graph

$= \frac{1}{2} \times 10 \times 30 + 20 \times 30$

$= 150 + 600$

= 750 m (***A***)

e.

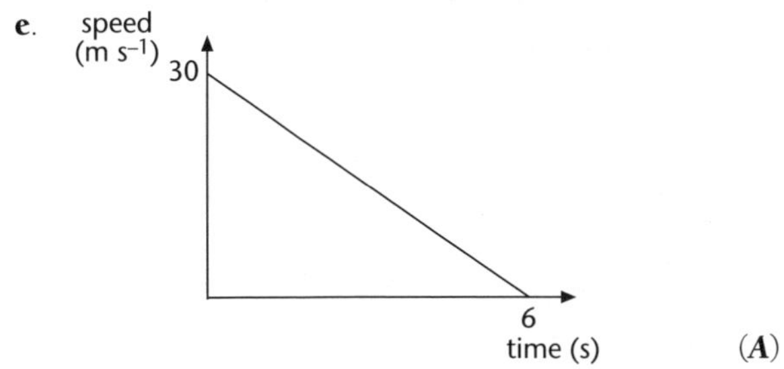

(***A***)

Unit 11.2 Activity 4C: Velocity–time graph (page 79)

Acceleration = 1 m/s^2

Unit 11.2 Activity 4D: Two-stage rocket (page 79)

a. 40 m/s^2 **b**. 20 m/s^2 **c**. –20 m/s^2

Unit 11.2 Activity 4E: Velocity–time graph – motion (page 82)

During interval A, the velocity was maintained constant/unchanged therefore, the acceleration was zero. During interval B, the particle/object was slowing down (velocity was decreasing, therefore, the acceleration was negative (deceleration). During interval C, the velocity was again constant/unchanged (acceleration is zero).

Unit 11.2 Activity 4F: The ticker timer (page 85)

1. **a**.

d-t graph for ticker-timer tape

(***A***)

b.

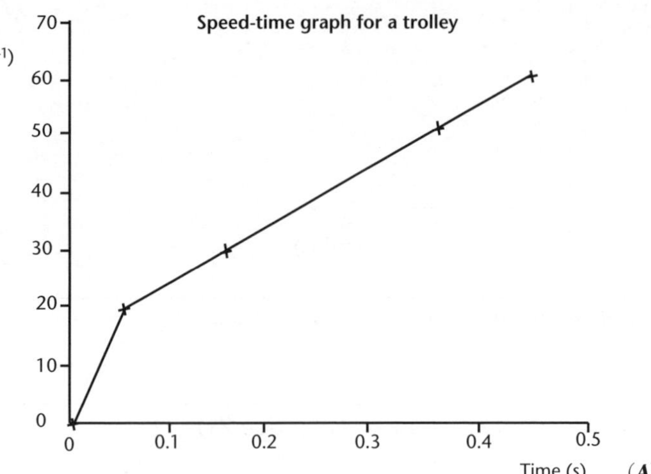

(**A**)

2. **a**. 0.2 s (**A**) **b**. 10.3 cm (**A**) **c**. $v = \frac{d}{t} = \frac{10.3}{0.2} = 51.5\ \text{cm s}^{-1}$ (**A**)

3. **a**.

Total distance travelled (cm)	Time (s)
0	0
1.5	0.2
4.6	0.4
10.4	0.6

(**A**)

b.

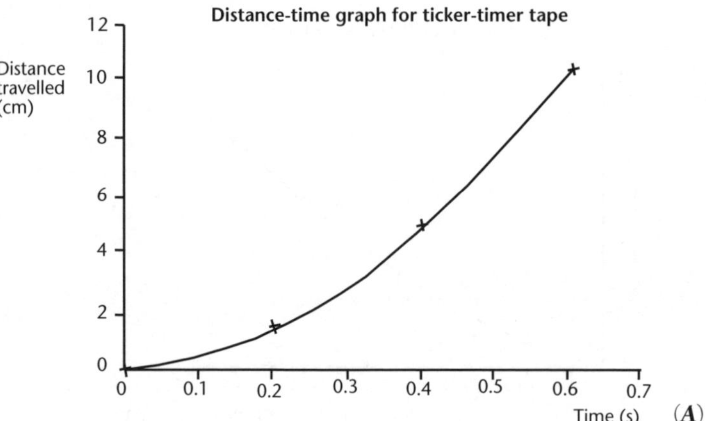

(**A**)

Topic 5: Equations of motion

Unit 11.2 Activity 5A: Graphs of motion (page 92)

1. **a.** 5 m s^{-2} (**A**) **b.** 90 m (**A**) **c.** 9 m s^{-1} (**A**)

2. **a.** 2 m s^{-2} (**A**) **b.** –3 m s^{-2} (**A**) **c.** 650 m (200 + 300 + 150) (**A**)

d. 45 m s^{-1} (**M**) **e.** Ahead of the car (distance = $0.5 \times 45 \times 30 = 675$ m). (**M**)

3. **a.** 5.6 m s^{-1} (**M**)

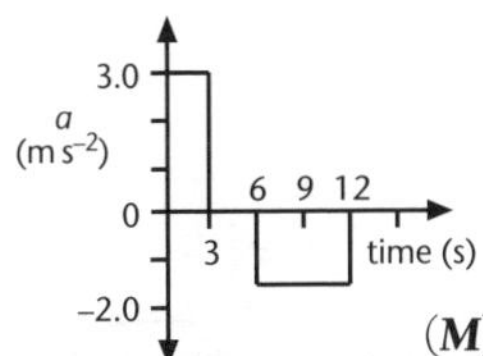

(**M**)

c.

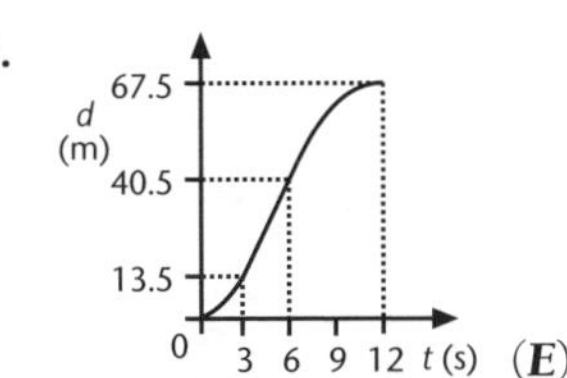

(**E**)

Unit 11.2 Activity 5B: Kinematic equations (page 94)

1. 6.0 m s^{-1} (**A**)

2. **a.** 350 m (**M**) **b.** **i.** 7.6 m s^{-1} (**M**) **ii.** 7.0 m s^{-1} (**A**)

3. 4.0 m (**A**)

4. **a.** 1 800 m (**M**) **b.** 22.4 m s^{-1} (**M**)

5. 30 s (**M**) **6.** 456 m (**M**)

7. **a.**

$$v_f = v_i + at$$

$$t = \frac{v_f - v_i}{a} \quad \text{[rearranging]}$$

$$d = \left(\frac{v_i + v_f}{2}\right)t$$

$$d = \left(\frac{v_i + v_f}{2}\right)\left(\frac{v_f - v_i}{a}\right) \quad \text{[substituting for } t\text{]}$$

$$2ad = (v_i + v_f)(v_f - v_i) \quad \text{[rearranging]}$$

$$= v_i v_f - v_i^2 + v_f^2 - v_f v_i$$

$$= v_f^2 - v_i^2 \quad \text{[simplifying]}$$

ie $v_f^2 = v_i^2 + 2ad$ (**E**)

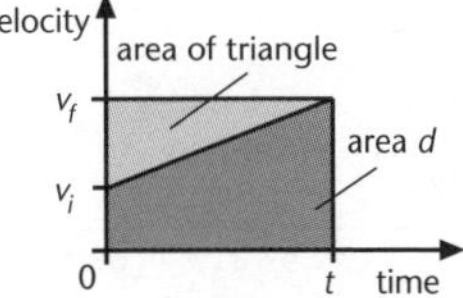

d = whole area – area of triangle

$$= v_f t - \frac{1}{2}(v_f - v_i)t \quad \text{[but } (v_f - v_i) = at\text{]}$$

$$= v_f t - \frac{1}{2}at^2$$ (**E**)

Unit 11.2 Activity 5C: Velocity (page 96)

1. A

2. B

3. False

4. True

5. False

Topic 6: Projectile motion

Unit 11.2 Activity 6A: Vertical motion under gravity (page 99)

1. 28 m s^{-1} (***M***)
2. **a.** (***M***)

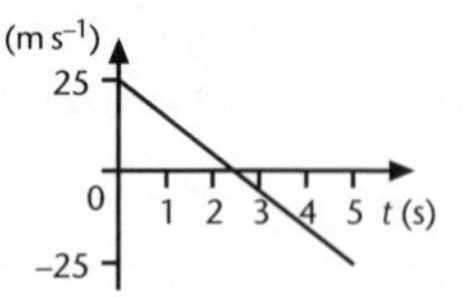

b. 2.5 seconds (***A***)

c. The force due to gravity is still acting on the ball at this moment. (***M***)

d.

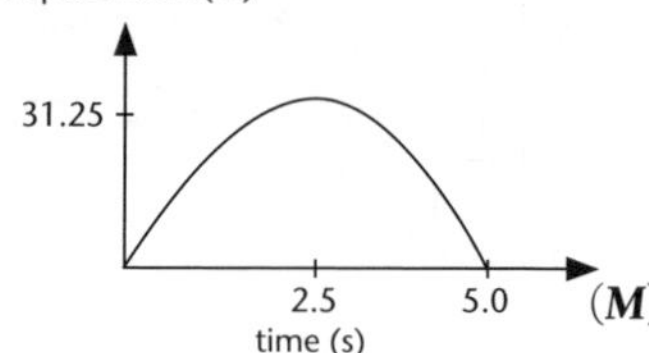

(***M***)

3. 1.55 s (***A***)
4. 80 m (***A***)

Unit 11.2 Activity 6B: Projectile motion (page 102)

1. **a.** The height of the ball increases to a maximum and then reduces to zero. (***A***)

   b. Initially the speed is a maximum and reduces during the upward path. On the return journey it increases back to the original speed. (***M***)

   c. At all points. (***A***)

   d.

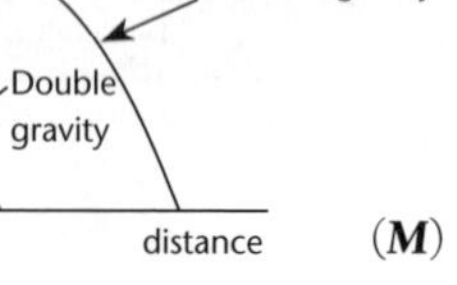

(***M***)

2. **a.** 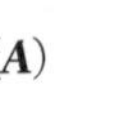 (***A***) **b.** (***A***) **c.** (***A***)

3. 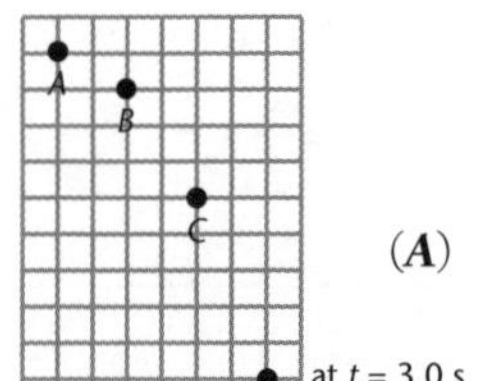

(***A***)

4. **a.** The engine should continue on with unchanged speed. (***A***)

   b.

5. 4.0 m (***M***)
6. **a.** Initial vertical velocity = 10 m s^{-1}; (***A***) initial horizontal velocity = 17.3 m s^{-1}. (***A***)

   b. 2.0 s (***A***) **c.** 35 m (***A***)

7. Time to reach goalposts = 2.5 s.

   Height at t = 2.5 s is 6.25 m so ball *does clear* the crossbar $\left(d = v_i t + \frac{1}{2}at^2\right)$. (***E***)

8. **a**. 20 m s^{-1}, 30 m s^{-1} (***A***) **b**. 6.0 s (***M***)
   c. 36 m s^{-1} at 56° to horizontal. (***E***) **d**. 25 m s^{-1} (***M***)
9. **a**. V_x is constant.
   b. V_y is initially positive, decreases to zero and then becomes progressively more negative.
   c. $a_x = 0$ throughout.
   d. $a_y = g$ throughout.
10. **a**. highest point **b**. lowest point
11. **a**. in your hand **b**. behind you **c**. in front of you
12. **a**. all tie **b**. all tie **c**. c, b, a **d**. c, b, a

Topic 7: Circular motion

Unit 11.2 Activity 7A: Circular motion (page 109)

1. 16 m s^{-1} (56.5 km h^{-1}) (***A***) 2. **a**. 1.3 s (***A***) **b**. 0.80 Hz (***A***) **c**. 48 rpm (***A***)
3. (***A***)
4. **a**. **i**. Correct. (***A***) **ii**. Correct. (***A***)
   iii. Correct. (***A***) **iv**. Incorrect. (***A***) **v**. Correct. (***A***)
   b. 0.28 Hz (***M***) **c**. 1.4 m s^{-1} (***A***)
5. **a**. 4 s (***M***) **b**. 0.79 m s^{-1} (***M***)
6. **a**. 3T (***M***) **b**. F/3 (***M***) **c**. 9F (***M***) **d**.
7. **a**. 2.1 s (***A***) **b**. 18 m s^{-2} (***A***)
   c. Bung will move in direction C. (***A***)
8. 2.5 m s^{-2} (***A***) 9. 0.60 m (***E***)
10. **a**. 10 m s^{-1} (***E***) **b**. 0.70 s (***M***) **c**. 7.3 m $\left(\sqrt{7.0^2 + 2.0^2}\right)$ (***E***)
11. B
12. B
13. C
14. C
15. D
16. False
17. False

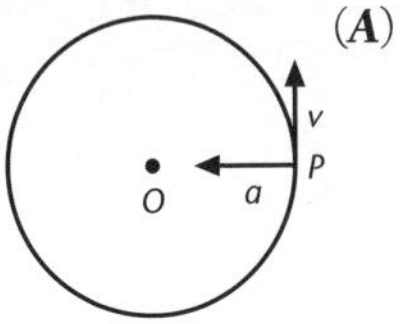

Unit 11.3 Force and Motion (Dynamics)

Topic 1: Force

Unit 11.3 Activity 1A: Using Newton's Laws (page 116)

1. **a**. 2.8 N (***A***) **b**. ↗ (***A***)
2. **a**. 1.20 m s^{-2} (***A***) **b**. 0.72 m s^{-2} (***A***)

3. **a**. 40 N (***A***) **b**. 10.0 kg (***A***) **c**. 4.0 m s^{-2} $\left(\frac{40\text{ N}}{10.0\text{ kg}}\right)$ (***M***)
   d. 2.0 m s^{-1} ($v_f^2 = v_i^2 + 2ad$) (***M***) **e**. 24 N ($F = ma = 6.0 \times 4.0$) (***M***)

4. **a**.

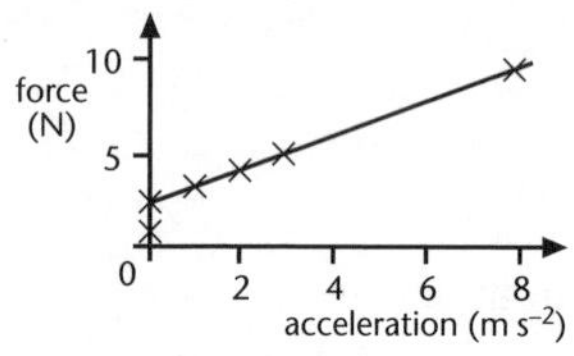

(***A***)

   b. gradient = $\left(\frac{10.0 - 2.0}{8.0 - 0}\right)$

   $= \frac{8.0}{8.0}$

   = 1.0 kg (***M***)

   c. 1.0 kg (***M***)
   d. Friction in the trolley's wheels and in the pulley. (***A***)
   e. The gradient of the graph should double to 2.0 kg. (***A***)

5. **a**. 1 000 N (***A***) **b**. 1 000 N (***A***) **c**. 22.5 m s^{-2} $\left(\frac{50 - 5.0}{2.0}\right)$ (***A***)
   d. 2 250 N ($F = ma = 100 \times 22.5$) (***A***) **e**. 1 000 N (***A***)
6. T = 5 N
7. F = 12 400 N
8. 15 N
9. $(5\sqrt{3})/8$ m/s^2
10. 10 N
11. 1 500 N
12. **a**. Magnitude of the normal force is equal to the weight, mg.
    b. It is greater than, because the acceleration is upward so the net force on the object (book) is up.
13. A bird flies forward by exerting a force on the air, but it is the air pushing back on the bird's wings that propels the bird forward: Newton's 3rd law of motion.
14. Because of the force exerted on it by the ground which is the reaction to the force exerted on the ground by the wheels.
15. A rocket accelerates because it exerts a strong force on the gases, expelling them. The gases exert an equal and opposite force on the rocket and it is this force that drives the rocket forward. Therefore, a space vehicle can maneuver in empty space just by firing its rockets in the direction opposite to that in which it wants to accelerate.
16. Gravitational pull (weight). Contact with another object (Normal force), Attachment to another object (tension).

Topic 2: Friction

Unit 11.3 Activity 2A: Force (page 121)

1. **a**. decreases
   b. stay the same
   c. decreases

Topic 3: Newton's laws

Unit 11.3 Activity 3A: Forces (page 126)

1. I and II (**M**)
2. Force 1 friction – opposes motion; Force 2 weight (gravity); Force 3 air friction – opposes motion. (**A**)
3. 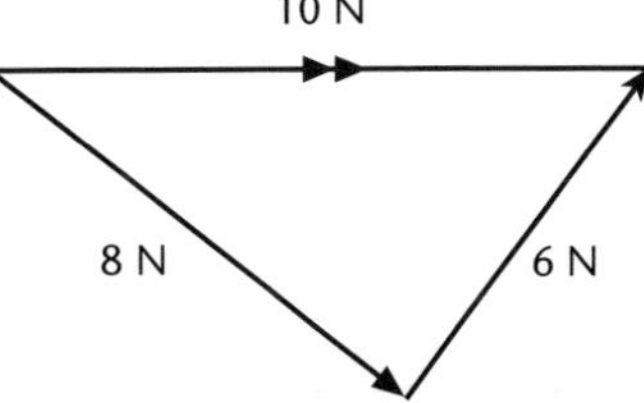

(**M**)
4. It stops accelerating because. $F = ma$; when $F = 0$ the acceleration is zero.
5. F sign = 15 g., a sign = 30 g. (**M**)
6. **a**. Engine force. (**A**)
   b. $W = 9\,000 + 6\,000 = 15\,000$ N (**A**)
   $m = \frac{W}{g} = \frac{15\,000}{10} = 1\,500$ kg (**A**)
   c. Since the car is travelling at constant speed, $F = 500$ N to left or –500 N. (**A**)
   d. New $F_{resultant} = 1\,000 - 500 = 500$ N to right. (**A**)
   e. $a = \frac{F}{m} = \frac{500}{15\,000} = 0.3$ m s^{-2} (**A**)
7. **a**. 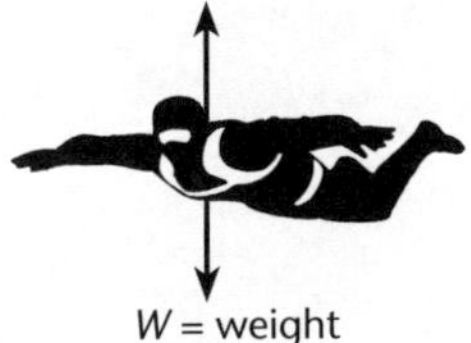

(**A**)
   b. The weight force remains constant but the air resistance increases. (**M**)
   c. Eventually, $F = R$ and the net force is zero. The skydiver starts moving at a constant speed called terminal velocity. (**E**)
   d. 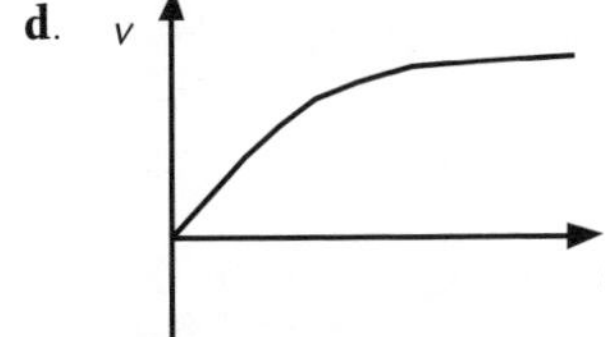

(**M**)
8. $F_{net} = ma = 2 \times 5 = 10$ N; $F_{friction} = 16 - 10 = 6$ N. (**E**)
9. 1 N → (one newton force to the right)
   [2 + 3 – 4 = 1] (**A**)

10 a. 16 N (***A***) **b**. 2.5 N (***A***) **c**. 20 000 000 N (***A***) **d**. 0.5 m s^{-2} (***A***)
e. 25 000 m s^{-2} (***A***) **f**. 40 m s^{-2} (***A***) **g**. 320 kg (***A***) **h**. 20 000 kg (***A***)

11. 3.6 N. (***A***)

12. Mass is the same, weight is greater. (***A***)

13. **a**. 20 m s^{-2} (***A***) **b**. 80 N (***A***) **c**. 40 m s^{-2} (***A***)

d.

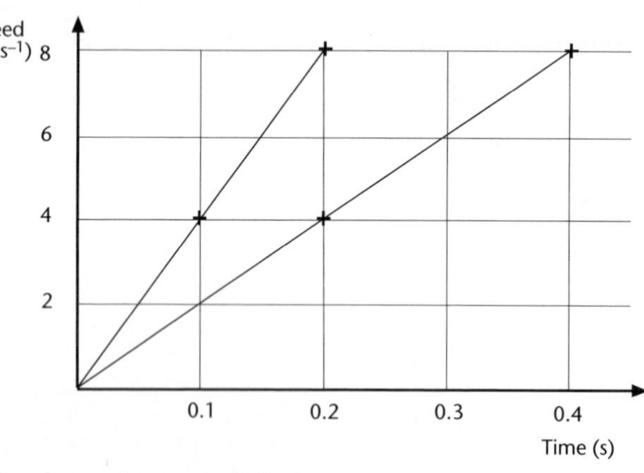

(***M***)

14. **a**. B (***A***) **b**. 20 m s^{-2} (***A***) **c**. 20 kg (***M***)

15. D (***M***)

16. **a**. 140 kg (***A***) **b**. Constant speed. (***A***) **c**. Speed reduces. (***M***)

17. **a**.

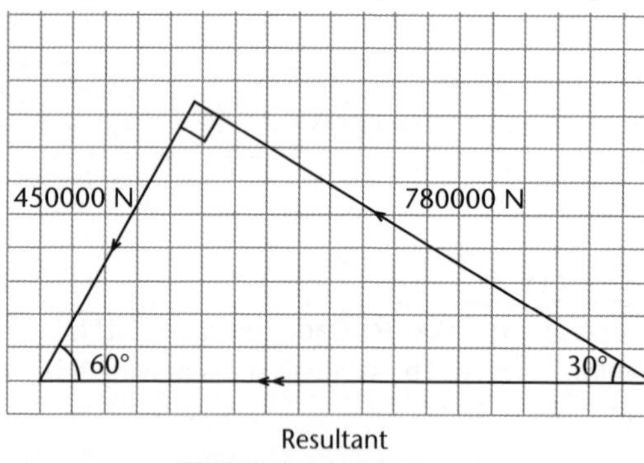

(***M***)

b. Resultant $= \sqrt{780\,000^2 + 450\,000^2}$
$= 900\,500$ N (***M***)

c. 900 500 N opposite to direction of travel (East) (***M***)

18. **a**.

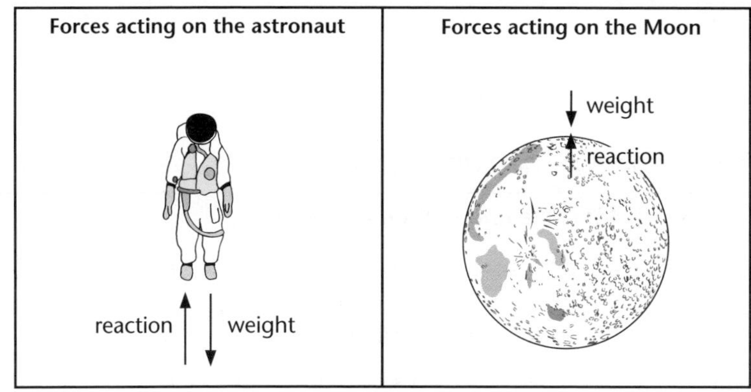

(***M***)

b. The two forces are equal and opposite. (***M***)

Unit 11.3 Activity 3B: Pressure and density (page 133)

1. $F = 6\,048$ N $l = 0.25$ m $b = 0.20$ m $A = ?$ $P = ?$

$A = l \times b = 0.25 \text{ m} \times 0.20 \text{ m}$ $\therefore A = 0.050 \text{ m}^2$

contact area $= 60\% \times 0.050 \text{ m}^2 = 0.030 \text{ m}^2$

$$P = \frac{F}{A} = \frac{6\,048 \text{ N}}{0.030 \text{ m}^2} \quad \therefore P = 201\,600 \text{ N m}^{-2}$$

Pressure = 201 600 Pa (pascal) *or* 201.6 kPa (***E***)

2. **a**. Pressure $= \frac{\text{Force}}{\text{Area}}$. Both Tom and Hone exert the same force, but Hone's force is exerted over a smaller area (ridges only in contact with ground), hence he exerts more pressure. (***E***)

b. $F = ma$ $m = 60$ kg $a = 10 \text{ N kg}^{-1}$

$\therefore F = 60 \text{ kg} \times 10 \text{ N kg}^{-1} = 600 \text{ N}$

Weight = 600 N (1 sf) (***M***)

c. $P = \frac{F}{A}$ $F = 600$ N $A = 1.6 \times 10^{-3} \text{ m}^2$

$\therefore P = \frac{600 \text{ N}}{1.6 \times 10^{-3} \text{ m}^2} = 375\,000$ Pa

Pressure = 400 000 Pa (1 sf)(***M***)

3. $P = \frac{F}{A} = \frac{75}{0.007} = 10\,714 \text{ N m}^{-2}$

$\therefore$ pressure $= 10\,000 \text{ N m}^{-2}$ (1 sf) (***M***)

4. 5 036 N

5. **a**. 750 kg
b. 270 kg

6. 75 km

7. 59.5 kg

8. 600 N

9. 52 000 KPa.

10. 20 000 Pa.

Topic 4: Rotational motion and equilibrium

Unit 11.3 Activity 4A: Force and motion (page 141)

1. **a**. The force required to accelerate the car around the curve is F = 3900 N, and F_{fr} = 5900 N; then the car will make the turn.
b. Here you find F_{fr} = 2000 N, the car will skid.

2. **a**. 60 m/s **b**. 3.2 rev/sec **c**. 0.32 seconds **d**. 1200 m/s^2

Unit 11.3 Activity 4B: Force, motion and equilibrium (page 146)

1. **a**.

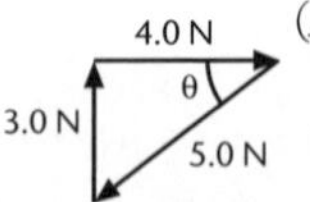

(**A**)

b. $\theta = \tan^{-1}\frac{3.0}{4.0} = 37°$ (**A**)
c. 5.0 N in the opposite direction to the 5.0 N force removed. (**A**)

2. **a**. 30 N (**A**) **b**.

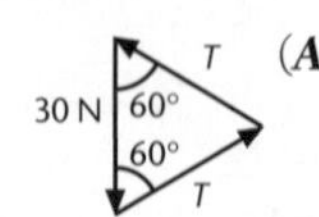

(**A**)

c. T = 30 N since forces form an equilateral triangle. (**M**)

3. **a**.

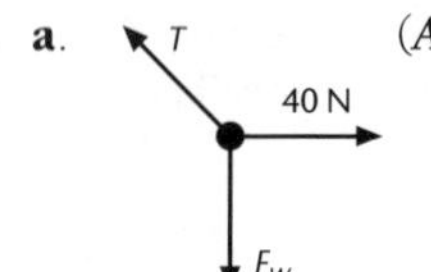

(**A**) **b**.

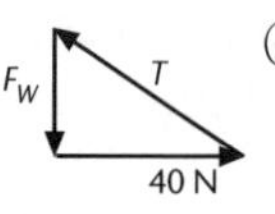

(**A**)

c. F_W = 30 N (g = 10 m s^{-2}) (**A**)
d. T = 50 N (3-4-5 triangle) (**M**)

4. The crank must be horizontal. (**A**)

5. 450 N (**A**)

6. **a**. 200 N (**M**) **b**. Beam would have to rotate until it was almost vertical. (**M**)

7. Between C and D. (**A**)

8. **a**. Resultant force = 1 N downwards (A) **b**. 1 N m clockwise. (M)
c. 1 N m anticlockwise. (**M**) **d**. 2 N m anticlockwise. (**M**)
e. It is not in equilibrium. (**A**)

9. **a**.

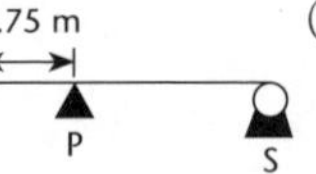

(**E**) **b**. 425 N (**E**)

10. **a**. 20 N (**M**) **b**. 80 N m (**M**)

11. 52 N (take moments about X). (**E**)

12. 1.1 m (**E**)

13. By applying the definition for the turning effect (moment or torque) of a force you can now summarise your results in a table as follows:

Magnitude of force	2F	F	3F
Distance from X	a	2a	4aSin40°
Anticlockwise moment about X	+2a	-2a	+7.71a

14. 1 Nm CCW
15. **a**. 460 Nm **b**. 345 Nm **c**. – 460 Nm **d**. 0
16. **a**. 40 Nm CCW **b**. 80 Nm CW

Topic 5: Momentum and impulse

Unit 11.3 Activity 5A: Momentum and impulse (page 155)

1. **a**. 3.5 kg m s^{-1} (***A***) **b**. 2×10^{-5} kg m s^{-1} (***A***) **c**. 2.55 kg m s^{-1} (***A***)
   d. 5×10^{4} kg m s^{-1} (***A***)
2. **a**. 0.4 kg m s^{-1} upwards. (***A***) **b**. 1.6 kg m s^{-1} upwards. (***M***)
3. (***M***) 4. **a**. (***A***) **b**. 20 kg m s^{-1} (***M***)

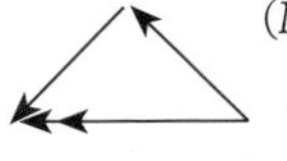

5. N s = kg m $s^{-2} \times$ s = kg m s^{-1} (***A***)
6. No, because impulse = 500 × 5.0 = 2 500 N s opposing the car's motion, and the car's initial momentum is p = 1 800 × 2.0 = 3 600 kg m s^{-1}.
   After 5.0 s, the car will have momentum p = 3 600 – 2 500 = 1 100 kg m s^{-1}, so the car will still be moving (at 0.61 m s^{-1}). (***E***)
7. **a**. 1.0 N s (***M***) **b**. –20 m s^{-1} (ie in opposite direction to initial velocity). (***E***)
8. **a**. 4.5 N (***M***) **b**. Most unlikely. (***A***)
9.

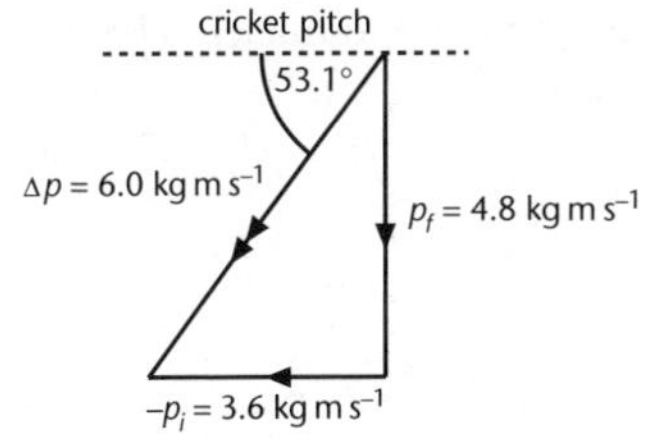

$$F = \frac{\Delta p}{\Delta t}$$

F = 60 N at 53.1° to cricket pitch (***E***)

Unit 11.3 Activity 5B: Collisions and explosions (page 160)

1. **a**. 2.25×10^{4} kg m s^{-1}, 3.2×10^{4} kg m s^{-1} (***A***)
   b. In the direction of the Mercedes Benz. (***A***)
   c. **i**. 4.0 m s^{-1} (***M***)
   ii. Friction acts to slow the wreck down. (***A***)
2. 4.5 kg m s^{-1} (***M***) **3**. 1.67 m s^{-1} (***M***)

4. a. 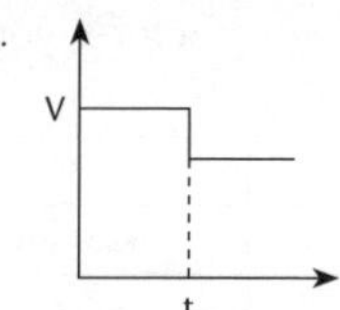

b. 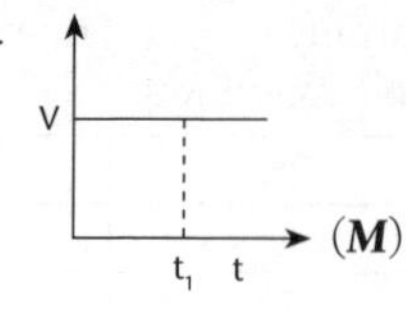

(**M**)

5. a. $p_i = 0$; both Bill and Ted are stopped;
$p_f = 58 \times -3.0 + 87 \times 2.0$ [right positive]
$= 0$ so momentum is conserved. (**M**)

b. i. 580 N (**M**) ii. 0.30 s (**A**) iii. 580 N (**A**)
iv. Forces equal and opposite. (**A**)

6. Southwest. (**M**)

7. a. Total momentum is 12 kg m s^{-1} angled 60° south of east (since equilateral triangle formed). (**M**)

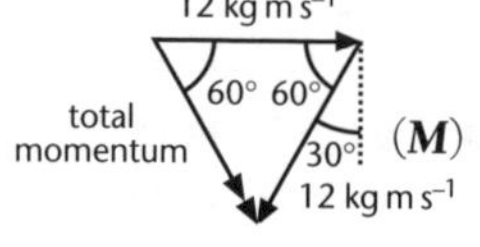

b. The same as the total before the collision, ie 12 kg m s^{-1} 60° south of east. (**A**)

c. 1.2 m s^{-1} $\left(v = \frac{P}{m} = \frac{12}{10}\right)$ (**M**)

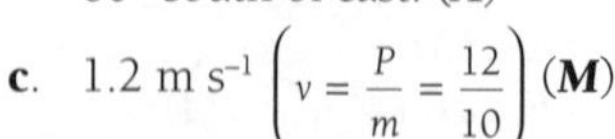

8. a. The 2.0 kg mass has the greater speed because it carries off an equal magnitude of momentum as the 6.0 kg mass (both angled at 60° to the original direction). (**M**)

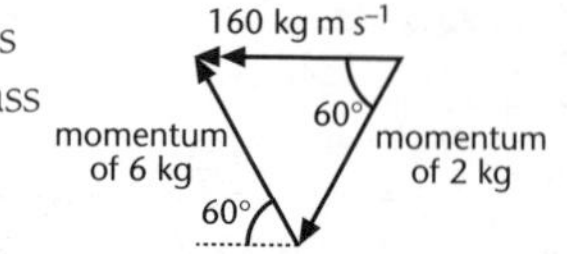

b. Smaller piece has speed 3v since 3 times smaller mass and the same momentum. (**M**)

c. Original momentum is 160 kg m s^{-1}; momentum of 6.0 kg = 160 kg m s^{-1} (since equilateral triangle).

$v = \frac{P}{m} = \frac{160}{6.0} = 27$ m s^{-1} (**E**)

9. 100 m s^{-1} at S 37° W (**E**)

10. a. ↓ (**A**) b. ↓ (**A**) c. 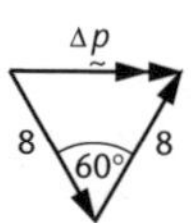

d. 8.0 kg m s^{-1} to the left. (**E**)

11. a. zero
b. positive
c. + y

12. a. unchanged
b. unchanged
c. decreased

Unit 11.4 Work, Power and Energy

Topic 1: Work and power

Unit 11.4 Activity 1A: Work (page 165)

1. **a**. 1 800 J (**A**) **b**. 0.325 J (**A**) **c**. 15 000 000 J (**A**) **d**. 100 N (**A**)
   e. 700 N (**A**) **f**. 30 m (**A**) **g**. 400 m (**A**)
2. **a**. 40 000 J (**A**) **b**. 2 500 N (**A**) **c**. 3 750 J (**A**) **d**. 43 750 J (**A**)
3. **a**. 1 500 N (**A**) **b**. 3 000 J (**A**) **c**. Increased. (**A**) **d**. 3 000 J (**A**)
4. **a**. 3 m (**A**) **b**. 600 N (**A**) **c**. 1 800 J (**A**) **d**. 900 W (**A**)

Unit 11.4 Activity 1B: Power (page 167)

1. **a**. 20 W (**A**) **b**. 0.2 W (**A**) **c**. 15 000 J (**A**) **d**. 200 000 J (**A**)
   e. 2 s (**A**) **f**. 0.001 s (**A**)
2. **a**. 8 000 N (**A**) **b**. 200 000 J (**A**) **c**. 10 000 W (**A**)
3. Electrical energy is converted into kinetic and potential energy. (**M**)
4. Thermal energy is dissipated to surroundings. (**M**)
5. $P = \frac{\Delta Ep}{t} = \frac{15\,000 \times 20}{10} = 30\,000$ W (**M**)
6. **a**. $W = F \times d$
   $\therefore W = mg \times \Delta h = 110 \times 10 \times 2.4 = 2\,640$ J (**E**)
   b. $P = \frac{W}{t} = \frac{2\,640}{4} = 660$ W (**M**)
7. **a**. $W = F \times d = 500 \times 12 = 6\,000$ J (**M**)
   b. $W = F \times d = 750 \times 10 = 7\,500$ J (**M**)
   c. Tom's speed is less than Marie's, but his power is greater than Marie's. (**M**)

Unit 11.4 Activity 1C: Power and speed (page 170)

1. F_{fr} = 800 N (resistance to motion)
2. Power of the engine = 1 110 KW
3. **a**. 37.5 m/s
   b. 19.5 m/s
   c. 461.5 m/s
4. Acceleration = 0.06 m/s^2
5. Tension = T = 147 200 N

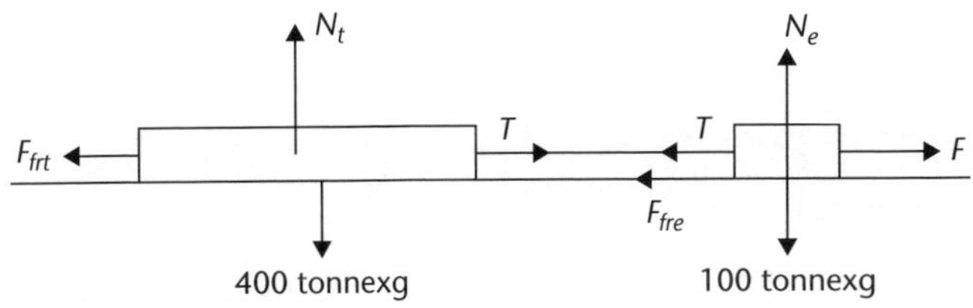

Topic 2: Energy

Unit 11.4 Activity 2A: Kinetic energy (page 173)

1. $Ek = \frac{1}{2} mv^2 = \frac{1}{2} \times 60 \times 3^2 = 270$ J (**A**)

2. Ek dog $= \frac{1}{2} \times 10 \times 10^2 = 500$ J; Ek child $= \frac{1}{2} \times 40 \times 6^2 = 720$ J (**A**)

Ek athlete $= \frac{1}{2} \times 80 \times 3^2 = 360$ J; Ek person $= \frac{1}{2} \times 120 \times 1^2 = 60$ J (**A**)

3. $Ek = \frac{1}{2} mv^2$

$= \frac{1}{2} \times 0.03 \times 400^2$

$= 2\,400$ J

4. Car Q travels at twice the speed; car Q has 4 × the kinetic energy of car P.

5. **a**. $Ek = \frac{1}{2} mv^2 = \frac{1}{2} \times 4 \times 4^2 = 32$ J (**A**) **b**. $\frac{1}{2} \times 6 \times 4^2 = 48$ J (**A**)

c. $\frac{1}{2} \times 2 \times 8^2 = 64$ J (**A**) **d**. $\frac{1}{2} \times 2 \times 12^2 = 144$ J (**A**)

Unit 11.4 Activity 2B: Gravitational potential energy (page 174)

1. $F = ma$; $35 - 5 = 60 \times a$

$30 = 60a$

$a = 0.5$ m s^{-2} (**M**)

2. **a**. $E_p = mgh = 10 \times 10 \times 2 = 200$ J (**A**) **b**. $E_p = 0.5 \times 10 \times 20 = 100$ J (**A**)

c. $m = \frac{Ep}{gh} = \frac{500}{10 \times 10} = 5$ kg (**A**) **d**. $h = \frac{Ep}{mg} = \frac{1\,500 \times 10^3}{500 \times 10} = 300$ m (**A**)

e. $m = \frac{Ep}{gh} = \frac{20 \times 10^3}{10 \times 0.5} = 4\,000$ kg = 4 tonne (**A**)

Unit 11.4 Activity 2C: Conservation of energy (page 175)

1. C (**A**)

2. **a**. **i**. Gravitational potential energy **ii**. kinetic energy (**A**)

b. **i**. Kinetic energy **ii**. electrical energy (**A**)

c. $w = mg$

$= 2\,000 \times 10$

$= 20\,000$ N (**A**)

3. **a**. $\Delta E_p = mg\Delta h$

$= 0.1 \times 10 \times 2$

$= 2$ J (**A**)

b. $E_k = 2$ J (energy is conserved). (**M**)

c. $\frac{1}{2}mv^2 = 2$

$\frac{1}{2} \times 0.1 \times v^2 = 2$

$$v = \sqrt{\frac{2}{0.1 \times 0.5}} = \sqrt{\frac{2}{0.05}}$$

$= 6.3 \text{ m s}^{-1}$ (***M***)

4. a. $m = 88$ kg $g = 10 \text{ m s}^{-2}$ $F = ?$

$F_w = mg = 88 \text{ kg} \times 10 \text{ m s}^{-2}$ $\therefore F_w = 880$ N downwards

$F_{support} = 880$ N upwards (***M***)

b. $m = 85$ kg $g = 10 \text{ m s}^{-2}$ $\Delta h = 18$ m $E_p = ?$

$\Delta E_p = mg\Delta h = 85 \text{ kg} \times 10 \text{ m s}^{-2} \times 18 \text{ m}$ $\therefore \Delta E_p = 15\,300$ J (*or* 15 kJ (2 sf)) (***A***)

c. gain in E_k = loss in E_p

$\frac{1}{2}mv^2 = mg\Delta h$

$\therefore 0.5 \times 85 \text{ kg} \times v^2 = 85 \text{ kg} \times 10 \text{ m s}^{-2} \times 6 \text{ m}$

$\therefore v^2 = 120 \text{ m}^2 \text{ s}^{-2}$ $\therefore v = \sqrt{120} \text{ m s}^{-1} = 10.95 \text{ m s}^{-1}$

$\therefore$ speed $= 11 \text{ m s}^{-1}$ (***E***)

d. The kinetic energy is converted into potential energy at X. Also kinetic energy is converted into heat (and sound) due to friction between the snowboard and the snow at right angles to the snowboard (perhaps pushing some of the snow into mounds or brushing it aside so work is done on moving it). (***M***)

5. U = 58.8 J; K = 24.0 J, therefore, total mechanical energy = 82.8 J

6. i. False
   ii. False
   iii. True
   iv. False

7. i. CD: decreasing
   ii. BC: decreasing
   iii. AB: increasing

Topic 3: Energy and mechanics

Unit 11.4 Activity 3A: Energy and springs (page 180)

1. a. 7.5 N (***A***) b. 0.56 J (***A***)
2. a. 300 N (***A***) b. 3 000 N m^{-1} (***A***) c. 15 J (***A***) d. 120 J (135 – 15) (***M***)
3. a. 5 J (***A***) b. 250 N m^{-1} (***A***) c.

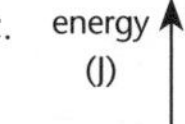

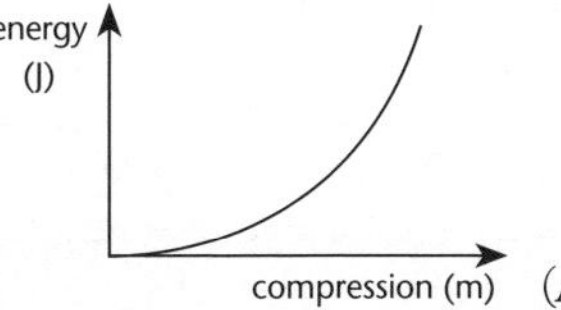

(***A***)

Unit 11.4 Activity 3B: Combined mechanics (page 184)

1. **a**. 3 000 J (**A**) **b**. 38 N (**A**) **c**. 0.63 m s^{-2} (**A**) **d**. 16 s (**A**)
2. **a**. 1.0 m s^{-1} (**E**) **b**. 0.5 m (**E**)
3. **a**. 0.5 m s^{-1} (**M**) **b**. 20 kg m s^{-1} (**M**) **c**. 20 kg m s^{-1} (**M**)
   d. 20 J (22.5 – 2.5) (**M**) **e**. No (girl gains only 5 J). (**M**)
4. **a**. Correct. (**A**) **b**. Correct. (**A**) **c**. Correct. (**A**) **d**. Correct. (**A**)
   e. Incorrect. (**A**)
5. **a**. 1.4 m s^{-1} (**E**) **b**. 10 cm (**E**)
6. **a**. 5 s (**A**) **b**. 4.0 kg (**M**) **c**. 3.0 N (**M**) **d**. 20.25 J (**A**)
   e. 15.75 m (**M**)
7. **a**. 1.41 : 1 ($\sqrt{2}$: 1) (**M**) **b**. 2 : 1 (**E**) **c**. 1 : 2.83 (1 : 2 $\sqrt{2}$) (**E**) **d**. 4 : 1 (**E**)
8. **a**. 2 393 J lost. (**M**) **b**. The 'lost' energy is converted to heat energy. (**A**)
9. **a**. 16.7 m below bridge. (**E**) **b**. 16.3 m s^{-1} (**E**)
10. **a**. 0.15 m s^{-1} (**A**) **b**. 1.1×10^{-3} m (**E**)
    c. Sphere B will return and collide with sphere A (stopping sphere B and lifting sphere A to the same height), and this pattern will continue indefinitely if no energy is lost through friction. (**M**)
    d. 0.075 m s^{-1} (**M**) **e**. 2.8×10^{-4} J (**M**)
    f. The 'lost' energy is converted to heat energy. (**A**)
    g. 2.8×10^{-4} m (**M**)
    h. The combined spheres will swing like a pendulum rising to the above height; this will continue indefinitely if no energy is lost through friction. (**M**)
11. $mgh - \frac{1}{2}mv^2$
12. 680 Joules
13. Case c is the minimum; Case b is the maximum dissipation

Topic 4: Simple machines

Unit 11.4 Activity 4A: Torque and equilibrium (page 191)

1. For equilibrium,

$$\begin{aligned} \Sigma F_{up} &= \Sigma F_{down} \\ F + 200 &= 600 + 200 \\ F &= 600 \text{ N} \quad (\mathbf{M}) \end{aligned}$$

2.

$$\begin{aligned} \text{Sum of torques about left support} &= 0 \\ 600 \times x + 200 \times 1.5 - 200 \times 3 &= 0 \\ 600\,x &= 600 - 300 \\ x &= \frac{300}{600} = 0.5 \text{ m} \quad (\mathbf{M}) \end{aligned}$$

3. Mass of A = 24 g, 2B = 24 g, B = 12 g, 3C = 12 g, C = 4 g, 4C = 2D, 16 = 2D, D = 8 g. (**M**)
4. **a**.

$$\begin{aligned} W &= mg \\ &= 45\,000 \times 10 \\ &= 450\,000 \text{ N} \quad (\mathbf{A}) \end{aligned}$$

b. $450\,000 \times 40 = F_B \times 60$

$F_B = 300\,000$ N (***A***)

c. $F_A + F_B = 450\,000$

$F_A = 450\,000 - 300\,000 = 150\,000$ N (***A***)

5. **a**. $\tau = Fd$

$= 40 \times 0.1$ (10 cm = 0.1 m)

$= 4$ N m anticlockwise (***A***)

b. $\tau_{cheese\ \&\ plank} = 10 \times 0.3 + 0.5 \times 0.7$

$= 3 + 0.35$

$= 3.35$ N m clockwise (***M***)

c. $\sum\tau_{clockwise} = \sum\tau_{anticlockwise}$

$3.35 + 1\,x = 4$

$x = 0.65$ m from pivot

Distance from support $= 0.20 + 0.65$

$= 0.85$ m (***M***)

6. **a**. $F \times 2 = 400 \times 1$

$F = \dfrac{400 \times 1}{2} = 200$ N (***A***)

b. $100 \times 5 = 2.5 \times F$

$F = \dfrac{100 \times 5}{2.5} = 200$ N (***A***)

7. **a**. Effort × 6 = 6 000 × 2

$\text{Effort} = \dfrac{12\,000}{6} = 2\,000$ N (***A***)

b. Effort × 1.5 = 500 × 0.3

$\text{Effort} = \dfrac{150}{5} = 100$ N (***A***)

8. **a**. $W = F \times d$

$= 500 \times 4$

$= 2\,000$ J (***A***)

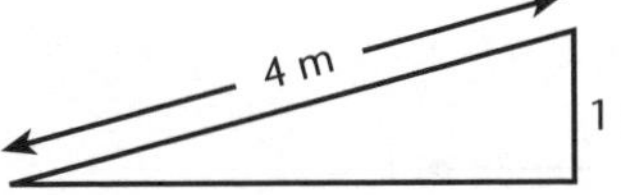

b. $W = mg\Delta h$

$= 500 \times 1$

$= 500$ J (***A***)

c. **i**. A smaller force can be used. (***M***)

ii. More work is required to raise the box. (***M***)

9. **a**. Because a larger torque is required. (***A***)

b. The handle acts as a lever. The longer the handle, the smaller the force required to operate the jack. (***M***)

c.

(***A***)

10. **Inclined plane machines**

Screw.
Ramp.
Wedge.
Coffee grinder.
Screw jack.

Lever type machines

Door handle.
Nut cracker.
Tyre lever.
Spanner.
Pliers.

(***A***)

11. **a.** Engine or propulsion force. (***A***)

b. $W = 9\,000 + 6\,000 = 15\,000$ N

$$m = \frac{W}{g} = \frac{15\,000}{10} = 1\,500 \text{ kg} \ (\boldsymbol{E})$$

c. $F = 500$ N (***M***)

d. $F_{\text{resultant}} = 500$ N (***M***)

e. $F = ma; \therefore a = \frac{F}{m} = \frac{500}{1500} = 0.3 \text{ m s}^{-2}$ (***M***)

12. **a.**

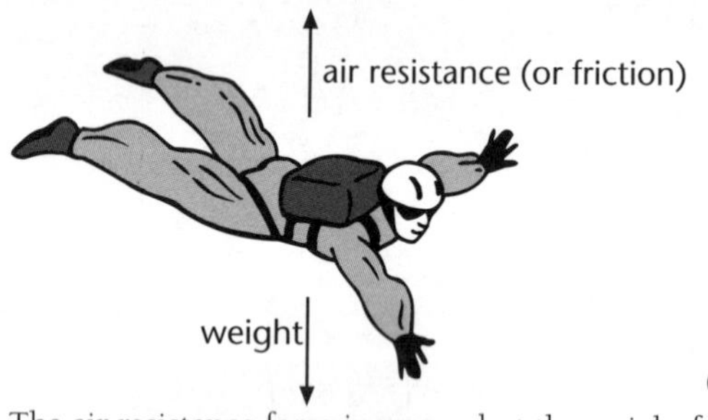

(***A***)

b. The air resistance force increases but the weight force stays the same. (***M***)

c. When the air resistance and weight forces are equal and opposite, the skydiver reaches terminal (constant) velocity. (***E***)

d.

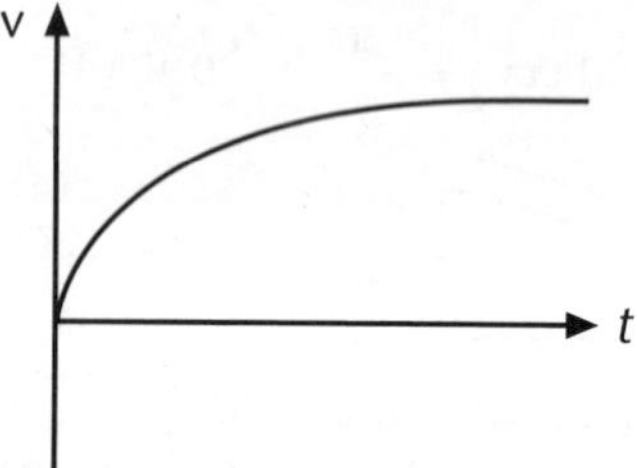

(***M***)

Unit 11.5 Electricity Principles

Topic 1: Electrostatics

Unit 11.5 Activity 1A: Attraction and repulsion (page 199)

1. **a.** Electrons will move to the bottom of the building. (***A***)

b. Positive. (***A***)

c. The electrons in the cloud will discharge to the roof of the building. (***A***)

d. A lightning conductor which is higher than the building and is connected to earth can safely conduct lightning to the ground. (***M***)

e. Length of spark $= \frac{1 \times 10^9}{20\,000} = 50\,000$ cm

$= 500$ m (0.5 km) (***A***)

f. $v = \frac{d}{t}$ $\quad d = v \times t$

$= 330 \times 3.5$

$= 1\,155$ m (1.16 km) (***A***)

2. **a**. Friction between the hairbrush and her hair has caused her hair to become charged. Like-charged hairs repel each other, and therefore stick out. (***M***)
   b. The hairbrush is oppositely charged to the hair and unlike charges attract. (***M***)
   c. On damp days the charges on the hair and brush leak to the water molecules in the air. (***E***)
3. **a**. Positive (***A***)
   b. Electrons travel from the jersey to the balloon because of friction between the balloon and jersey. (***M***)
   c. Description: stream is bent.
   Explanation: stream is attracted towards balloon because water molecules have an uneven charge distribution. (***M***)
   d.

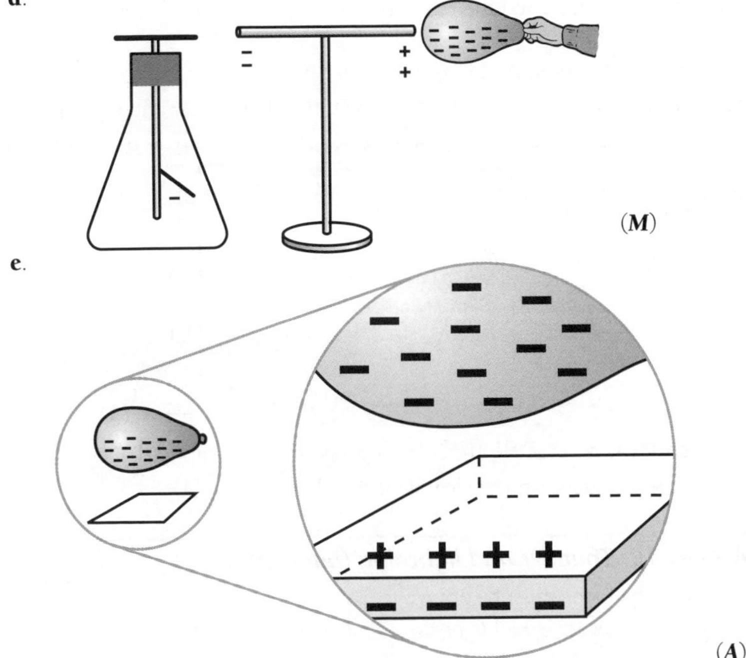

(***M***)

   e.

(***A***)

   f. Description: Paper is lifted and touches balloon then falls.
   Explanation: Positive charges on top surface are attracted to balloon. Then electrons travel from balloon to paper, paper falls because like charges repel. (***E***)

Unit 11.5 Activity 1B: The electroscope (page 202)

1. **a**. No current will flow. (***A***)
   b. No current will flow. (***A***)
   c. Electrons will flow from the negatively charged electroscope. (***A***)

2. The rod must be positively charged. Electrons are attracted from the leaf and stem of the electroscope towards the cap. This leaves a greater positive charge on the leaf and stem and spreads the leaf further. (**M**)
3. The rod must be a neutral uncharged insulator. (**A**)
4. **a**. Ball A will become negatively charged by induction when it is brought close to the cap of the positively charged electroscope; the leaf will go down. (**M**)
   b. Ball A must be negatively charged. (**A**)
   c. The electroscope leaf goes up. (**A**)
   d. Ball B must be positively charged. (**A**)
5. Electrons are transferred from the rod to the silk and the rod becomes positively charged. (**M**)
6. The cap is negatively charged and the leaf is positively charged because electrons are attracted towards the cap.
7. **a**. Because of friction between the air and the car. (**A**)
   b. Because he provides a conducting path between the charged car and the earth. (**A**)
   c. Attach a conducting strip that touches the metal of the car body to the earth. (**M**)
8. **a**.

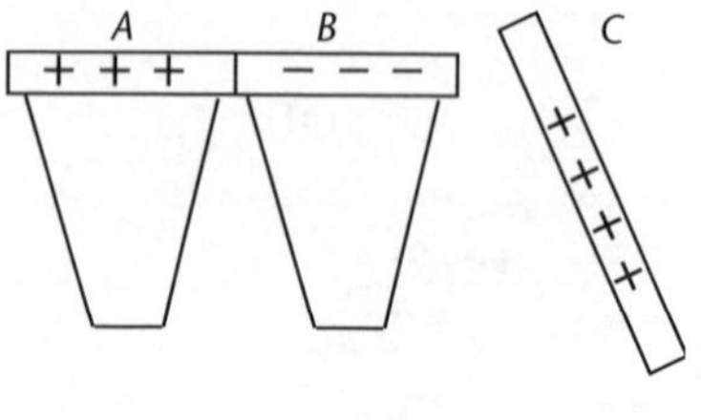

(**A**)

b.

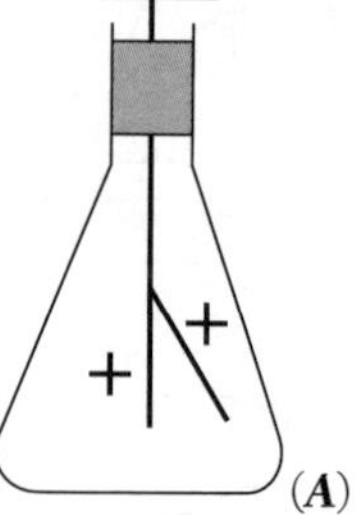

(**A**)

   c. The leaf goes down because electrons are repelled from the cap to the leaf. (**A**)
9. The cap is negatively charged and the leaf is positively charged. (**A**)

Unit 11.5 Activity 1C: Thunder and lightning (page 206)

1. The charges on the balloons must be opposite. (**A**)
2. Ball A positive; ball B negative; ball C positive; ball D negative. (**A**)
3. **a**. The silk became negatively charged. (**A**)
   b. Charge an electroscope positively with the glass rod. Bringing the silk cloth near the cap should make the leaf go down. (**M**)
4.

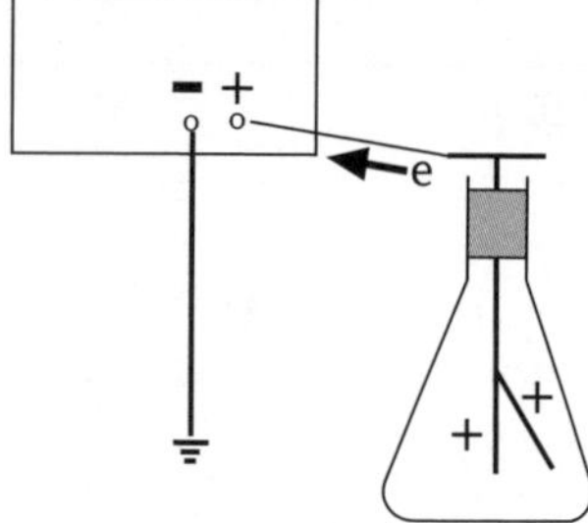

(**A**)

5. 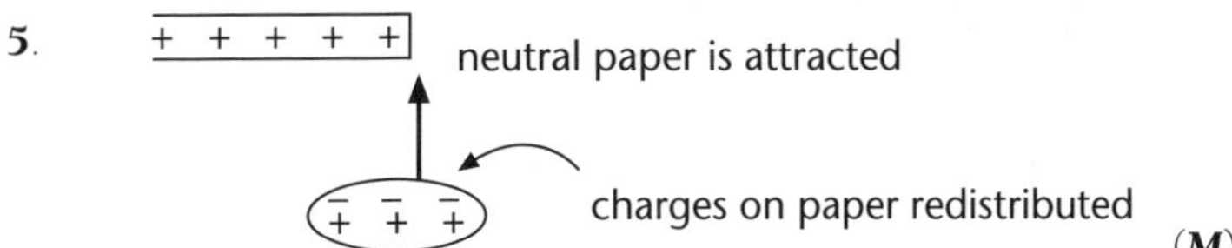

(***M***)

6. **a**. Stream of water is bent. (***A***)
   b. The pen becomes charged because electrons are transferred between the pen and the jersey; the charged pen repels the stream of water because water molecules have an uneven charge distribution. (***M***)
7. Because a thunderbolt might discharge electrons from the cloud to the tree and give the person seeking shelter an electric shock. (***M***)
8. The positively charged ball is repelled by the positive rod and attracted to the negatively charged rod. It will move in direction B. (***M***)
9. Electrons are attracted from Q to P leaving P negatively charged and Q positively charged. (***M***)
10. No charge on P and a positive charge on Q because electrons are discharged from P to earth. (***M***)
11. **a**. **b**.

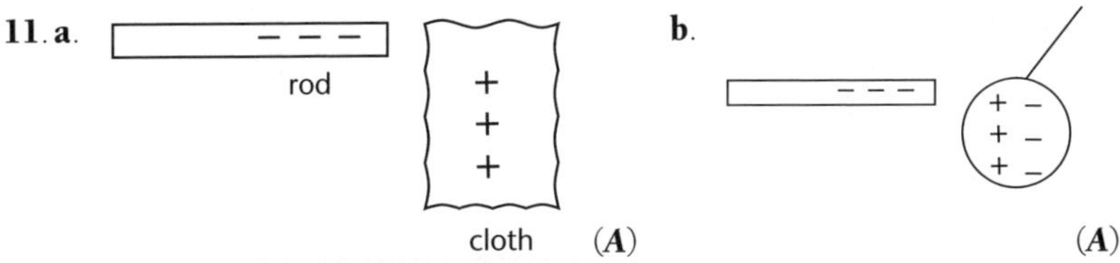

(***A***) (***A***)

   c. Electrons are discharged from the ball to his finger and the ball becomes positively charged. (***M***)
   d. Induction. (***A***)
   e.

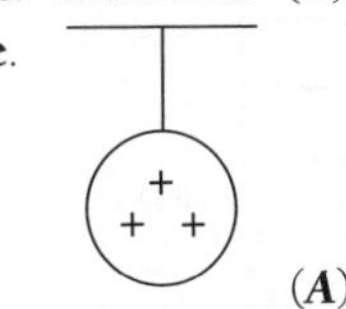

(***A***)

12. **a**. The plane becomes charged because of the friction between the fuselage and air during flight. (***A***)
   b. The plane needs to be earthed. (***A***)
   c. A spark may discharge from the plane to the fuel tanker and ignite it, causing an explosion. (***M***)

Topic 2: Current electricity

Unit 11.5 Activity 2A: Circuits (page 215)

1. **a**. 0.23 A (***A***) **b**. 23 000 V (***A***) **c**. 8 500 mA (***A***)
2. **a**. 1.4 A (***A***) **b**. 6 mA (***A***) **c**. 6 A (***A***) **d**. 13 mA (***A***)
   e. 0 µA (***A***) **f**. 1.4 mV (***A***) **g**. 3.5 V (***A***) **h**. 0.56 A (***A***)

3. a.

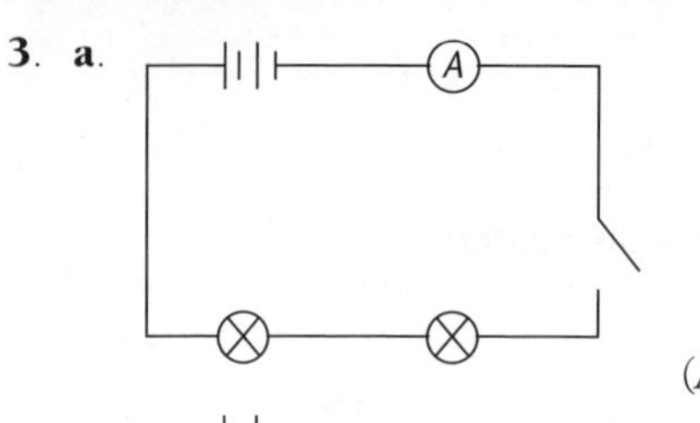

(***A***)

b.

(***A***)

c.

(***A***)

d.

(***A***)

4.

12 V

(***M***)

5.

(***M***)

6.

(***A***)

7.

Meter	Type of meter
1	voltmeter
2	ammeter
3	voltmeter
4	ammeter

(***M***)

8. 5 A. (***A***)

9. The current at B is 0.5 A, the current at A is 1 A and the current at E is 1 A. (***A***)

10. Current 2 is greater than current 3. Current 1 is greater than currents 2 and 3. (***M***)

11. **a**. A and B are both off. (***A***) **b**. A and B are both on. (***A***) **c**. A is on but B is off. (***A***)

12.

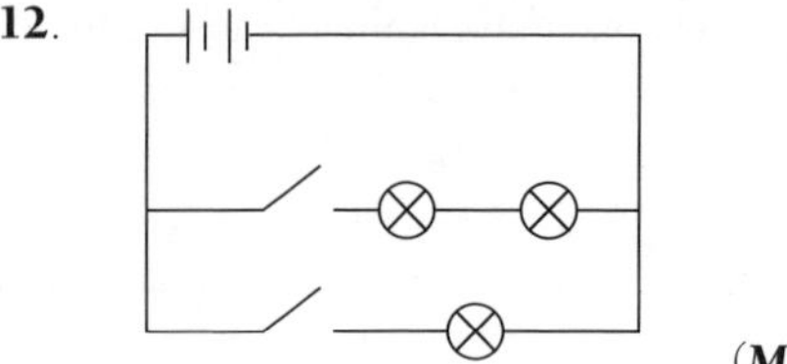

(***M***)

13. a. $A_3 = 1.5$ A, since $A_3 = A_2$.
$A_1 = 1.5 + 1.5 = 3$ A.
$A_4 = A_1 = 1.5$ A. (**A**)

b.

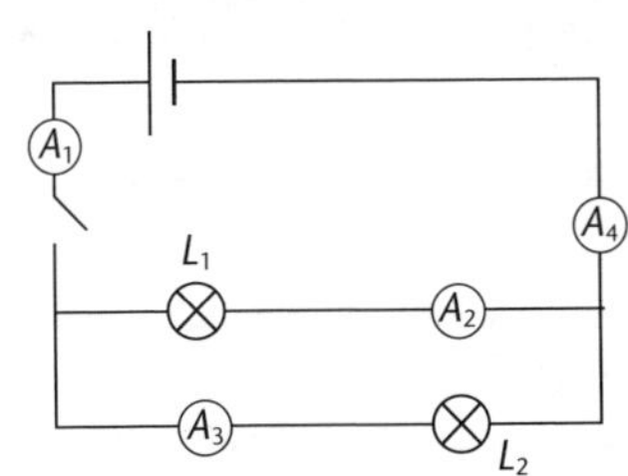

(**A**)

c.

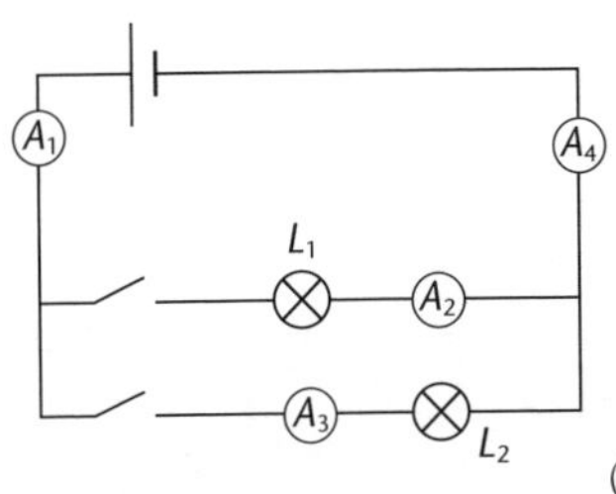

(**A**)

d.

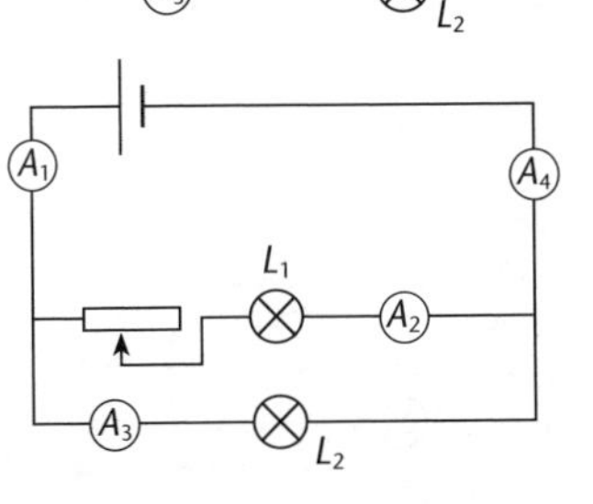

(**A**)

14. Light 1 and 2 are equally bright. Light 3 is brighter than 1 and 2. (**M**)

15. Light 1 becomes brighter. Light 2 is turned off. Light 3 becomes dimmer. (**M**)

16. Connection 1 = 3 V. Connection 2 = 1.5 V. Connection 3 = 3 V. (**A**)

17. A

18. a. $V_1 = V_2 = V_3$ (**A**)

b. $A_1 = A_2 + A_3$ (**A**)

19.

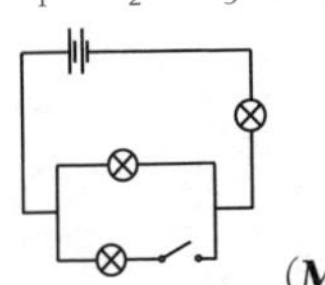

(**M**)

20. a.

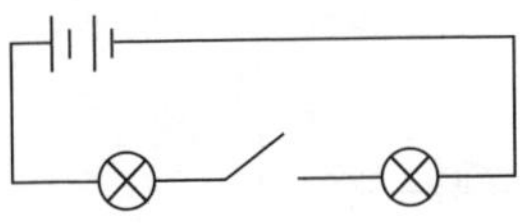

(**A**)

b.

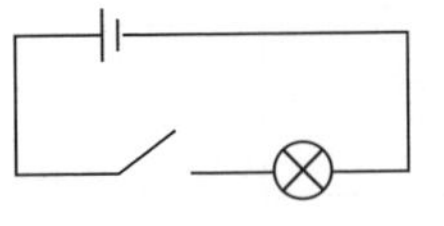

(**A**)

c.

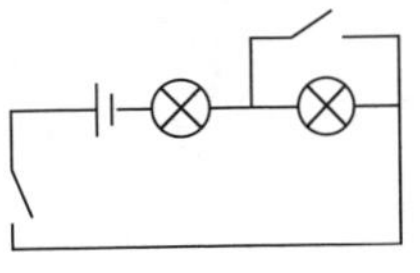

(**A**)

d.

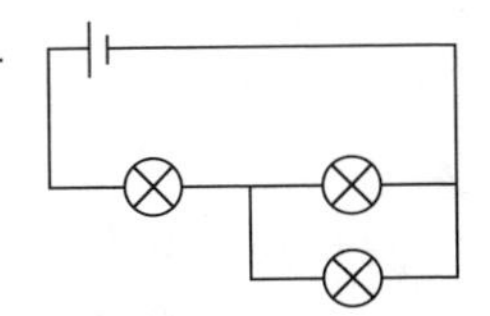

(**A**)

e.

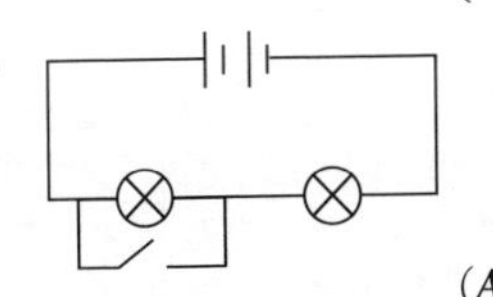

(**A**)

21. a. S2 and S4. (***A***) **b.** S2 and S3. (***A***) **c.** S2 and S3 and S4. (***M***)

22. a. ——▷|—— (***A***)

b. An LED emits light when current flows through it, but does so only when current flows in one particular direction. (***A***)

c.
- Protection for an ammeter to stop current through the meter when the meter is wrongly connected.
- Used as a small indicator light to show a circuit is switched on. (***A***)

23. The only LED that lights up is c. (***A***)

24. 2.5×10^{-3} A = 2.5 mA (***A***)

25. Voltage. (***A***)

26. a.

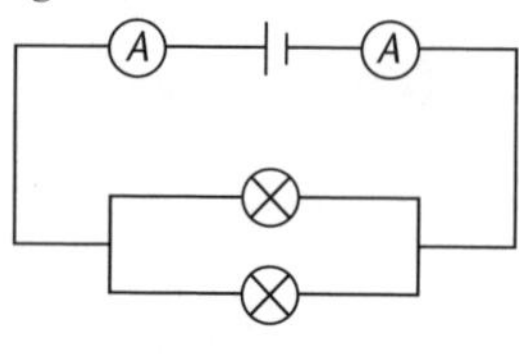

(***A***)

b. Reading on A_2 is 1.0 A. (***A***)

c. i. The lamps are connected in parallel. (***A***)

ii. The circuit resistance increases and the current decreases; reading on A_1 decreases. (***M***)

27. a.

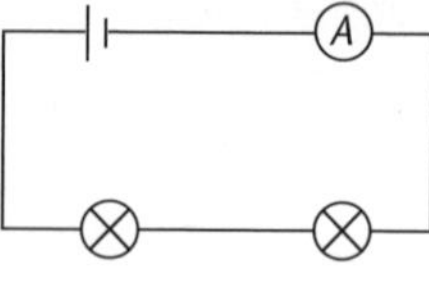

(***A***)

b.

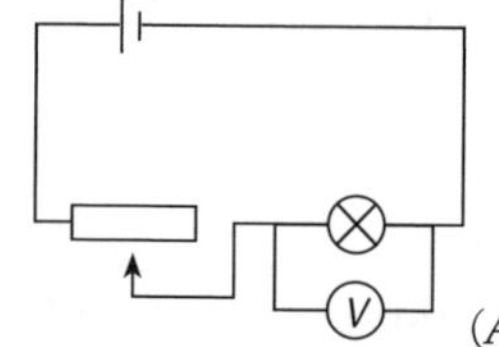

(***A***)

c.

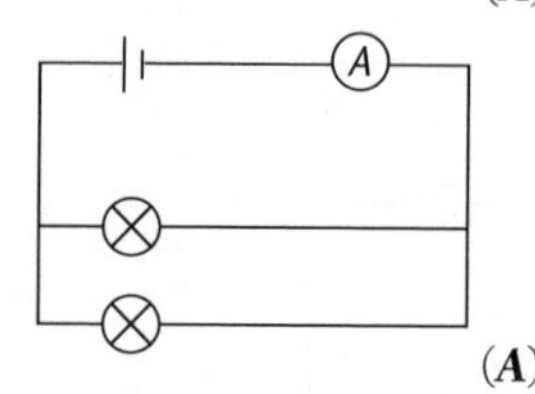

(***A***)

d.

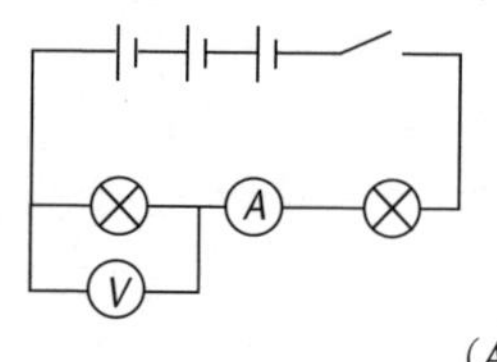

(***A***)

28. $A_3 = 5 - 2 = 3$ A (***A***)

29. a. $R_T = R_1 + R_2$ $\therefore R_T = (4.0 + 8.0)\ \Omega$ $\therefore R_T = 12\ \Omega$ (***A***)

b. $V = IR$ where $V = 12$ V, $R = 8.0\ \Omega$ $\therefore I = \frac{V}{R}$ $\therefore I = \frac{12}{8.0}$ A $\therefore I = 1.5$ A (***A***)

c. $I = 1.5$ A (***A***)

d. $E = VIt$ where $V = 6$ V, $I = 1.5$ A, $t = 13$ s $\therefore E = 6 \times 1.5 \times 13$ J

$\therefore E = 117$ J (or 120 J (2 sf)) (***M***)

e. Voltage = 18 *V* (***A***)

f. $P = VI$ where $V = 18$ V, $I = 0.9$ A

$\therefore P = 18 \times 0.9$ W

$\therefore P = 16.2$ W (or 20 W (1 sf)) (***M***)

g. $E = VIt$ or $E = Pt$ where $t = 15$ s

$\therefore E = 16.2 \times 15$ J

$\therefore E = 243$ J (or 240 J (2 sf)) (***M***)

h. $I = 1.5$ A (check $I = \frac{V}{R} = \frac{18 \text{ V}}{12\ \Omega} = 1.5$ A) (***M***)

i. Total resistance = $\frac{\text{total voltage}}{\text{total current}}$

$\therefore \text{RT} = \frac{18 \text{ V}}{1.5 \text{ A} + 0.9 \text{ A}}$

$\therefore \text{RT} = \frac{18 \text{ V}}{2.4 \text{ A}}$

$\therefore \text{RT} = 7.5\ \Omega$ (***M***)

j. **i**.

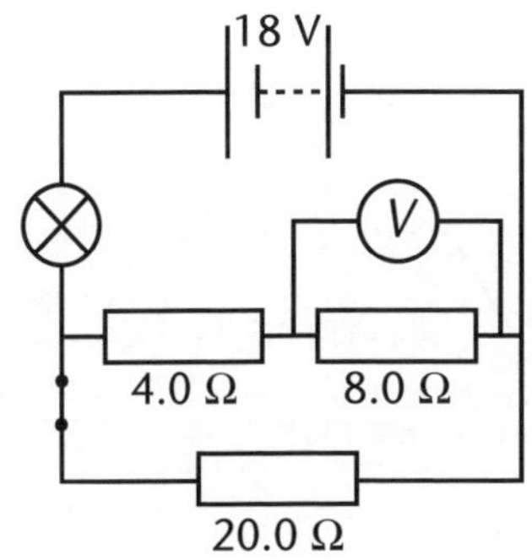

(***M***)

ii. In the position shown ALL the current passes through the bulb. (***M***)

Topic 3: Resistance

Unit 11.5 Activity 3A: Ohm's law and resistance (page 234)

1. **a**.

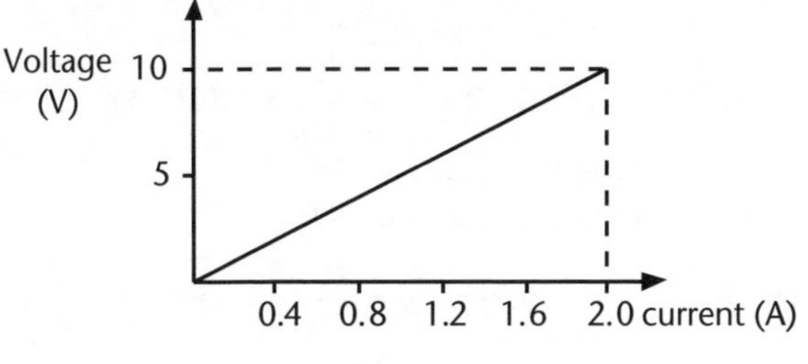

(***A***)

b. $R = \frac{V}{I} = \frac{10}{2} = 5\ \Omega$ = gradient (***A***)

c. $R = 5\ \Omega$ (***A***)

d. When $V = 12$ V, I will be 2.4 A. (***A***)

2. **a**. 4 A (***A***) **b**. 1.2 Ω (***A***) **c**. 0.05 V (***A***) **d**. 15.3 A (***A***) **e**. 0.4 V (***A***)
f. 600 V (***A***) **g**. 8 Ω (***A***) **h**. 3.6 mA (***A***) **i**. 23 Ω (***A***)

3. $I = \frac{V}{R} = \frac{12}{10} = 1.2$ A (***A***)

4. $V = IR = 10 \times 10^{-3} \times 10 \times 10^{3} = 100$ V (***A***)

5. $R = \frac{V}{I} = \frac{230}{5} = 46\ \Omega$ (***A***)

6. $R_T = R_1 + R_2 + R_3 = 10 + 15 + 5 = 30\ \Omega$ (***A***)

7. a. $R_T = 8 + 4 + 2 = 14\ \Omega$ (**A**)

$I = \frac{V}{R_T} = \frac{28}{14} = 2\text{ A}$

b. $I = \frac{V}{R} = \frac{12}{1\,000} = 0.012\text{ A} = 12\text{ mA}$ (**A**)

c. $R_T = 100 + 150 + 50 = 300\ \Omega$

$I = \frac{V}{R_T} = \frac{100}{300} = 0.33\text{ A}$ (**M**)

d. $I = \frac{V}{R} = \frac{10}{0.1} = 100\text{ A}$ (**A**)

e. $R_T = 12 + 31 + 41 + 27 = 111\ \Omega$

$I = \frac{V}{R_T} = \frac{115}{111} = 1.04\text{ A}$ (**M**)

8. a.

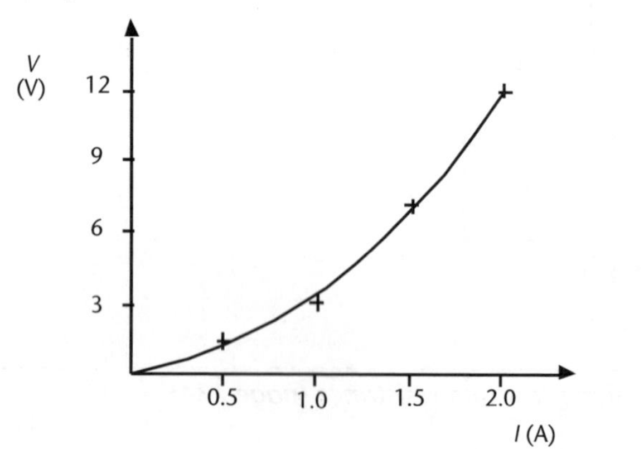

(**A**)

b. i. $R = \frac{V}{I} = \frac{1}{0.5} = 2\ \Omega$ (**A**)

ii. $R = \frac{V}{I} = \frac{7}{1.5} = 4.7\ \Omega$ (**A**)

c. As the filament gets hotter its resistance increases. (**A**)

d. No; an ohmic conductor has a constant resistance. (**M**)

9. a. $R_T = 5 + 2.5 = 7.5\ \Omega$ (**A**)

b. $I = \frac{V}{R} = \frac{10}{7.5} = 1.33\text{ A}$ (**A**)

c. $V = IR = 1.33 \times 5 = 6.66\text{ V}$ (**A**)

d. $I = \frac{1.33}{2} = 0.67\text{ A}$ (**A**)

10. a.

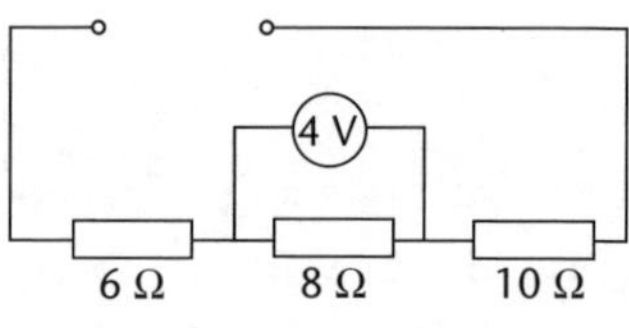

(**A**)

b. $I = \frac{V}{R} = \frac{4}{8} = 0.5\text{ A}$; the current is the same everywhere in the circuit. (**A**)

c. $V = IR = 0.5 \times 10 = 5\text{ V}$ (**A**)

d. V across $6\ \Omega$ resistor $= IR = 0.5 \times 6 = 3\text{ V}$

$V = 5 + 4 + 3 = 12\text{ V}$ (**M**)

e. $V_{supply} = V$ across all these resistors

$= 12\text{ V}$ (**A**)

11. a.

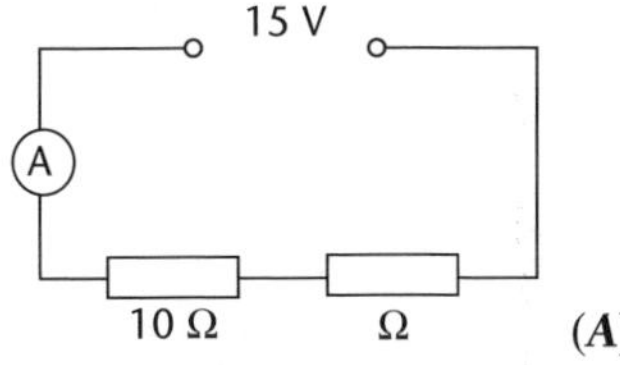

(***A***)

b. $R_r = \frac{V}{I} = \frac{15}{0.5} = 30\ \Omega$

$R = 30 - 10 = 20\ \Omega$ (***A***)

c. $V = IR = 0.5 \times 10 = 5$ V (***A***)

d. $V = 15 - 5 = 10$ V (***A***)

12. a. $I = \frac{V}{R} = \frac{10}{4} = 2.5$ A (***A***)

b. $I = 2.5$ A (***A***)

c. $V = IR = 2.5 \times 3 = 7.5$ V (***A***)

d. $V_{battery} = 10 + 7.5 = 17.5$ V (***A***)

13. a. $R_T = 2 + 3 = 5\ \Omega$ (***A***)

b. $I = \frac{V}{R_T} = \frac{12}{5} = 2.4$ A (***A***)

c. $V = IR = 2.4 \times 2 = 4.8$ V (***A***)

d. $V = IR = 2.4 \times 3 = 7.2$ V (***A***)

14. a. $I = \frac{V}{R} = \frac{8}{50} = 0.16$ A (***A***)

b. $I = \frac{V}{R} = \frac{6}{10\,000} = 6 \times 10^{-4}$ A (***A***)

c. $I = \frac{V}{R} = \frac{12}{50 \times 10^6} = 2.4 \times 10^{-7}$ A (***A***)

15. a. $I = 2$ A (***A***)

b. $V = 3$ V (***A***)

c. $R = \frac{V}{I} = \frac{3}{2} = 1.5\ \Omega$ (***A***)

16. a. $V_{across\ 1\ \Omega\ resistor} = IR = 3 \times 1 = 3$ V (***A***)

$V_{across\ 5\ \Omega\ resistor} = 3 \times 5 = 15$ V (***A***)

$V_{across\ 12\ \Omega\ resistor} = 3 \times 12 = 36$ V (***A***)

b. $V_{battery} = 3 + 15 + 36 = 54$ V (***M***)

17. a. $R = \frac{V}{I} = \frac{3}{2} = 6\ \Omega$ (***A***)

b. $R = \frac{V}{I} = \frac{230}{0.5} = 460\ \Omega$ (***A***)

c. $R = \frac{V}{I} = \frac{6}{150 \times 10^{-3}} = 40\ \Omega$ (***A***)

18. The same as the current in the 1000 Ω resistor. The current in a series circuit is the same at all points in the circuit. (***A***)

19. a. $V_{across}\ R_2 = IR = 0.2 \times 6 = 1.2$ V (***A***)

b. I through $A_3 = \frac{V}{R} = \frac{1.2}{4} = 0.3$ A (***A***)

c. Reading on A_1, = reading on A_2 + reading on A_3

$= 0.2 + 0.3$

$= 0.5$ A (***M***)

d. $V_{across}\ R_1 = IR = 0.5 \times 3.6 = 1.8$ V (***A***)

e. $V_{battery} = 1.8 + 1.2 = 3$ V (***A***)

f. $E = VIt = I^2Rt$

$= 0.2^2 \times 6 \times 60$

$= 14.4$ J (***A***)

g. Anticlockwise. [Away from the positive terminal of the battery towards the negative terminal of the battery.] (***A***)

h. **i**.

ii. $R_T = R_1 + R_3$
$= 3.6 + 4$
$= 7.6\ \Omega$ (***A***)

20. $\frac{20}{60} \times 12\ \text{V} = 4\ \text{V}$ (***M***)

21. A $R_T = R_1 + R_2 = 10\ \Omega$; (***A***)
B $R_T = 9 + 1 = 10\ \Omega$; (***A***)
C $R_T = 4 + 3.33 = 7.3\ \Omega$; (***A***)
D $\frac{I}{R_T} = \frac{1}{20} + \frac{1}{20} = \frac{2}{20}$; $R_T = \frac{2}{20} = 10\ \Omega$. (***M***)

22. **a**.

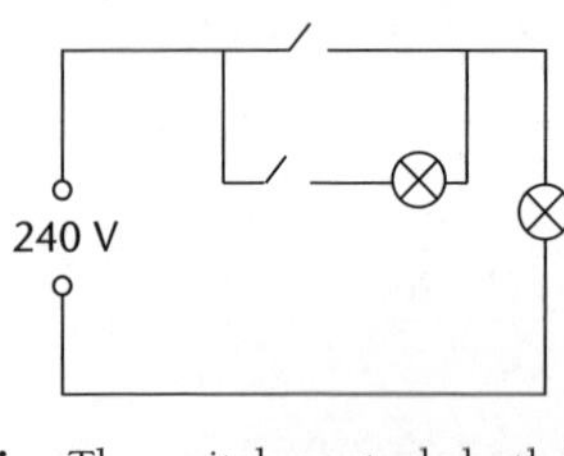

(***A***)

b. **i**. The switch controls both lamps; ii. the variable resistor changes the current for both lamps. (***A***)

ii.

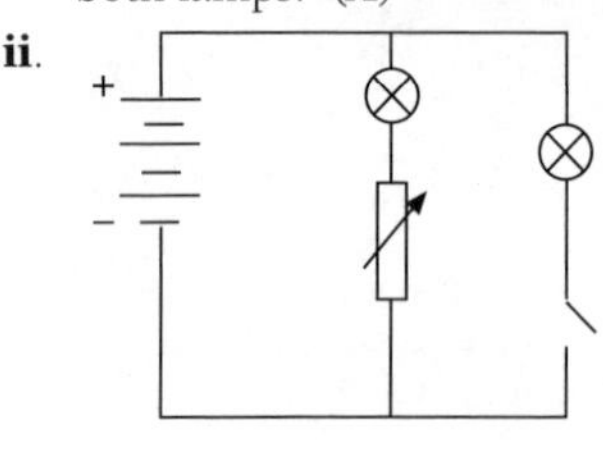

(***A***)

23. **a**.

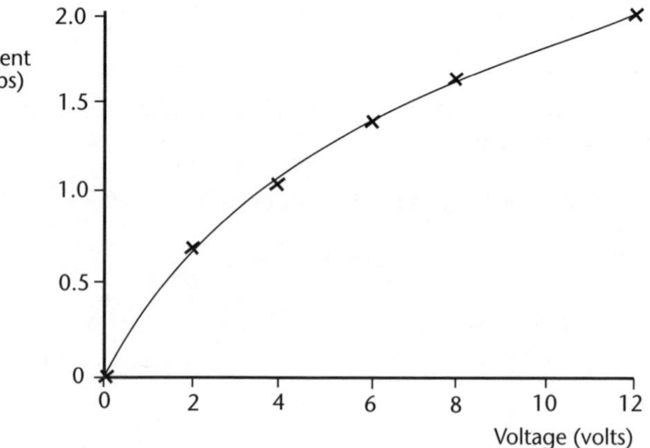

(***A***)

b. $I = 1.8\ \text{A}$ (***A***)

c. $R = \frac{V}{I} = \frac{12}{2} = 6\ \Omega$

i. For two bulbs:

$R = 12\ \Omega$

$I = \frac{V}{R} = \frac{12}{12} = 1\text{ A}$ (***M***)

ii. For three bulbs:

$R = 18\ \Omega$

$I = \frac{V}{R} = \frac{12}{18} = 0.66\text{ A}$ (***M***)

24. **a**. $V_1 = IR = 0.01 \times 10 = 0.1\text{ V}$ (***A***)

$V_2 = IR = 0.01 \times 100 = 1\text{ V}$ (***A***)

$V_3 = IR = 0.01 \times 632 = 6.32\text{ V}$ (***A***)

b. $V_4 = IR = 0.3 \times 25 = 7.5\text{ V}$ (***A***)

$V_5 = IR = 0.3 \times (5 + 20 + 1\,000 + 200) = 367.5\text{ V}$ (***A***)

c. $V_6 = IR = 2 \times 10^{-3} \times 48 \times 10^3 = 96\text{ V}$ (***A***)

$V_7 = IR = 2 \times 10^{-3} \times (33 \times 10^3 + 68 \times 10^3) = 202\text{ V}$ (***A***)

25. $\frac{I}{R_T} = \frac{1}{20} + \frac{1}{20} = \frac{2}{20}$ (for parallel combination)

$R_r = \frac{20}{2} = 10\ \Omega$

$R_T = 20 + 10$ (for complete circuit)

$R_T = 30\ \Omega$

V across parallel combination = 4 V (***M***)

26. $\frac{I}{R_T} = \frac{1}{60} + \frac{1}{120}$ (for parallel combination)

$= \frac{2}{120} + \frac{1}{120}$

$= \frac{3}{120}$

$R_r = \frac{120}{3} = 40\ \Omega$ (for parallel combination)

$R_T = 20 + 40 = 60\ \Omega$ (for circuit)

$I = \frac{V}{R_T} = \frac{36}{60} = 0.6\text{ A}$ (through 20 Ω resistor)

$I = 0.6 - 0.2 = 0.4\text{ A}$ (through 60 Ω resistor) (***E***)

27. **a**.

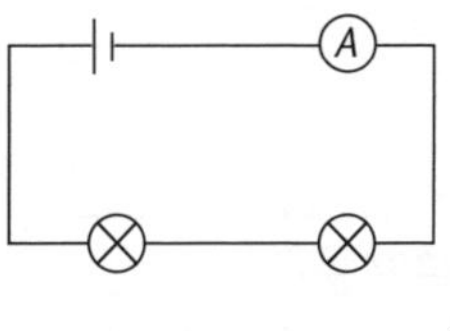

(***A***)

b.

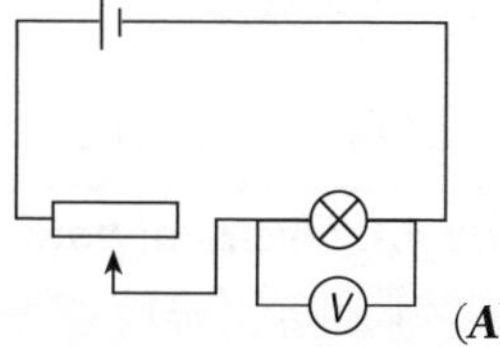

(***A***)

c. 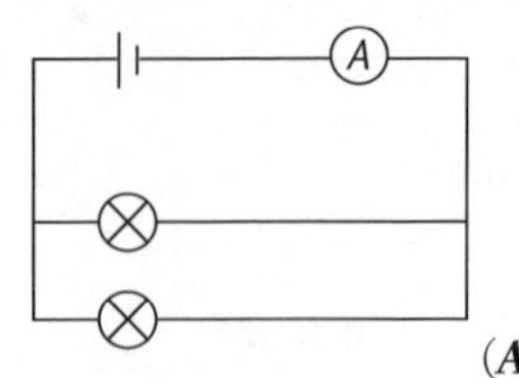

(**A**)

d. 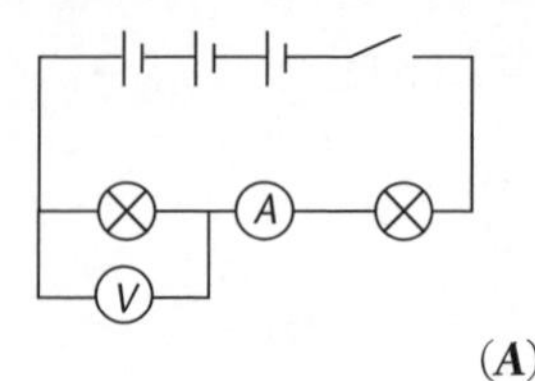

(**A**)

28. $A_3 = 5 - 2 = 3$ A (**A**)

29. a. $R_T = R_1 + R_2 \quad \therefore R_T = (4.0 + 8.0)\ \Omega \quad \therefore R_T = 12\ \Omega$ (**A**)

b. $V = IR$ where $V = 12$ V, $R = 8.0\ \Omega \quad \therefore I = \frac{V}{R} \quad \therefore I = \frac{12}{8.0}$ A $\quad \therefore I = 1.5$ A (**A**)

c. $I = 1.5$ A (**A**)

d. $E = VIt$ where $V = 6$ V, $I = 1.5$ A, $t = 13$ s $\quad \therefore E = 6 \times 1.5 \times 13$ J

$\therefore E = 117$ J (or 120 J (2 sf)) (**M**)

e. Voltage = 18 V (**A**)

f. $P = VI$ where $V = 18$ V, $I = 0.9$ A

$\therefore P = 18 \times 0.9$ W

$\therefore P = 16.2$ W (or 20 W (1 sf)) (**M**)

g. $E = VIt$ or $E = Pt$ where $t = 15$ s

$\therefore E = 16.2 \times 15$ J

$\therefore E = 243$ J (or 240 J (2 sf)) (**M**)

h. $I = 1.5$ A (check $I = \frac{V}{R} = \frac{18\text{ V}}{12\ \Omega} = 1.5$ A) (**M**)

i. Total resistance $= \frac{\text{total voltage}}{\text{total current}}$

$\therefore RT = \frac{18\text{ V}}{1.5\text{ A} + 0.9\text{ A}}$

$\therefore RT = \frac{18\text{ V}}{2.4\text{ A}}$

$\therefore RT = 7.5\ \Omega$ (**M**)

j. i. 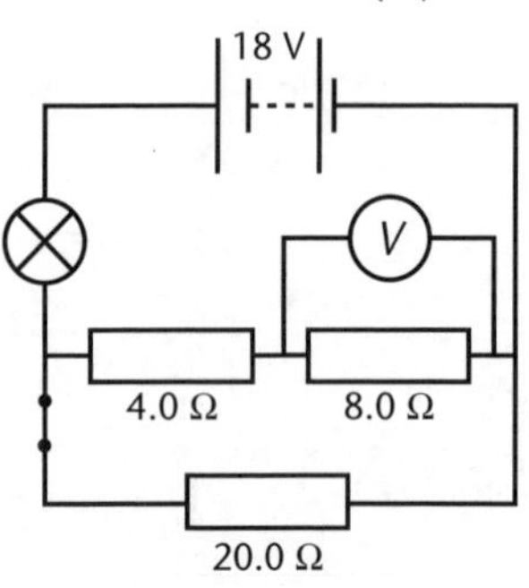

(**M**)

ii. In the position shown ALL the current passes through the bulb. (**M**)

Topic 4: Electrical power and energy

Unit 11.5 Activity 4A: Electrical power and energy (page 245)

1. a. $E = Pt = 150 \times 1 = 150$ J (**A**)

b. $E = Pt = 150 \times 8 \times 60 \times 60 = 4.3$ MJ (**A**)

c. $E = Pt = 150 \times 365 \times 8 \times 60 \times 60 = 15.8 \times 10^8$ J (**A**)

2. $P = VI; I = \frac{P}{V}\ \frac{2.4 \times 10_3}{230} = 10.4$ A (**A**)

3. $I = \frac{P}{V} = \frac{3\,000}{230} = 13$ A (***A***)

4. Total power $= 5 \times 60 = 300$ W $= 0.3$ kW
   Total kW h $= 0.3 \times 4.5 = 1.35$
   Total cost $= 1.35 \times 0.13$
   $= 0.176$ cents (***M***)

5. Total power $= 5 \times 100 \times 4 \times 60$
   $= 500 + 240$
   $= 740$

   $P = VI$; $I = \frac{P}{V} = \frac{740}{230} = 3.2$ A; hence a 5 A fuse is sufficient. (***E***)

6. **a**. Total units $= 3 \times 3 = 9$ kW h (***A***) **b**. Total cost $= 9 \times \$0.12 = \$1.08$ (***A***)

7. **a**. In parallel; if they were in series they would have to be used all at the same time. (***M***)
   b. Total power $= 3 + 2 \times 4 \times 0.5 = 7$ kW
   Total cost $= 7 \times 1.5 \times \$0.12 = \$1.26$ (***M***)
   c. $P = VI$ $I = \frac{P}{V} = \frac{7\,000}{230} = 30.4$ A (***A***)
   d. A 30 A fuse would not be sufficient if all parts of the stove were used at the same time; the fuse would blow. (***M***)

8. Maximum power $= VI = 230 \times 5 = 1\,150$ W;
   number of lamps $= \frac{1150}{150} = 7.6$; ie 7 lamps (8 lamps would blow fuse) (***E***)

9. **a**. $\frac{I}{R_T} = \frac{1}{10} + \frac{1}{2} = \frac{1}{10} + \frac{5}{10} = \frac{6}{10}$

   R_T for parallel combination $= \frac{10}{6} = 1.67\ \Omega$

   R_T for circuit $= 1.67 + 5 = 6.67\ \Omega$

   $I = \frac{V}{R_T} = \frac{12}{6.67} = 1.8$ A (***E***)

   b. $V_1 = IR$
   $= 1.8 \times 1.67$
   $= 3$ V (***A***)

   c. $P = VI = I^2R$
   $= 1.8^2 \times 5$
   $= 16$ W (***A***)

   d. $V_2 = IR = 1.8 \times 5$
   $= 9$ V (***A***)

   e. Current in 10 Ω resistor $= \frac{V}{R} = \frac{3}{10} = 0.33$ A
   $P = VI = 3 \times 0.3 = 1$ W (***M***)

10. **a**. $P = I^2R = 1^2 \times 1 = 1$ W (***A***)
   b. $P = I^2R = (1 \times 10^{-3})^2 \times 5 \times 10^3 = 5 \times 10^{-3}$ W (***A***)
   c. $P = I^2R = (0.2)^2 \times 60 = 2.4$ W (***A***)
   d. $P = I^2R = (0.2)^2 \times 1\,000 = 40$ W (***A***)
   e. $P = I^2R = 1^2 \times 0.7 \times 10^3 = 0.7 \times 10^3$ W (700 W) (***A***)

11. **a**. $P = \frac{V^2}{R} = \frac{36}{8} = 4.5$ W (***A***) **b**. $P = \frac{V^2}{R} = \frac{230^2}{80} = 661$ W (***A***)

c. $P = \frac{V^2}{R} = \frac{1.5^2}{1\,000} = 2.25 \times 10^{-3}$ W (***A***)

12. Number of units used = 17 084 – 16 110 = 974
Total cost = 974 × 12.19 cents
= \$118.73 (***M***)

13. **a**. Five 100 W lamps dissipate 0.5 kW for 10 hours.
Total cost = 0.5 × 10 × \$0.12
= \$0.60 (***M***)

b. Total cost = 5 × 4 × \$0.12
= \$2.40 (***M***)

14. **a**. Meter 1 = 10 000 + 4 000 + 600 + 6 = 14 606 kWhr
Meter 2 = 10 000 + 5 000 + 300 + 60 + 8 = 15 368 kWhr (***M***)

b. Electricity used = 15 368 – 14 606 = 762 units (***A***)

c. Cost = 762 × \$0.12 = \$91.44 (***A***)

15. Power of first lamp = $\frac{V^2}{R} = \frac{36}{4} = 9$ W (***A***)

Resistance of second lamp = $\frac{V^2}{P} = \frac{12^2}{9} = 16\ \Omega$ (***A***)

16. **a**.

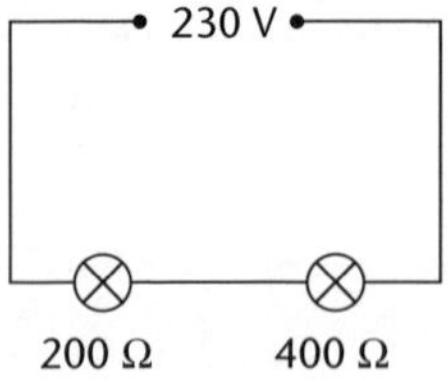

(***A***)

b. $R_T = 200 + 400 = 600\ \Omega$

$I = \frac{V}{R_T} = \frac{230}{600} = 0.38$ A (***M***)

c. For 200 Ω bulb:

$P = I^2R = 0.38^2 \times 200 = 28.9$ W (***A***)

For 400 Ω bulb:

$P = I^2R = 0.38^2 \times 400 = 57.8$ W (***A***)

17. **a**. $I = \frac{P}{V} = \frac{2\,400}{230} = 10.4$ A (***A***) **b**. $R = \frac{V}{I} = \frac{230}{10.4} = 22.1\ \Omega$ (***A***)

c. $Q = mc\Delta T$ = 1.5 × 4 200 × (100 – 16)
= 529 200 J
= 0.53 MJ (***M***)

d. $P = \frac{E}{t}$; $t = \frac{E}{P} = \frac{529\,200}{2\,400}$ = 221 s = 0.06 hours (***M***)

e. $R_T = R1 + R2$
= 22.1 + 12
= 34.1 Ω (***A***)

f. $P = \frac{V^2}{R} = \frac{230^2}{34.1}$ = 1 551 W = 1.6 kW (***A***)

g. i. It takes *more* time. (***A***)

ii. $Q = mc\Delta t$
$= 1.5 \times 4\,200 \times 84$
$= 529\,200$

$t = \frac{E}{P} = \frac{529\,200}{1\,551} = 341$ s; extra time = 341 – 221 = 120 s (***E***)

h. Insulating material base; shiny outer surface. (***M***)

18. a. $Q = mc\Delta T$
$= 180 \times 4\,200 \times (65 - 18)$
$= 35\,532\,000$ J
$= 3.6 \times 10^7$ J (***A***)

Topic 6: Alternative current (AC) circuits and effective values

Unit 11.5 Activity 6A: AC circuits (page 263)

1. For the inductive circuit, the current amplitude decreases and for the capacitive circuit, the current amplitude increases.

2. D. and E.

3. **a.** more
b. same
c. same
d. same (zero)

4. **a.** 0.60 Amp
b. 0.60 Amp.

5. **a.** 4.60 Khz
b. 26.6 nf
c. $X_L = 2.60$ KiloOhm, $X_C = 0.650$ KiloOhm

Unit 11.6 Electronics

Topic 2: Solid state devices – diode and transistor

Unit 11.6 Activity 2A: Diodes (page 274)

1. **a.** 0.6 volts
b. approximately 9 v. The diode is in reverse bias and does not exhibit the characteristic voltage drop.

2. The diode is in forward bias and will put a short-circuit across the battery. Both the diode and the battery can be damaged if you do this.

3. The voltage across each diode will add to get 1.2 v. They are in a forward bias condition.

Unit 11.6 Activity 2B: The transistor (page 282)

6 mA

Unit 11.6 Activity 2C: Transistors (page 285)

1. Emitter follower
2. Saturated transistor
3. 0.6 volts
4. Approximately 0.15 volts
5. Transistor in CUTOFF
6. 50 mA

Topic 3: Digital electronics

Unit 11.6 Activity 3A: Binary numbers (page 291)

1. a. 1010110
   b. 1000000
2. 46
3. 1010111_2
4. 83
5. 01010011
6. 4610
7. 56_8 and 123_8

Unit 11.6 Activity 3B: Truth table (page 296)

1. Write a truth table for this combination of gates, showing all the possible states of A, B, C, and Q.
2. What must the states of inputs A and B be for the output to be HIGH?

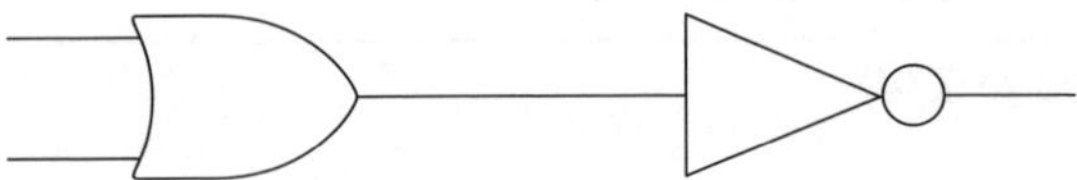

Glossary/Index

accelerate (48): when something changes its velocity.
acceleration (47): symbol **a** – the rate of change of velocity per second.
acceleration due to gravity (86): rate at which objects change their velocity when they move under the influence of gravity.
accurate (32): when a measurement is close to the actual value.
action (96): force acting on an object.
air resistance (86, 119): friction caused when an object moves through air; also called 'drag'.
alternating current (84, 212): electric current that changes its direction periodically.
ammeter (28, 212): device used to measure electrical current in a circuit.
ampere (212): symbol **A** – the SI unit of current (1 A = 1 C s^{-1}).
amplitude (255): maximum distance a particle in a medium moves from its undisturbed (rest) position.
atom (197): smallest particle of matter that cannot be broken down chemically.
average speed (48): symbol v – total distance travelled divided by total time taken.
average velocity (52): total displacement divided by the total time taken.
axes (57): the two perpendicular directions used in a reference frame, or the vertical and horizontal lines drawn on a graph.

bar graphs (41): graph used to display the variation of discrete variables.
battery (171, 211): power source that converts chemical potential energy to electrical energy; two or more cells in series.
bearing (58): direction measured clockwise from North.

calibration error (32): an error in a measuring instrument when it is compared with standard measurements.
centre of mass (143): point at which all the mass of an object can be considered to be; all the weight can be thought to act down from this point.
centrifugal force (108): apparent outwards force associated with circular motion.
centripetal acceleration (107): inwards acceleration of an object as it moves in a circle.
centripetal force (107): inwards force necessary to keep an object moving in a circle.
changing units (18): swapping from one SI unit to another – dividing or multiplying by 1 000.
chemical energy (177): energy stored in the chemical bonds in substances.
circuit (23): a conducting path for conventional current to flow from the positive side of a cell around to the negative side of a cell.
circuit diagram (213): diagram that uses electrical symbols to represent components and how they are joined together.
collisions (151): when two or more objects interact.
commutator (253): parts of a DC electric motor that connect the rotating coil to the outside circuit.
compass (57): device that can be used to show the direction of a magnetic field; consists of a small pivoted magnet that can rotate freely.
components (62, 211): vectors – two vectors at right angles which when added together are equivalent to a single vector; electricity – the different parts of an electrical circuit.
compression (181): where particles in a medium are bunched together due to longitudinal wave motion.
computer spreadsheet (2): computer-generated table that can be used to electronically calculate and graph data.
conduction (249): transfer of heat energy by particle vibration.

conductor (203): a substance that allows an electric current (*electrical energy*) or heat (*thermal energy*) to travel through it.

conservation of momentum (157): momentum before a collision or explosion same as momentum after, assuming no external forces are acting.

conserved (157, 178): when something stays constant.

constant speed (48): time during which an object does not change it speed throughout its journey.

contact force (113): force that makes contact with the object it acts on.

continuous variables (41): variables that can take on any value inside a range of values.

'conventional' current (212): describing the movement of charge from positive to negative around a circuit (electrons actually travel the other way in wires).

couple (142): two equal and opposite forces that act at a perpendicular distance apart to cause rotation.

current (7, 212): rate of movement of charge in a closed circuit; amount of charge passing a point in a circuit per second.

deceleration (48): symbol **a** – slowing down, a negative acceleration.

density (10): symbol ρ – mass per unit volume.

dependent variable (12): variable that is plotted on the y-axis of a graph; changes in response to changes in the independent variable.

digital multimeter (23): device used to measure a range of electric qualities.

diode (270): electric component that allows current to flow in one direction only.

direct current (212): electric current that always moves in the same direction around a circuit (away from the positive terminal of the supply towards the negative terminal) – this is in the opposite direction to electron flow.

discrete (41): variables that are only able to take integer values (eg number of students in a classroom).

displacement (15, 47): distance and direction from a known origin.

dissipated (186): spread out into the surrounding environment.

distance (47): length between two positions.

distance versus time graph (71): graph of distance (y-axis) versus time (x-axis).

dry cell (211): cell that produces a DC voltage; electrolyte in the cell is a paste.

dynamic equilibrium (187): occurs when all the forces and torques acting on an object are balanced and the object is travelling at constant velocity.

elastic (183): kinetic energy is conserved.

elastic potential energy (178): energy stored in an extended or compressed spring.

electric circuit (211): unbroken pathway through which electricity can flow.

electric field (114): region in which a charged object experiences a force.

electromagnet (84): coil of wire that becomes a magnet when a current passes through it.

electromagnetic spectrum (171): family of waves that travel at the speed of light (3×10^8 m s^{-1}).

electromotive force (224): emf; alternative term for induced voltage.

electron (114): small negatively-charged particle that orbits the nucleus of an atom.

electroscope (201): device for measuring the type and quantity of electric charge.

electrostatic force (114): a field force – force a charged particle has on it when it enters an electric field.

emf (224): electromotive force; a source of voltage.

energy (57): symbol **E** – capacity to do work.

equilibrium (142): situation when an object is at rest or moving uniformly as described in Newton's First Law.

error (24): word used to describe how a measurement could differ from the true value.
explosion (151): when an object breaks up into two or more parts violently.
extrapolation (40): process of estimating data values outside the range of known data.

fair test (1): investigation where one variable is changed (independent variable) to determine the effect on another variable (dependent variable) while all other variables are kept constant.
field force (113): force that acts through a force field, eg gravitational, electrostatic and magnetic forces.
force (7, 57): symbol **F** – a push or a pull.
free body force diagrams (115): diagrams drawn seperately with only the relevant forces indicated.
free electrons (265): electrons that are only loosely associated with atoms, so are able to move through a conducting substance.
frequency (106): number of revolutions, vibrations or events that occur in one second; number of waves that pass a point per second.
friction (113): force produced when two surfaces come in contact or slide past each other.
fundamental units (7): the seven units in the SI System from which all others can be derived.
fuse (245): wire of low melting point and high resistance that melts when excessive current flows in a circuit.

galvanometer (249): very sensitive ammeter for measuring small currents.
gradient (42): slope.
graph (12): visual way of displaying data using axes.
gravitation (97): force of attraction between two objects due to their mass.
gravitational potential energy (172): symbol **Ep** – energy possessed by an object because of its position in a gravitational field.

head (57): end point of a vector.
heat (113): energy transferred when an object changes temperature or state.
hertz (106): SI unit of frequency.
hydroelectric power (172): electrical power generated by turbines which are driven by falling water.
hypothesis (1): a proposed explanation that can guide an investigation.

illumination (276): describing how intense a light source is at a distance away from it.
impulse (153): change in momentum produced by a force acting for a length of time.
independent variable (12): variable plotted on the x-axis of a graph; the variable is changed in increments to investigate changes in the dependent variable.
inelastic (183): kinetic energy is not conserved.
instantaneous speed (48): speed at an instant in time.
instantaneous velocity (52): velocity at a particular instant in time.
insulator (203): substance that does not permit electrons (electrical energy) or heat (thermal energy) to flow through.
interpolation (40): process of estimating data values within the range of known values.
inverse (44): opposite mathematical operation, eg division by 3.6 instead of multiplication by 3.6.

joule (163): symbol **J** – SI unit of energy and work.

kilogram (18): symbol **kg** – SI unit of mass.

kilowatt (178): 1 000 watts.
kilowatt-hour (243): amount of energy transferred when 1 000 watts of power is used for 1 hour.
kinematic equations of motion (89): set of formulas used to describe motion mathematically.
kinematics (47): the study of motion.
kinetic energy (157, 172): symbol **Ek** – energy possessed by a moving object.

leverage (141): alternative name for torque.
light (178): narrow band of frequencies of electromagnetic radiation visible to our eyes.
light-emitting diodes (275): LED; special diodes that give off light when current passes through them.
line graph (41): graph used to display two continuous variables.
linear (42): forming a straight line.
lubricant (113): substance used to reduce friction between moving surfaces, eg oil or grease.

magnet (114): an object that can attract iron or steel.
magnetic field (114): region around a magnet where iron or other magnets experience a force.
magnetic field lines (253): lines drawn to graphically represent the strength and direction of a magnetic field.
magnetic force (114): a field force – for a magnetic pole has on it when it enters magnetic field.
mass (7): symbol **m** – amount of matter in an object.
mechanical energy (172): kinetic or potential energy.
medium (208): material through which waves or light can pass, eg, water, air.
metric (7): related in multiples of ten.
micrometer screw gauge (22): device for accurate measurement of small distances.
milliwatt (178): $\frac{1}{1\,000}$ watt.
mistake (33): an error, such as misreading a measurement.
moment (141): the turning effect of a force about a pivot (*see torque*).
momentum (151): physical quantity defined as the mass times the velocity of an object.
motion (47): movement.

net (123): another word for resultant.
neutral (197): situation where the number of positive and negative charges are equal.
newton (7, 125): symbol **N** – SI unit of force.
nichrome (229): alloy of nickel, chromium and iron that is used to make wire of high resistance.
node (255): place of zero displacement in a standing wave.
non-ohmic resistor (230): resistor that does not have a constant resistance; has a curved V versus I graph.
nuclear force (97): force which keeps the particles in the nucleus together.
nuclear reactor (171): device to obtain energy from controlled nuclear chain reactions.
nucleus (171): collection of protons and neutrons at the centre of an atom; tiny, dense and positively charged.

ohm (230): symbol Ω – SI unit of electrical resistance.
ohmic conductors (231): conductors that obey Ohm's law so their graph of voltage against current is a straight line.
ohmic resistor (230): resistor that has constant resistance; obeys Ohm's Law.
Ohm's Law (229): relationship between voltage, resistance, and current; $V = IR$.
oscilloscope (250): electronic device that displays vibrations in graphical form on a screen.

parabola (99): shape of a projectile's path.
parabolic (100): U-shaped, like a parabola.
parallax (24): apparent movement of two objects due to the movement of the observer.
parallax error (32): error produced by reading a scale from an angle.
parallel (169): connected 'side by side' in a circuit.
parallel circuit (214): an electrical circuit in which there is more than one pathway for the electric current.
pascal (131): symbol **Pa** – SI unit of pressure; 1 Pa = 1 N m^{-2}.
period (105): time taken for one vibration or for one wave to pass a point; the time taken for one revolution or event.
periodic (261): wave motion that repeats regularly.
phase (256): measure of how much one wave is out of step with another.
pivot (141): point about which the torque can act or an object can revolve.
potential difference or voltage (205): symbol **V** – the energy per coulomb gained or lost between two points in an electrical circuit.
potential divider (296): two resistors connected in series that divide up the voltage across them.
power (166): rate of doing work, defined as work divided by time.
power pack (211): a variable DC or AC power supply that is connected to the mains supply.
precise (31): when measurements are closely grouped together.
pressure (131): symbol **P** – the force per unit area.
projectile (99): an object moving through the air only under the force of gravity.
proportional (43): when two quantities are related by a constant ratio.
proportionality constant (42): value of the ratio between two quantities.
proton (197): a positively-charged particle found in the nucleus of an atom.

quantum theory (6): idea that electrons have definite amounts of energy as they orbit an atom's nucleus.

radioactive decay (13, 40): nucleus that breaks apart and releases an α-particle, β-particle or γ-radiation.
random error (31): error that is equally likely to be more or less than the true value.
ratio (19): the quotient of two quantities.
reaction (120): name given to the force that is equal in size and opposite in direction to an action force.
reference frame (57): pair of directions at right angles used to describe the direction of a vector.
reflection (25): wave motion that 'bounces back' at a boundary.
relationships (1): how different quantities or ideas are connected.
relative velocity (65): velocity of one object in relation to another.
relay (296): a magnetic switch for switching a circuit on or off.
resistance (5): quality that describes the ability of a conductor to conduct electricity; the lower the resistance, the better it conducts.
resistor (232): electrical component that converts electrical energy to heat.
resolving (63, 101): process of finding vector components.
resultant (16): vector obtained by the addition or subtraction of other vectors.
revolution (105): motion once around a circle.
revolutions per minute (106): number of rotations in one minute.

rheostat (229): a variable resistor.
rotation (105): motion once around a circle.
rounded off (35): if a digit is 0, 1, 2, 3, 4, the preceding digit stays the same, whereas 5, 6, 7, 8 rounds up the preceding digit by one.
rounding error (36): error introduced into a calculation caused by using partial results that have been rounded.

scalar (15): a quantity that has size (magnitude) only.
scientific notation (8, 33): way of writing any number between 1 and 10 multiplied by an approriate power of 10, eg 1 260 = 1.26×10^3.
sense (141): anticlockwise or clockwise direction.
series (211): connected 'one after the other' in a circuit.
series circuit (214): electrical circuit in which there is only one pathway for the current.
SI system (7): international system of units.
significant figures (9): digits in a number or measurement that are not being used as place holders.
slope (42): quantity used to describe how steep a straight line is – vertical change divided by the horizontal change.
specific heat capacity (247): symbol **c** – amount of energy required to raise the temperature of 1 kg of a substance by 1 °C.
speed (7): symbol **s** – rate of change of distance.
speed versus time graph (74): a graph of speed (in m s^{-1}) on y-axis, time (in s) on x-axis.
spring constant (179): force required to extend or compress a spring by one metre.
standard form (33): alternative name for scientific notation.
static equilibrium (187): occurs when an object is at rest and all the forces and torques acting on it are balanced.
systematic error (32): an error consistently larger (or smaller) than the true value.

tail (57): starting point of a vector.
temperature (28, 229): symbol **T** – measure of the average kinetic energy of the particles in a substance.
tension (107, 115): force in connecting strings and ropes that tries to stretch them.
terminal velocity (188): constant velocity attained when the air resistance force is equal and opposite to the gravitational force acting on a falling object.
thermometer (28): device for measuring temperature.
thrust (121): a force, usually from an engine.
ticker tape (84): paper tape pulled through a ticker timer; records dots.
ticker timer (84): device for recording distance and time for a moving object.
tolerance (233): expected range of variation in the value of a manufactured component such as a resistor.
torque (141): turning or twisting effect about a pivot; defined as the product of force and perpendicular distance to the pivot.
transducer (171): device for transforming energy from one form into another, eg a loudspeaker converts electrical energy into sound energy.
transparent (266): a substance that can be clearly seen through.
turning effect (141): another term for torque.

unit (7): reference standard for measuring a quantity.

van der Graaf generator (114, 204): device for generating electrical charge.

variables (1, 40): physical quantities that can have a range of values and can be drawn on a graph.
vector (15): quantity that possesses both size (magnitude) and direction.
vector diagram (60): scale diagram showing vector magnitudes and directions.
velocity (47): symbol **v** – speed in a stated direction.
Vernier caliper (21): device for accurate measurement of distances.
volt (211): symbol **V** – SI unit of potential difference.
voltage (205): common term for potential difference.
voltmeter (28, 211): electrical meter used to measure potential difference.

watt (166): symbol **W** – SI unit of power; 1 W = 1 J s^{-1}.
wavelength (278): distance between adjacent corresponding points in a wave that are vibrating in the same manner.
weight (125): symbol **W** – force of gravity acting on an object.
work (163): process of transforming energy from one form to another, defined as force times distance moved in the direction of the force.

zero error (32): error in a measuring instrument where the measuring scale does not read zero for zero measurement.